Cambridge IGCSE™ and O Level

Additional Mathematics

Second edition

Val Hanrahan, Jeanette Powell,
Stephen Wrigley
Series editor: Roger Porkess

Cambridge International copyright material in this publication is reproduced under licence and remains the intellectual property of Cambridge Assessment International Education.

Endorsement indicates that a resource has passed Cambridge International's rigorous quality-assurance process and is suitable to support the delivery of a Cambridge International syllabus. However, endorsed resources are not the only suitable materials available to support teaching and learning, and are not essential to be used to achieve the qualification. Resource lists found on the Cambridge International website will include this resource and other endorsed resources.

Any example answers to questions taken from past question papers, practice questions, accompanying marks and mark schemes included in this resource have been written by the authors and are for guidance only. They do not replicate examination papers. In examinations the way marks are awarded may be different. Any references to assessment and/or assessment preparation are the publisher's interpretation of the syllabus requirements. Examiners will not use endorsed resources as a source of material for any assessment set by Cambridge International.

While the publishers have made every attempt to ensure that advice on the qualification and its assessment is accurate, the official syllabus, specimen assessment materials and any associated assessment guidance materials produced by the awarding body are the only authoritative source of information and should always be referred to for definitive guidance. Cambridge International recommends that teachers consider using a range of teaching and learning resources based on their own professional judgement of their students' needs.

Cambridge International has not paid for the production of this resource, nor does Cambridge International receive any royalties from its sale. For more information about the endorsement process, please visit www.cambridgeinternational.org/endorsed-resources

Questions from Cambridge IGCSE™ and O Level Mathematics past papers are reproduced by permission of Cambridge Assessment International Education. Unless otherwise acknowledged, the questions, example answers and comments that appear in this book were written by the authors. In examinations, the way marks are awarded may be different. Cambridge Assessment International Education bears no responsibility for the example answers to questions taken from its past question papers which are contained in this publication.

Photo credits

p.3 © Yulia Grogoryeva – 123RF.com; p.20 © The Granger Collection/Alamy Stock Photo; p.41 © FabrikaSimf/Shutterstock.com; p.55 © SCIENCE PHOTO LIBRARY; p.77 © Bonma Suriya/Shutterstock.com; p.89 © Monkey Business – stock.adobe.com; p.118 © Vitalii Nesterchuk/Shutterstock.com; p.136 © DenPhoto – stock.adobe.com; p.157 © siraphol – stock.adobe.com; p.166 © vazhdaev – stock.adobe.com; p.216 © white78 – stock.adobe.com; p.256 *left* © marcel – stock.adobe.com; p.256 *right* © Mariusz Blach – stock.adobe.com; p.289 © Peter Bernik – stock.adobe.com; p.312 © NASA/JPL

Every effort has been made to trace all copyright holders, but if any have been inadvertently overlooked, the Publishers will be pleased to make the necessary arrangements at the first opportunity.

Although every effort has been made to ensure that website addresses are correct at time of going to press, Hodder Education cannot be held responsible for the content of any website mentioned in this book. It is sometimes possible to find a relocated web page by typing in the address of the home page for a website in the URL window of your browser.

Hachette UK's policy is to use papers that are natural, renewable and recyclable products and made from wood grown in well-managed forests and other controlled sources. The logging and manufacturing processes are expected to conform to the environmental regulations of the country of origin.

Orders: please contact Hachette UK Distribution, Hely Hutchinson Centre, Milton Road, Didcot, Oxfordshire, OX11 7HH. Telephone: +44 (0)1235 827827. Email education@hachette.co.uk Lines are open from 9 a.m. to 5 p.m., Monday to Friday. You can also order through our website: www.hoddereducation.com

© Val Hanrahan, Jeanette Powell and Stephen Wrigley 2023

First published in 2018
This edition published in 2023
by Hodder Education (a trading division of Hodder & Stoughton Limited), an Hachette UK Company, Carmelite House, 50 Victoria Embankment, London EC4Y 0DZ

www.hoddereducation.com

The authorised representative in the EEA is Hachette Ireland,
8 Castlecourt Centre, Dublin 15, D15 XTP3, Ireland (email: info@hbgi.ie)

Impression number	10 9 8 7 6 5 4 3
Year	2027 2026 2025

All rights reserved. Apart from any use permitted under UK copyright law, no part of this publication may be reproduced or transmitted in any form or by any means, electronic or mechanical, including photocopying and recording, or held within any information storage and retrieval system, without permission in writing from the publisher or under licence from the Copyright Licensing Agency Limited. Further details of such licences (for reprographic reproduction) may be obtained from the Copyright Licensing Agency Limited, www.cla.co.uk

Cover photo © mimadeo – stock.adobe.com

Illustrations by Integra Software Services Pvt. Ltd., Pondicherry, India

Typeset in Times Ten LT Std 10/12 by Integra Software Services Pvt. Ltd., Pondicherry, India

Printed and Bound in Great Britain by Bell & Bain Ltd, Glasgow

A catalogue record for this title is available from the British Library.

ISBN: 978 1 3983 7395 2

Contents

	Introduction	iv
	Review chapter	1
CHAPTER 1	Functions	3
CHAPTER 2	Quadratic functions	20
CHAPTER 3	Factors of polynomials	41
	Review exercise 1	53
CHAPTER 4	Equations, inequalities and graphs	55
CHAPTER 5	Simultaneous equations	77
CHAPTER 6	Logarithmic and exponential functions	89
	Review exercise 2	116
CHAPTER 7	Straight line graphs	118
CHAPTER 8	Coordinate geometry of the circle	136
CHAPTER 9	Circular measure	157
CHAPTER 10	Trigonometry	166
	Review exercise 3	204
CHAPTER 11	Permutations and combinations	206
CHAPTER 12	Series	216
CHAPTER 13	Vectors in two dimensions	241
	Review exercise 4	254
CHAPTER 14	Differentiation	256
CHAPTER 15	Integration	289
CHAPTER 16	Kinematics	312
	Review exercise 5	325
	Mathematical notation	327
	Answers	329
	Glossary	378
	Index	381

Introduction

This book has been written for all students studying the Cambridge IGCSE™ and O Level Additional Mathematics syllabuses (0606/4037) for first examination from 2025. The book carefully and precisely follows the syllabus. It provides the detail, guidance and practice needed to support and encourage learners as they advance their mathematical reasoning, skills and communication.

This is the second edition of this book, comprehensively updated to cover the revised syllabus. Teachers and learners using the first edition have provided invaluable feedback, and their suggestions are incorporated into this edition. There are a few instances where the book goes beyond the syllabus to provide additional context to the topic for the benefit of students' deeper understanding. It is clearly indicated wherever this is the case.

Organisation of content

Where possible, the chapter titles and chapter section headings match those of the syllabus; however, the long, final section on calculus is split into three chapters: Differentiation, Integration and Kinematics, so that it is easily manageable for students.

The content of every chapter is split into several short sections. Numerous worked examples are included to illustrate every aspect of the topic, as well as Exercises that provide ample opportunity to reinforce learning. Exercise questions increase in difficulty, from those that are very straightforward through to others that provide greater challenge.

There are also five summative Review exercises distributed throughout the book. These contain practice questions, including past paper questions, based on the topics in the preceding chapters. Each question is mapped to the chapters that students should have worked through before attempting the question. This means that students can complete the questions at the appropriate stage of the course even if they are not following the chapter order. Alternatively, they can be used at the end of the course for revision. Indicative marks are assigned to each question.

Prior knowledge

Throughout this book, it is assumed that readers are competent and fluent in the basic algebra that is covered in Cambridge IGCSE™ / O Level Mathematics:

- working with expressions and formulae, simplifying and collecting like terms
- substituting numbers into algebraic expressions
- linear and quadratic factorisation and the use of brackets
- solving simple, simultaneous and quadratic equations
- working with inequalities
- changing the subject of a formula
- plotting and sketching graphs.

The book opens with a Review chapter of 20 multiple choice questions providing readers with an opportunity to check that they are still familiar with these topics.

Assessment

For both Cambridge IGCSE™ and O Level Additional Mathematics you will take two examination papers, Paper 1 (Non-calculator) and Paper 2 (Calculator):

- 2 hours each
- 50% each.

The information in this section is taken from the Cambridge International syllabus. You should always refer to the appropriate syllabus document for the year of examination to confirm the details and for more information. The syllabus document is available on the Cambridge International website at www.cambridgeinternational.org

From the authors

We very much hope you enjoy this book. It introduces you to some of the exciting ideas of mathematics. These will broaden your understanding of the subject and prove really helpful when you go on to further study. They include topics such as identities, vectors and particularly calculus; all of these are covered in the later chapters of the book. In order to handle such topics confidently, you will need to be fluent in algebra and numerical work and be able to communicate the mathematics you are doing. The early chapters are designed to build on your previous experience in a way that develops these essential skills and at the same time expands the techniques you are able to use.

First edition authors
Val Hanrahan
Jeanette Powell
Roger Porkess

Second edition authors
Stephen Wrigley
Roger Porkess

INTRODUCTION

How to use this book

To make your study of IGCSE and O Level Additional Mathematics as rewarding and successful as possible, this endorsed textbook offers the following important features:

Approach

Each chapter is broken down into several sections, with each section covering a single topic. Topics are introduced through clear explanations, with key terms picked out in bold type.

The modulus function

The **modulus** of a number is its positive value even when the number itself is negative.

The modulus is denoted by a vertical line on each side of the number and is sometimes called the **magnitude** of the quantity.

Worked examples

The worked examples cover important techniques and question styles. They are designed to reinforce the explanations, and give you step-by-step help for solving problems.

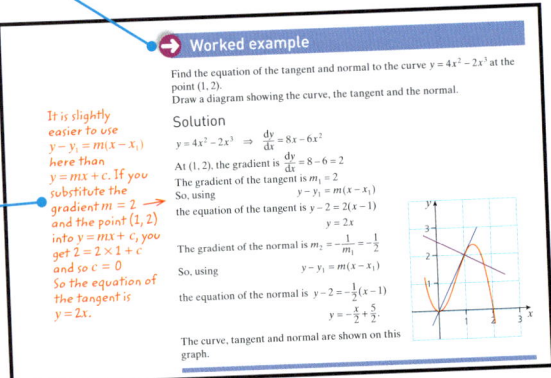

Commentaries

The commentaries provide additional explanations and encourage full understanding of mathematical principles.

Discussion points

These are points you should discuss in class with your teacher or fellow students, to encourage deeper exploration and mathematical communication.

Discussion point

Consider the example above where $y = 4x^2 - 2x^3$. At the point (1, 2), if the value of x increases by 0.001, what is the corresponding increase in y? What is the connection with the gradient at (1, 2)? What about at the points **i** (2, 0) and **ii** (0, 0)?

Note

The Note feature contains useful information; for example, on the differences between how calculators may display information. Explanations encourage full understanding of mathematical principles.

Note

The gradient at a particular point can be used to find the approximate change in y corresponding to a small change in x.

How to use this book

Exercises

These appear throughout the text and provide ample and varied opportunities to practice and apply what you've learned.

Exercise 1.2
1 Given that $f(x) = 3x + 2$, $g(x) = x^2$ and $h(x) = 2x$, find:
   a) $fg(2)$ b) $fg(x)$ c) $gh(x)$ d) $fgh(x)$
2 Given that $f(x) = \sqrt{2x+1}$ and $g(x) = 4 - x$, find:
   a) $fg(-4)$ b) $gf(12)$ c) $fg(x)$ d) $gf(x)$
3 Given that $f(x) = x + 4$, $g(x) = 2x^2$ and $h(x) = \frac{1}{2x+1}$, find:
   a) $f^2(x)$ b) $g^2(x)$ c) $h^2(x)$ d) $hgf(x)$
4 For each function, find the inverse and sketch the graphs of $y = f(x)$ and $y = f^{-1}(x)$ on the same axes. Use the same scale on both axes.
   a) $f(x) = 3x - 1$ b) $f(x) = x^3$, $x > 0$

Learning outcomes

Each chapter ends with a summary of the learning outcomes and a list of key points to confirm what you should have learned and understood.

Now you should be able to:
★ recognise the difference between permutations and combinations and know when each should be used
★ know and use the notation $n!$ and the expressions for permutations and combinations of n items taken r at a time
★ answer problems on arrangement and selection using permutations or combinations.

Key points

✔ The number of ways of arranging n different objects in a line is $n!$ This is read as n factorial.
✔ $n! = n \times (n-1) \times (n-2) \ldots \times 3 \times 2 \times 1$ where n is a positive integer.
✔ By convention, $0! = 1$.
✔ The number of permutations of r objects from n is $^nP_r = \frac{n!}{(n-r)!}$
✔ The number of combinations of r objects from n is $^nC_r = \frac{n!}{(n-r)!r!}$
✔ The order matters for permutations, but not for combinations.

Review exercises

After Chapters 3, 6, 10, 13 and 16, you will find Review exercises that cover the concepts learnt in previous chapters. You can work through these exercises during the course to summarise your learning or at the end of the course as revision. Next to each question are the chapters that you should have studied before attempting the question. Marks for each question are shown in square brackets.

 These icons highlight questions where a calculator should not be used.

Review exercise 3

Ch 7 1 Solutions to this question by accurate drawing will not be accepted.
The points $A(3, 2)$, $B(7, -4)$, $C(2, -3)$ and $D(k, 3)$ are such that CD is perpendicular to AB.
Find the equation of the perpendicular bisector of CD. [6]
Cambridge O Level Additional Mathematics (4037)
Paper 22 Q5, February/March 2019
Cambridge IGCSE Additional Mathematics (0606)
Paper 22 Q5, February/March 2019

Ch 7 2 It is thought that the relationship $y = ax^n$, where a and n are constants, connects the variables x and y. An experiment was carried out recording the values of y for certain values of x.
a) Transform the relationship $y = ax^n$ into straight line form. [2]
The values of $\ln x$ and $\ln y$ were plotted and a line of best fit was drawn. It is given that the line of best fit crosses through the points with coordinates (1.35, 4.81) and (5.55, 2.29).
b) Calculate the constants a and n. [4]

Ch2, 5, 7, 8 3 The diagram shows the circle $x^2 + y^2 - 4x + 4y - 17 = 0$ and the lines l_1, $y = x + 1$, and l_2. The line l_1 intersects the circle at points P and Q and the line l_2 intersects the circle at points $R(5, 2)$ and $S(7, -2)$. The lines intersect at point T.

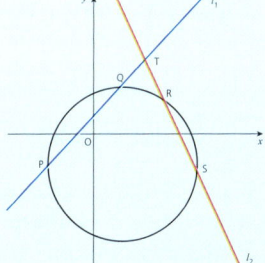

a) Find the coordinates of the point of intersection of l_1 and l_2. [5]
b) Give the coordinates of the points P and Q. [4]
c) Find the area of the triangle PST. [2]

Ch 8 4 Two circles with equations $x^2 + y^2 + 6x - 8y + 9 = 0$ and $x^2 + y^2 - 2x - 15 = 0$ intersect at points A and B.

vii

INTRODUCTION

Command words

Command words are used to tell you how to answer a specific question. The list below contains the command words for this syllabus. The definitions explain what the words are asking you to do.

Command word	What it means
Calculate	Work out from given facts, figures or information
Describe	State the points of a topic / give characteristics and main features
Determine	Establish with certainty
Explain	Set out purposes or reasons / make the relationships between things clear / say why and/or how and support with relevant evidence
Give	Produce an answer from a given source or recall/memory
Plot	Mark point(s) on a graph
Show (that)	Provide structured evidence that leads to a given result
Sketch	Make a simple freehand drawing showing the key features
State	Express in clear terms
Verify	Confirm a given statement/result is true
Work out	Calculate from given facts, figures or information with or without the use of a calculator
Write	Give an answer in a specific form
Write down	Give an answer without significant working

The information in this section is taken from the Cambridge International syllabus. You should always refer to the appropriate syllabus document for the year of examination to confirm the details and for more information. The syllabus document is available on the Cambridge International website at www.cambridgeinternational.org

Explore the book cover: how are pine cones mathematical?

The Fibonacci sequence begins 0, 1, 1, 2, 3, 5, 8, …. Each number in the sequence is equal to the sum of the preceding two numbers. Pine cones contain seed pods arranged in two spirals that twist in opposite directions. The number of steps in each spiral usually matches a pair of consecutive Fibonacci numbers. Research 'Fibonacci in pine cones' to discover more.

Review chapter

These questions are multiple choice.

1. Work out $5 \times 4^2 - 6 \div 2$.
   - A 397
   - B 197
   - C 77
   - D 37

2. Work out $\dfrac{22 - 6 + 9}{5 + 5 \times 7}$.
   - A $\dfrac{5}{14}$
   - B $\dfrac{5}{8}$
   - C $\dfrac{7}{8}$
   - D $\dfrac{7}{14}$

3. Evaluate $2\dfrac{5}{7} - 1\dfrac{11}{14}$, giving your answer as a fraction in its simplest form.
   - A $\dfrac{13}{14}$
   - B $1\dfrac{6}{7}$
   - C $\dfrac{9}{14}$
   - D $4\dfrac{1}{2}$

4. Evaluate $2\dfrac{2}{11} \div \dfrac{5}{8}$, giving your answer as a fraction in its simplest form.
   - A $1\dfrac{16}{55}$
   - B $1\dfrac{4}{11}$
   - C $1\dfrac{49}{88}$
   - D $3\dfrac{27}{55}$

5. Work out the exact value of $2\sqrt{6} \times 5\sqrt{10}$. Write your answer as simply as possible.
   - A $20\sqrt{15}$
   - B $\sqrt{600}$
   - C $7\sqrt{60}$
   - D $10\sqrt{16}$

6. Work out the exact value of $\sqrt{24} + 2\sqrt{6}$. Write your answer as simply as possible.
   - A $2\sqrt{30}$
   - B $2\sqrt{24}$
   - C $4\sqrt{6}$
   - D 6

7. Evaluate $3^5 \times 3^{-3}$, giving your answer in index form.
   - A 9^2
   - B 9^{-15}
   - C 3^{-15}
   - D 3^2

8. Simplify $(4x^6)^2$.
   - A $8x^8$
   - B $16x^8$
   - C $8x^{12}$
   - D $16x^{12}$

9. Work out $x^2 + y^2 - 4x - 6y$ when $x = 3$ and $y = -2$.
   - A 13
   - B 5
   - C -19
   - D 37

10. Expand and simplify $4(5p + 3) - 3(2 - 4p)$.
    - A $16p - 3$
    - B $8p + 6$
    - C $32p + 6$
    - D $24p - 3$

11. Expand and simplify $(x + 2)(x - 2)(x + 4)$.
    - A $x^3 - 16$
    - B $x^3 + 4x^2 - 4x - 16$
    - C $x^3 - 20x - 16$
    - D $x^3 - 4x - 16$

12. Fully factorise $15x^2y + 25xy$.
    - A $5xy(3x + 5)$
    - B $5(3x^2y + 5xy)$
    - C $xy(15x + 25)$
    - D $40x^3y^2$

REVIEW CHAPTER

13 Solve $2(10 - 4x) = 8(1 + x)$.

  A $x = -\frac{3}{4}$

  B $x = \frac{3}{4}$

  C $x = \frac{4}{3}$

  D $x = \frac{7}{4}$

14 Solve the inequality $4(2x - 3) < 3(1 + 2x)$.

  A $x < \frac{15}{14}$

  B $x > -\frac{15}{2}$

  C $x < \frac{1}{6}$

  D $x < \frac{15}{2}$

15 Rearrange $\frac{1}{2}ax^2 - b = c$ to make x the subject.

  A $x = \frac{\pm\sqrt{2c+b}}{a}$

  B $x = \pm\sqrt{\frac{2(c+b)}{a}}$

  C $x = \frac{\pm\sqrt{2(c+b)}}{a}$

  D $x = \pm\sqrt{\frac{2c+b}{a}}$

16 Rearrange $y = \frac{5x}{x-5}$ to make x the subject.

  A $x = \frac{y(x-5)}{5}$

  B $x = \frac{y-5}{5y}$

  C $x = 5y(y-5)$

  D $x = \frac{5y}{y-5}$

17 Work out $\frac{2x+5}{4x} \div \frac{4x^2+10x}{2}$. Give your answer as a fraction in its simplest form.

  A $4x^2$

  B $\frac{2(2x+5)}{4x(4x^2+10x)}$

  C $\frac{1}{4x^2}$

  D $\frac{7}{8x^2+20x}$

18 Use Pythagoras' Theorem to calculate the length of the side marked x.

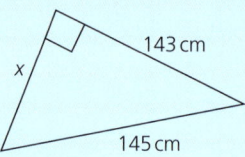

  A 204 cm
  B 576 cm
  C 143 cm
  D 24 cm

19 An isosceles triangle, shown below, has a base of length 10 cm and an area of 60 cm². Work out the perimeter of the triangle.

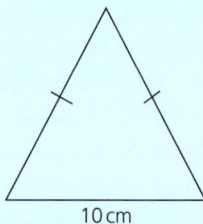

  A 36 cm
  B 34 cm
  C 30 cm
  D 22 cm

20 The bearing of Zurich from London is 126°. Find the bearing of London from Zurich.

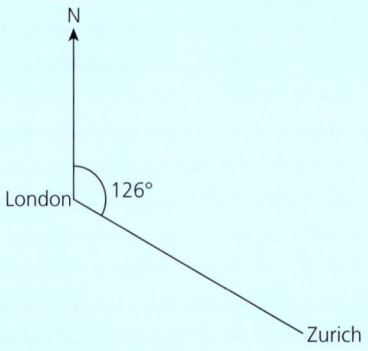

  A 54°
  B 126°
  C 234°
  D 306°

1 Functions

If A equals success, then the formula is A equals X plus Y plus Z, with X being work, Y play, and Z keeping your mouth shut.

Albert Einstein (1879–1955)

Discussion point

Look at the display on this fuel pump. One of the quantities is measured and one is calculated from it. Which is which?

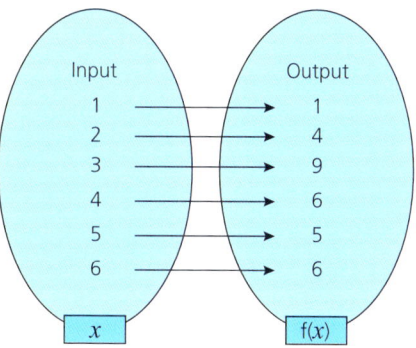

Discussion point

Which digits will never appear in the output set of the previous example?

1 FUNCTIONS

A **function** is a rule that associates each element of one set (the **input**) with only one element of a second set (the **output**). It is possible for more than one input to have the same output, as shown above.

You can use a **flow chart** (or **number machine**) to express a function.

This flow chart shows a function, f, with two operations. The first operation is ×2 and the second operation is +3.

Input ⟶ ×2 ⟶ +3 ⟶ Output

You can write the equation of a line in the form $y = 2x + 3$ using **function notation**.

$$f(x) = 2x + 3$$

or $f: x \mapsto 2x + 3$

Read this as 'f of x equals two x plus three'.

Read this as 'f maps x onto two x plus three'.

Using this notation, you can write, for example:

$$f(4) = 2 \times 4 + 3 = 11$$

or $f: (-5) \mapsto 2 \times (-5) + 3 = -7$

The domain and range

The **domain** of a function $f(x)$ is the set of all possible inputs. This is the set of values of x that the function operates on. In the first mapping diagram of the next worked example, the domain is the first five positive odd numbers. If no domain is given, it is assumed to be all real values of x. This is often denoted by the letter $\mathbb{R}$.

Real numbers are all of the rational and irrational numbers.

The **range** of the function $f(x)$ is all the possible output values, i.e. the corresponding values of $f(x)$. It is sometimes called the **image set** and is controlled by the domain.

In certain functions one or more values must be excluded from the domain, as shown in the following example.

➡ Worked example

For the function $f(x) = \dfrac{1}{2x+1}$:

a) Draw a mapping diagram showing the outputs for the set of inputs odd numbers from 1 to 9 inclusive.

b) Draw a mapping diagram showing the outputs for the set of inputs even numbers from 2 to 10 inclusive.

c) Which number cannot be an input for this function?

Solution

a)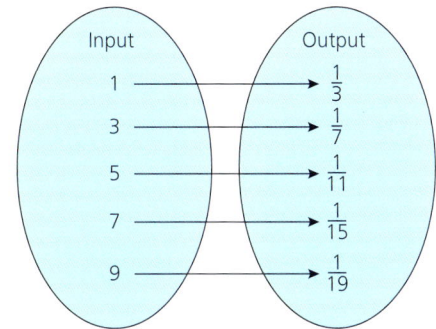

b)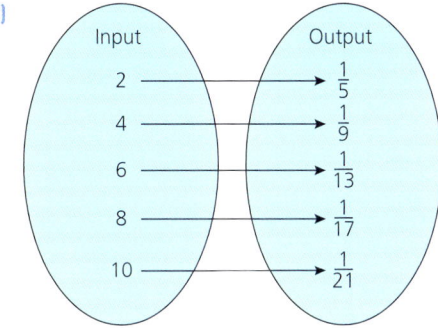

c) A fraction cannot have a denominator of 0, so $2x + 1 \neq 0$
$\Rightarrow x = -\frac{1}{2}$ must be excluded.

Mappings

A mapping is the process of going from an object to its image.

For example, this mapping diagram shows the function $f(x) = x^2 + 1$ when the domain is the set of integers $-2 \leq x \leq 2$.

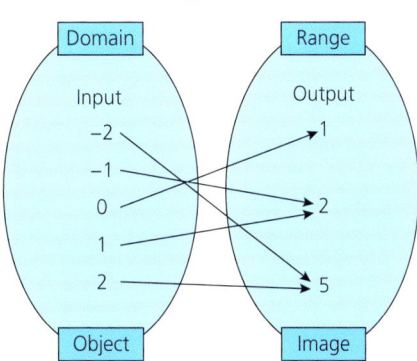

A mapping diagram is one way to illustrate a function.

There are four different types of mappings.

1 FUNCTIONS

One-one

Every object has a unique image and every image comes from only one object.

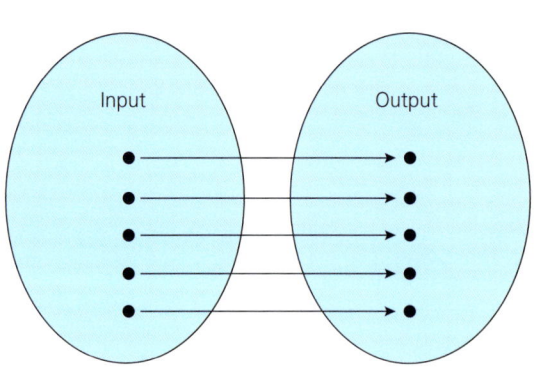

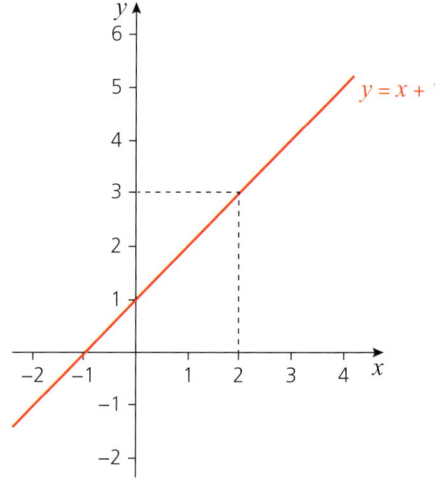

Many-one

Every object has a unique image but at least one image corresponds to more than one object.

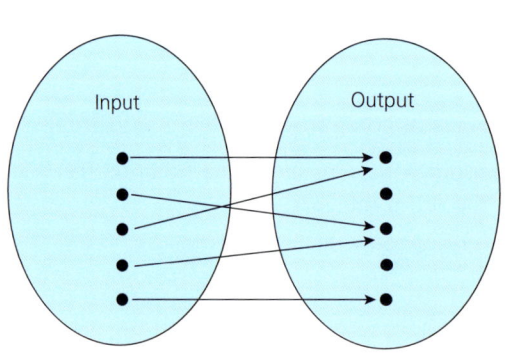

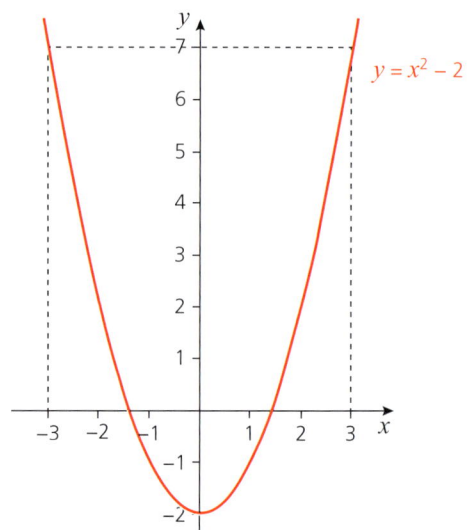

Functions

One-many
There is at least one object that has more than one image but every image comes from only one object.

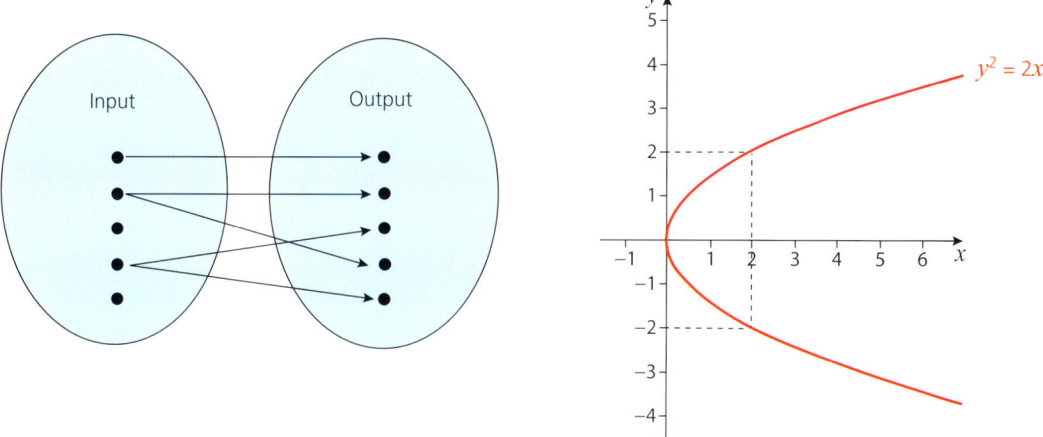

Many-many
There is at least one object that has more than one image and at least one image that corresponds to more than one object.

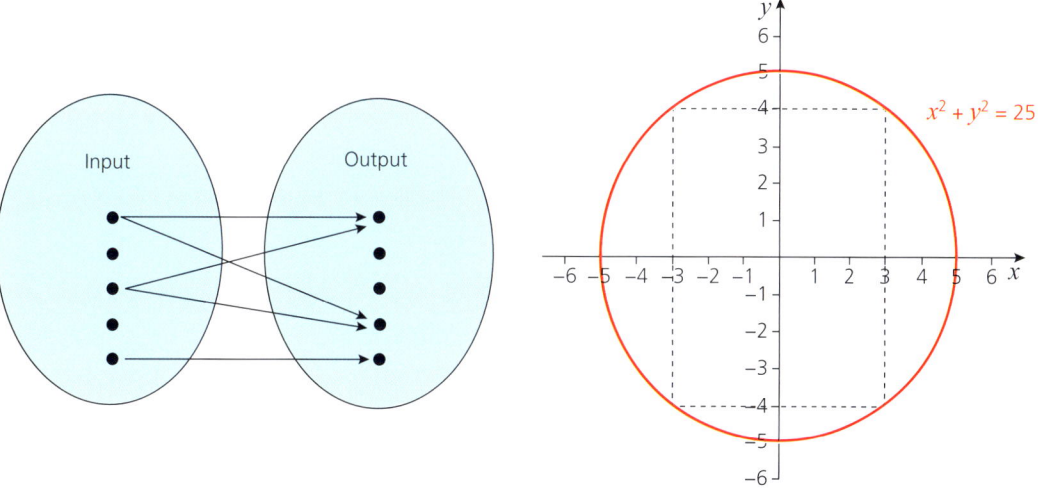

1 FUNCTIONS

Types of function

A function is a mapping that is either one-one or many-one.

For a one-one function, the graph of y against x doesn't 'double back' on itself.

Below are some examples of one-one functions.
- All straight lines that are not parallel to either axis.
- Functions of the form $y = x^{2n+1}$ for integer values of n.
- Functions of the form $y = a^x$ for $a > 0$.
- $y = \cos x$ for $0° \leq x \leq 180°$.

These are examples of many-one functions:
- all quadratic curves,
- cubic equations with two turning points.

Worked example

Sketch each function and state whether it is one-one or many-one.

a) $y = x + 3$ b) $y = x^2 - 1$

Solution

a) $y = x + 3$ is a straight line.

When $x = 0$, $y = 3$, so the point $(0, 3)$ is on the line.

When $y = 0$, $x = -3$, so the point $(-3, 0)$ is on the line.

$y = x + 3$ is a one-one function.

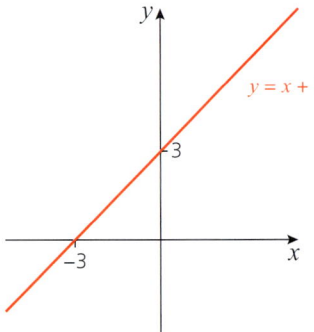

b) $y = x^2$ is a ∪-shaped curve through the origin.

$y = x^2 - 1$ is the same shape, but has been moved down one unit so crosses the y-axis at $(0, -1)$.

$y = x^2 - 1$ factorises to $y = (x + 1)(x - 1)$

$\Rightarrow$ When $y = 0$, $x = 1$ or $x = -1$.

$y = x^2 - 1$ is a many-one function since, for example, $y = 0$ corresponds to both $x = 1$ and $x = -1$.

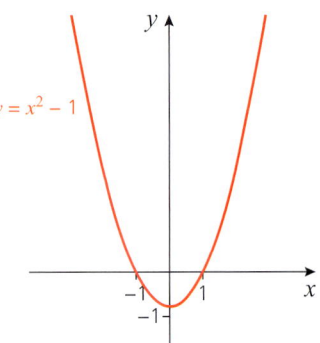

Inverse function

The inverse function reverses the effect of the function. For example, if the function says 'double', the inverse says 'halve'; if the function says 'add 2', the inverse says 'subtract 2'. All one-one functions have an inverse; many-one functions do not.

➡ Worked example

a) Use a flow chart to find the inverse of the function $f(x) = \dfrac{3x+2}{2}$.

b) Sketch the graphs of $y = f(x)$ and $y = f^{-1}(x)$ on the same axes. Use the same scale on both axes.

c) What do you notice?

Solution

a) For $f(x) = \dfrac{3x+2}{2}$:

Input $\rightarrow$ $\boxed{\times 3}$ $\rightarrow$ $\boxed{+2}$ $\rightarrow$ $\boxed{\div 2}$ $\rightarrow$ Output

x $\rightarrow$ $3x$ $\rightarrow$ $3x+2$ $\rightarrow$ $\dfrac{3x+2}{2}$ $\rightarrow$ $f(x)$

Reversing these operations gives the inverse function.

Output $\leftarrow$ $\boxed{\div 3}$ $\leftarrow$ $\boxed{-2}$ $\leftarrow$ $\boxed{\times 2}$ $\leftarrow$ Input

$f^{-1}(x)$ $\leftarrow$ $\dfrac{2x-2}{3}$ $\leftarrow$ $2x-2$ $\leftarrow$ $2x$ $\leftarrow$ x

b)

Reflecting in the line $y = x$ has the effect of switching the x- and y-coordinates.

c) The graphs of $y = f(x)$ and $y = f^{-1}(x)$ are reflections of each other in the line $y = x$.

1 FUNCTIONS

An alternative method is to interchange the coordinates, since this gives a reflection in the line $y = x$, and then use an algebraic method to find the inverse as shown in the next example.

➡ Worked example

a) Find $g^{-1}(x)$ when $g(x) = \frac{x}{3} + 4$.

b) Sketch $y = g(x)$ and $y = g^{-1}(x)$ on the same axes. Use the same scale on both axes.

Solution

a) Let $y = \frac{x}{3} + 4$.

Interchange x and y. $\quad x = \frac{y}{3} + 4$

Rearrange to make y the subject. $\quad x - 4 = \frac{y}{3}$

$\Rightarrow y = 3(x - 4)$

Rearranging and interchanging x and y can be done in either order.

The inverse function is given by $g^{-1}(x) = 3(x - 4)$.

b)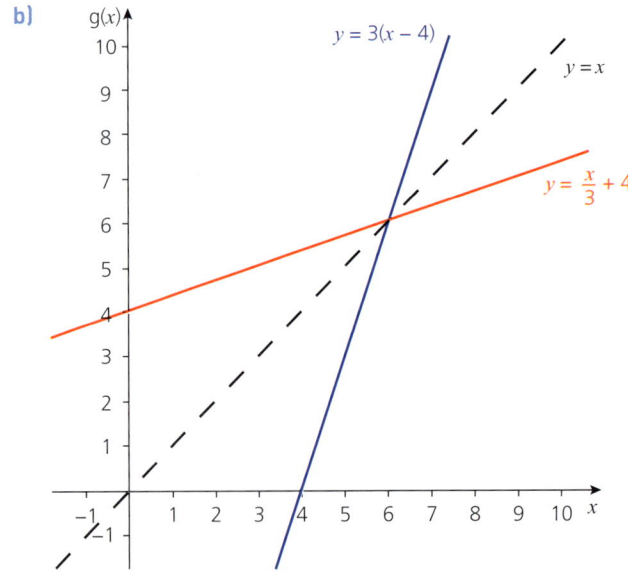

➡ Worked example

a) Sketch the graph of the function $f(x) = x^2$ for $-4 \leq x \leq 4$.

b) Explain, using an example, why $f(x)$ does not have an inverse with $-4 \leq x \leq 4$ as its domain.

c) Suggest a suitable domain for $f(x)$ so that an inverse can be found.

Inverse function

Solution

a)

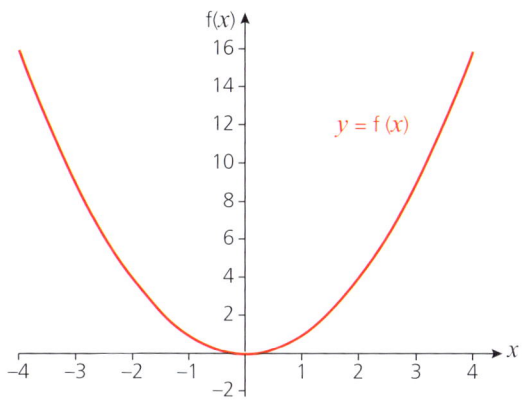

f(x) is a many-one function so it does not have an inverse.

b) The function does not have an inverse with $-4 \leq x \leq 4$ as its domain because, for example, f(2) and f(−2) both equal 4. This means that if the function were reversed, there would be no unique value for 4 to return to. In other words, $f(x) = x^2$ is not a one-one function for $-4 \leq x \leq 4$.

c) Any domain in which the function is one-one, for example, $0 \leq x \leq 4$.

> **Note**
> - The domain of f(x) is the same as the range of $f^{-1}(x)$.
> - The range of f(x) is the same as the domain of $f^{-1}(x)$.

→ Worked example

a) State the range of the function $f(x) = \sqrt{x-1}$ for $x \geq 1$.

b) State the domain and range of the inverse function $f^{-1}(x)$.

Solution

a)

You may find it useful to draw a sketch.

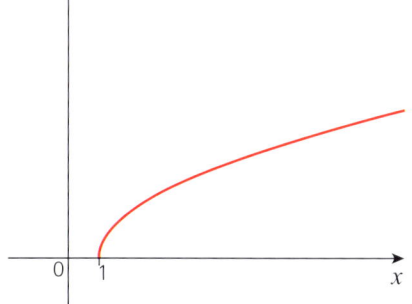

So, the range of f(x) is $f(x) \geq 0$

11

1 FUNCTIONS

b) The range of f(x) is f(x) ⩾ 0 ⇒ the domain of f⁻¹(x) is x ⩾ 0
The domain of f(x) is x ⩾ 1 ⇒ the range of f⁻¹(x) is f⁻¹(x) ⩾ 1

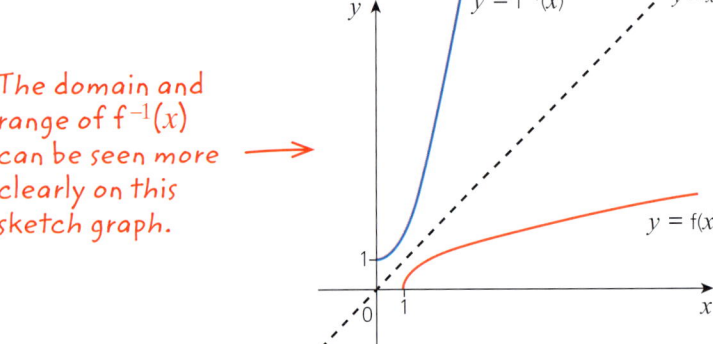

The domain and range of f⁻¹(x) can be seen more clearly on this sketch graph.

Exercise 1.1

1 For the function f(x) = 3x + 4, find:
  a) f(3) b) f(−2) c) f(0) d) f($\frac{1}{2}$)

2 For the function g(x) = (x + 2)², find:
  a) g(4) b) g(−6) c) g(0) d) g($\frac{1}{2}$)

3 For the function h: x → 3x² + 1, find:
  a) h(2) b) h(−3) c) h(0) d) h($\frac{1}{3}$)

4 For the function f: x → $\frac{2x+6}{3}$, find:
  a) f(3) b) f(−6) c) f(0) d) f($\frac{1}{4}$)

√2x+1 is the notation for the positive square root of 2x+1

5 For the function f(x) = $\sqrt{2x+1}$:
  a) Draw a mapping diagram to show the outputs when the set of inputs is the odd numbers from 1 to 9 inclusive.
  b) Draw a mapping diagram to show the outputs when the set of inputs is the even numbers from 2 to 10 inclusive.
  c) Which number must be excluded as an input?

6 Find the range of each function:
  a) f(x) = 3x − 2; domain {1, 2, 3, 4, 5}
  b) g(x) = $\frac{x-4}{2}$; domain {−2, −1, 0, 1, 2}
  c) h(x) = 2x²; domain x∈ℝ
  d) f: x → x² + 6; domain x∈ℝ

7 Which value(s) must be excluded from the domain of these functions?
  a) f(x) = $\frac{1}{x}$
  b) f(x) = $\sqrt{x-1}$
  c) f(x) = $\frac{3}{2x-3}$
  d) f(x) = $\sqrt{2-x^2}$

8 Find the inverse of each function:
  a) f(x) = 7x − 2
  b) g(x) = $\frac{3x+4}{2}$
  c) h(x) = (x − 1)² for x ⩾ 1
  d) f(x) = x² + 4 for x ⩾ 0

9 a) Find the inverse of the function f(x) = 3x − 4.
  b) Sketch f(x), f⁻¹(x) and the line y = x on the same axes. Use the same scale on both axes.

Exercise 1.1 (cont)

Plot: Start with a table of values.
Sketch: Show the main features of the curve.

10 a) Plot the graph of the function $f(x) = 4 - x^2$ for values of x such that $0 \leq x \leq 3$. Use the same scale on both axes.
b) Find the values of $f^{-1}(-5)$, $f^{-1}(0)$, $f^{-1}(3)$ and $f^{-1}(4)$.
c) Sketch $y = f(x)$, $y = f^{-1}(x)$ and $y = x$ on the same axes. Use the range -6 to $+6$ for both axes.
d) Find the domain and range of $f^{-1}(x)$.

Composition of functions

When two functions are used one after the other, the single equivalent function is called the **composite function**.

For example, if $f(x) = 3x + 2$ and $g(x) = 2x - 3$, then the composite function $gf(x)$ is obtained by applying f first and then applying g to the result.

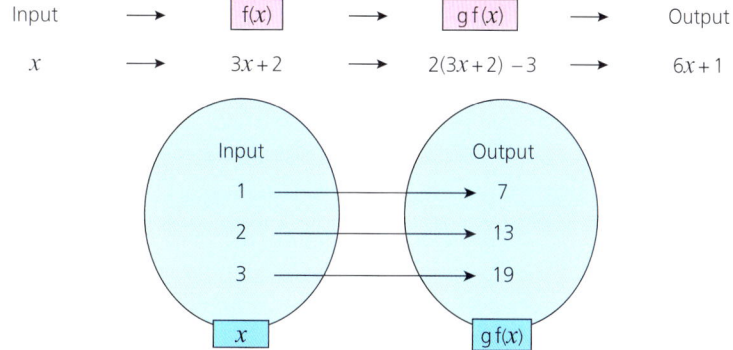

Think of two functions $f(x)$ and $g(x)$ such that the combined function $gf(x)$ exists. The function $f(x)$ is applied first and so the domain of the combined function $gf(x)$ must be contained in the domain of $f(x)$. The domain of $gf(x)$ cannot include any stray elements that $f(x)$ cannot act on because they are outside its domain. So the domain of $gf(x)$ is either the same as that of $f(x)$ or a subset of it. This is written **domain of $gf(x) \subseteq$ domain of $f(x)$**.

Similarly, the function $g(x)$ is applied second and so any element in the outcome from the combined function $gf(x)$ must be an element that is a possible outcome from the function $g(x)$. This is written **range of $gf(x) \subseteq$ range of $g(x)$**.

$f^2(x)$ is the same as $f(f(x))$ and means that you apply the same function twice.

The order in which these operations are applied is important, as shown below.

➡ Worked example

Given that $f(x) = 2x$, $g(x) = x^2$ and $h(x) = \frac{1}{x}$, find:
a) $fg(x)$
b) $gf(x)$
c) $h^2(x)$
d) $fgh(x)$
e) $hgf(x)$

Solution

a) $fg(x) = f(x^2)$
$= 2x^2$

b) $gf(x) = g(2x)$
$= (2x)^2$
$= 4x^2$

c) $h^2(x) = h[h(x)]$
$= h\left(\frac{1}{x}\right)$
$= 1 \div \frac{1}{x}$
$= x$

d) $fgh(x) = fg\left(\frac{1}{x}\right)$
$= f\left[\left(\frac{1}{x}\right)^2\right]$
$= f\left(\frac{1}{x^2}\right)$
$= \frac{2}{x^2}$

e) $hgf(x) = hg(2x)$
$= h((2x)^2)$
$= h(4x^2)$
$= \frac{1}{4x^2}$

➡ Worked example

a) Find $f^{-1}(x)$ when $f(x) = \frac{2x-1}{4}$.

b) Find $f[f^{-1}(x)]$.

c) Find $f^{-1}[f(x)]$.

d) What do you notice?

Solution

a) Write $f(x)$ as $y = \frac{2x-1}{4}$
Interchange x and y. $\quad x = \frac{2y-1}{4}$
$\Rightarrow 4x = 2y - 1$
$\Rightarrow 2y = 4x + 1$
$\Rightarrow y = \frac{4x+1}{2}$
$\Rightarrow f^{-1}(x) = \frac{4x+1}{2}$

b) $f[f^{-1}(x)] = f\left[\frac{4x+1}{2}\right]$
$= \frac{2\left(\frac{4x+1}{2}\right) - 1}{4}$
$= \frac{(4x+1) - 1}{4}$
$= \frac{4x}{4}$
$= x$

c) $f^{-1}[f(x)] = f^{-1}\left(\frac{2x-1}{4}\right)$
$= \frac{4\left(\frac{2x-1}{4}\right) + 1}{2}$
$= \frac{(2x-1) + 1}{2}$
$= \frac{2x}{2}$
$= x$

d) Questions **a** and **b** show that $f[f^{-1}(x)] = f^{-1}[f(x)] = x$.

This result is true for all functions that have an inverse.

The examples above show that applying a function and its inverse in either order leaves the original quantity unchanged, which is what the notation $f(f^{-1})$ or $f^{-1}(f)$ implies.

➡ Worked example

Using the functions $f(x) = \sin x$ and $g(x) = x^2$, express the following as functions of x:

a) $fg(x)$ b) $gf(x)$ c) $f^2(x)$

Solution

a) $fg(x) = f[g(x)]$
   $= \sin(x^2)$

b) $gf(x) = g[f(x)]$
   $= (\sin x)^2$

c) $f^2(x) = f[f(x)]$
   $= \sin(\sin x)$

Notice that $\sin(x^2)$ is not the same as $(\sin x)^2$ or $\sin(\sin x)$.

The modulus function

The **modulus** of a number is its positive value even when the number itself is negative.

The modulus is denoted by a vertical line on each side of the number and is sometimes called the **magnitude** of the quantity.

For example, $|28| = 28$ and $|-28| = 28$

$|x| = x$ when $x \geq 0$ and $|x| = -x$ when $x < 0$

Therefore for the graph of the modulus function $y = |f(x)|$, any part of the corresponding graph of $y = f(x)$ where $y < 0$, is reflected in the x-axis.

➡ Worked example

For each of the following, sketch $y = f(x)$ and $y = |f(x)|$ on separate axes:

a) $y = x - 2$; $-2 \leq x \leq 6$

b) $y = x^2 - 2$; $-3 \leq x \leq 3$

c) $y = \cos x$; $0° \leq x \leq 180°$

1 FUNCTIONS

Solution

a)

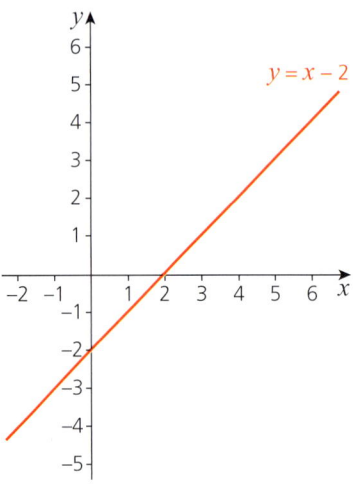

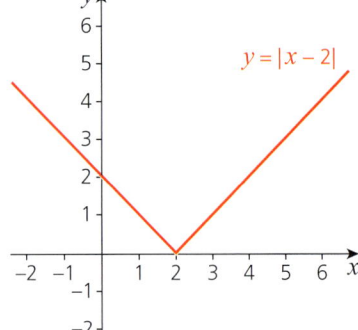

b)

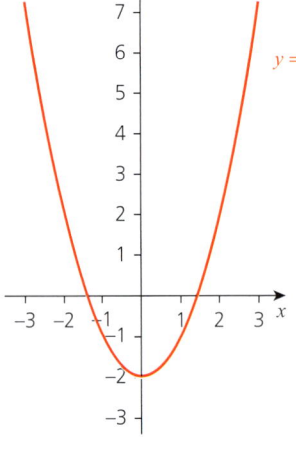

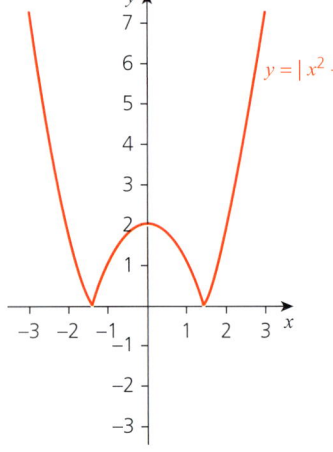

Notice the sharp change of gradient from negative to positive, where part of the graph is reflected. This point is called a 'cusp'.

c)

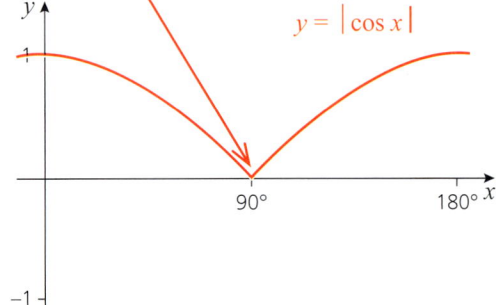

The modulus function

Exercise 1.2

1. Given that $f(x) = 3x + 2$, $g(x) = x^2$ and $h(x) = 2x$, find:
   a) $fg(2)$
   b) $fg(x)$
   c) $gh(x)$
   d) $fgh(x)$

2. Given that $f(x) = \sqrt{2x+1}$ and $g(x) = 4 - x$, find:
   a) $fg(-4)$
   b) $gf(12)$
   c) $fg(x)$
   d) $gf(x)$

3. Given that $f(x) = x + 4$, $g(x) = 2x^2$ and $h(x) = \dfrac{1}{2x+1}$, find:
   a) $f^2(x)$
   b) $g^2(x)$
   c) $h^2(x)$
   d) $hgf(x)$

4. For each function, find the inverse and sketch the graphs of $y = f(x)$ and $y = f^{-1}(x)$ on the same axes. Use the same scale on both axes.
   a) $f(x) = 3x - 1$
   b) $f(x) = x^3, x > 0$

5. Solve the following equations:
   a) $|x - 3| = 4$
   b) $|2x + 1| = 7$
   c) $|3x - 2| = 5$
   d) $|x + 2| = 2$

6. Sketch the graph of each function:
   a) $y = x + 2$
   b) $y = |x + 2|$
   c) $y = |x + 2| + 3$

7. Sketch these graphs for $0° \leq x \leq 360°$:
   a) $y = \cos x$
   b) $y = \cos x + 1$
   c) $y = |\cos x|$
   d) $y = |\cos x| + 1$

8. Graph 1 represents the line $y = 2x - 1$. Graph 2 is related to graph 1 and graph 3 is related to graph 2.
   Write down the equations of graph 2 and graph 3.

Note

In Chapter 10, you will look again at the relationship between $y = f(x)$ and $y = |f(x)|$, where $f(x)$ is trigonometric.

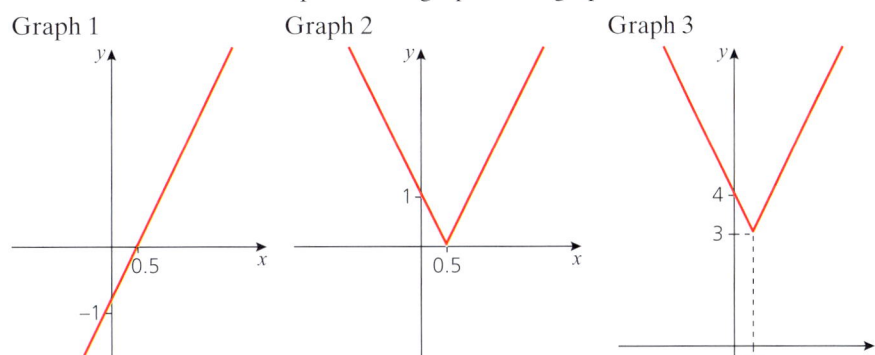

Graph 1   Graph 2   Graph 3

9. The graph shows part of a quadratic curve and its inverse.
   a) What is the equation of the curve?
   b) What is the equation of the inverse?

10. a) Sketch the graphs of these functions:
    i) $y = 1 - 2x$
    ii) $y = |1 - 2x|$
    iii) $y = -|1 - 2x|$
    iv) $y = 3 - |1 - 2x|$
    b) Use a series of transformations to sketch the graph of $y = |3x + 1| - 2$.

11. For each part:
    a) Sketch both graphs on the same axes.

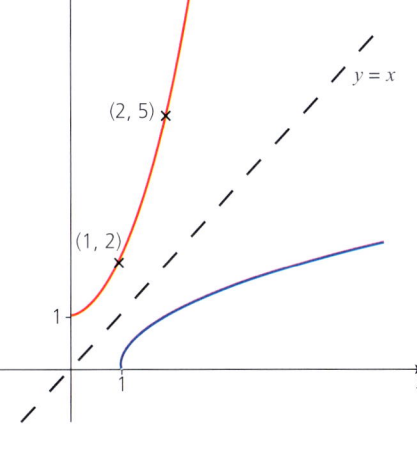

1 FUNCTIONS

b) Write down the coordinates of their points of intersection.
i) $y = |x|$ and $y = 1 - |x|$
ii) $y = 2|x|$ and $y = 2 - |x|$
iii) $y = 3|x|$ and $y = 3 - |x|$

Past-paper questions

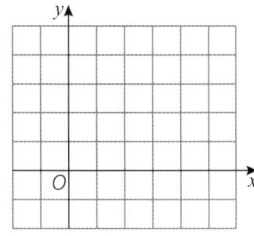

1 The functions f and g are defined by

$$f(x) = \frac{2x}{x+1} \text{ for } x > 0,$$

$$g(x) = \sqrt{x+1} \text{ for } x > -1$$

(i) Find fg(8). [2]
(ii) Find an expression for $f^2(x)$, giving your answer in the form $\frac{ax}{bx+c}$, where a, b and c are integers to be found. [3]
(iii) Find an expression for $g^{-1}(x)$, stating its domain and range. [4]
(iv) On axes like the ones shown, sketch the graphs of $y = g(x)$ and $y = g^{-1}(x)$, indicating the geometrical relationship between the graphs. [3]

Cambridge O Level Additional Mathematics (4037)
Paper 21 Q12, June 2014
Cambridge IGCSE Additional Mathematics (0606)
Paper 21 Q12, June 2014

2 (i) Sketch the graph of $y = |3x - 5|$, for $-2 \leq x \leq 3$, showing the coordinates of the points where the graph meets the axes. [3]
(ii) On the same diagram, sketch the graph of $y = 8x$. [1]
(iii) Solve the equation $8x = |3x - 5|$. [3]

Cambridge O Level Additional Mathematics (4037)
Paper 13 Q7, November 2010
Cambridge IGCSE Additional Mathematics (0606)
Paper 13 Q7, November 2010

Now you should be able to:

★ understand the terms: function, domain, range (image set), one-one function, many-one function, inverse function, and composition of functions
★ find the domain and range of functions
★ recognise and use function notations
★ understand the relationship between $y = f(x)$ and $y = |f(x)|$, where $f(x)$ may be linear, quadratic, cubic or trigonometric
★ explain in words why a given function does not have an inverse
★ find the inverse of a one-one function
★ form and use composite functions
★ use sketch graphs to show the relationship between a function and its inverse.

The modulus function

Key points

- A **mapping** is a rule for changing one number into another number or numbers.
- A **function**, f(x), is a rule that maps one number onto another single number.
- The **graph of a function** has only one value of y for each value of x. However, two or more values of x may give the same value of y.
- A **flow chart** can be used to show the individual operations within a function in the order in which they are applied.
- The **domain** of a function is the set of **input values**, or **objects**, that the function is operating on.
- The **range** or **image set** of a function is the corresponding set of **output values** or **images**, f(x).
- A **mapping diagram** can be used to illustrate a function. It is best used when the domain contains only a small number of values.
- In a **one-one function** there is a unique value of y for every value of x and a unique value of x for every value of y.
- In a **many-one** function two or more values of x correspond to the same value of y.
- In a **one-many** function one value of x corresponds to two or more values of y.
- In a **many-many** function two or more values of x correspond to the same value of y and two or more values of y correspond to the same value of x.
- The **inverse** of a function reverses the effect of the function. Only one-one functions have inverses.
- The term **composition of functions** is used to describe the application of one function followed by another function(s). The notation fg(x) means that the function g is applied first, then f is applied to the result.
- The **modulus** of a number or a function is always a positive value. $|x| = x$ if $x \geq 0$ and $|x| = -x$ if $x < 0$.
- The modulus of a function $y = f(x)$ is denoted by $|f(x)|$ and is illustrated by reflecting any part of the graph where $y < 0$ in the x-axis.

2 Quadratic functions

Algebra is but written geometry, and geometry is but figured algebra.

Sophia Germain (1776–1831)

Early mathematics focused principally on arithmetic and geometry. However, in the sixteenth century a French mathematician, François Viète, started work on 'new algebra'. He was a lawyer by trade and served as a privy councillor to both Henry III and Henry IV of France. His innovative use of letters and parameters in equations was an important step towards modern algebra.

François Viète (1540–1603)

Discussion point

Viète presented methods of solving equations of second, third and fourth degrees and discovered the connection between the positive roots of an equation and the coefficients of different powers of the unknown quantity. Another of Viète's remarkable achievements was to prove that claims that a circle could be squared, an angle trisected and the cube doubled were untrue. He achieved all this, and much more, using only a ruler and compasses, without the use of either tables or a calculator! In order to appreciate the challenges Viète faced, try to solve the quadratic equation $2x^2 - 8x + 5 = 0$ without using a calculator. Give your answers correct to two decimal places.

This chapter is about quadratic functions and covers a number of related themes.

The graph below illustrates these themes:

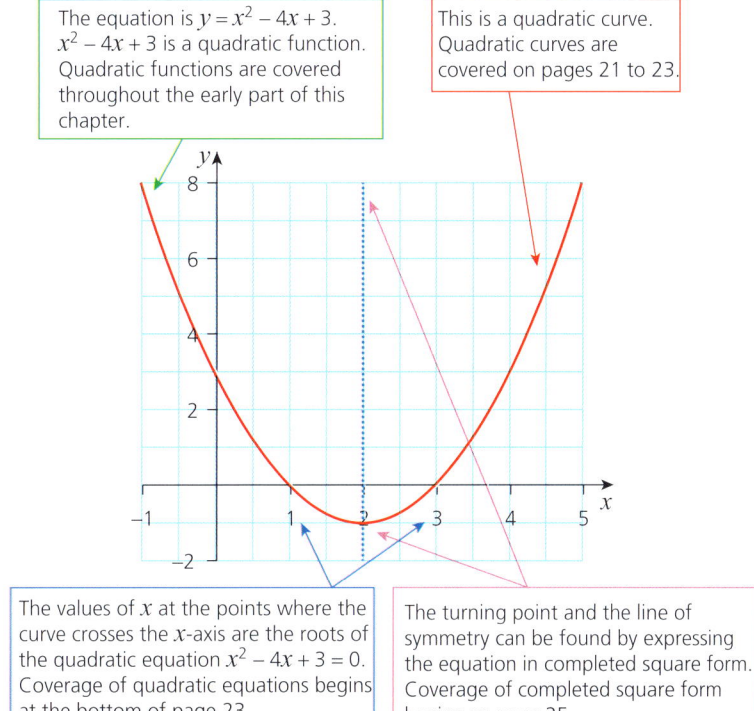

The equation is $y = x^2 - 4x + 3$. $x^2 - 4x + 3$ is a quadratic function. Quadratic functions are covered throughout the early part of this chapter.

This is a quadratic curve. Quadratic curves are covered on pages 21 to 23.

The values of x at the points where the curve crosses the x-axis are the roots of the quadratic equation $x^2 - 4x + 3 = 0$. Coverage of quadratic equations begins at the bottom of page 23.

The turning point and the line of symmetry can be found by expressing the equation in completed square form. Coverage of completed square form begins on page 25.

Maximum and minimum values

A **polynomial** is an expression in which, with the exception of a constant, the terms are positive integer powers of a variable. The highest power is the **order** of the polynomial.

A **quadratic function** or expression is a polynomial of order 2. $x^2 + 3$, a^2 and $2y^2 - 3y + 5$ are all quadratic expressions. Each expression contains only one variable (letter), and the highest power of that variable is 2.

The graph of a quadratic function is either ∪-shaped or ∩-shaped. Think about the expression $x^2 + 3x + 2$. When the value of x is very large and positive, or very large and negative, the x^2 term dominates the expression, resulting in large positive values. Therefore the graph of the function is ∪-shaped.

Similarly, the $-2x^2$ term dominates the expression $5 - 4x - 2x^2$ for both large positive and large negative values of x giving negative values of the expression for both. Therefore the graph of this function is ∩-shaped.

Although many of the quadratic equations that you will meet will have three terms, you will also meet quadratic equations with only two, or even one term. These fall into two main categories.

1. Equations with no constant term, for example, $2x^2 - 5x = 0$.

   This has x as a common factor so factorises to $x(2x - 5) = 0$
   $\Rightarrow x = 0$ or $2x - 5 = 0$
   $\Rightarrow x = 0$ or $x = 2.5$

2. Equations with no 'middle' term, which come into two categories:
   i) The sign of the constant term is negative, for example, $a^2 - 9 = 0$ and $2a^2 - 7 = 0$.

   $a^2 - 9 = 0$ factorises to $(a + 3)(a - 3) = 0$
   $\qquad\qquad\qquad\quad \Rightarrow a = -3$ or $a = 3$
   $2a^2 - 7 = 0 \Rightarrow a^2 = 3.5$
   $\qquad\qquad\quad \Rightarrow a = \pm\sqrt{3.5}$

   ii) The sign of the constant term is positive, for example, $p^2 + 4 = 0$.
   $p^2 + 4 = 0 \Rightarrow p^2 = -4$, so there is no real-valued solution.

> **Note**
>
> Depending on the calculator you are using, $\sqrt{(-4)}$ may be displayed as 'Math error' or '2i', where i is used to denote $\sqrt{(-1)}$. This is a **complex number** or **imaginary number** which you will meet if you study Further Mathematics at Advanced Level.

The vertical line of symmetry

Graphs of all quadratic functions have a vertical line of symmetry. You can use this to find the maximum or minimum value of the function. If the graph crosses the horizontal axis, then the line of symmetry is halfway between the two points of intersection. The maximum or minimum value lies on this line of symmetry.

Maximum and minimum values

Worked example

a) Plot the graph of $y = x^2 - 4x - 5$ for values of x from -2 to $+6$.

b) Identify the values of x where the curve intersects the horizontal axis.

c) Hence find the coordinates of the maximum or minimum point.

Solution

First create a table of values for $-2 \leq x \leq 6$.

a)
x	−2	−1	0	1	2	3	4	5	6
y	7	0	−5	−8	−9	−8	−5	0	7

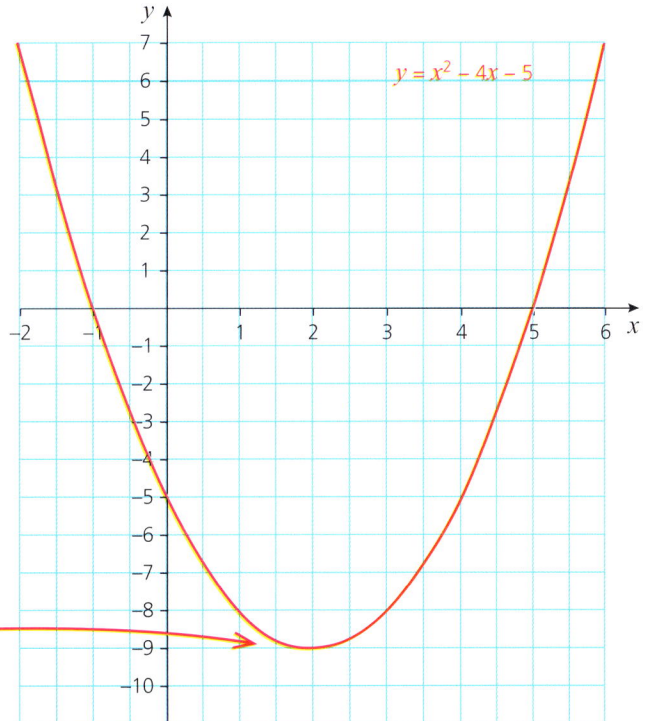

*This point is often referred to as the **turning point** of the curve.*

This is also shown in the table.

*The line $x = 2$ passes through the turning point. It is a **vertical line of symmetry** for the curve.*

b) The graph intersects the horizontal axis when $x = -1$ and when $x = 5$.

c) The graph shows that the curve has a minimum turning point halfway between $x = -1$ and $x = 5$. The table shows that the coordinates of this point are $(2, -9)$.

Factorising

Drawing graphs by hand to find maximum or minimum values can be time-consuming. The following example shows you how to use algebra to find these values.

2 QUADRATIC FUNCTIONS

→ Worked example

Find the coordinates of the turning point of the curve $y = x^2 + x - 6$. State whether the turning point is a maximum or a minimum value.

Solution

Find two integers (whole numbers) that multiply together to give the **constant** term, -6.

Possible pairs of numbers with a product of -6 are: 6 and -1, 1 and -6, 3 and -2, 2 and -3.

Identify any of the pairs of numbers that can be added together to give the coefficient of x (1). 3 and -2 are the only pair with a sum of 1, so use this pair to split up the x term.

$$x^2 + x - 6 = x^2 + 3x - 2x - 6$$
$$= x(x + 3) - 2(x + 3)$$
$$= (x + 3)(x - 2)$$

The first step is to factorise the expression. One method of factorising is shown, but if you are confident using a different method then continue to use it.

Both expressions in the brackets must be the same. Notice the sign change due to the negative sign in front of the 2.

Note

You would get the same result if you used $3x$ and $-2x$ in the opposite order:

$$x^2 + x - 6 = x^2 - 2x + 3x - 6$$
$$= x(x - 2) + 3(x - 2)$$
$$= (x - 2)(x + 3)$$

The graph of $y = x^2 + x - 6$ crosses the x-axis when $(x + 3)(x - 2) = 0$, i.e. when $x = -3$ and when $x = 2$.

The x-coordinate of the turning point is halfway between these two values, so:

$$x = \frac{-3 + 2}{2}$$
$$= -0.5$$

Substituting this value into the equation of the curve gives:

$$y = (-0.5)^2 + (-0.5) - 6$$
$$= -6.25$$

The equation of the curve has a positive x^2 term so its graph is ∪-shaped. Therefore the minimum value is at $(-0.5, -6.25)$.

The method shown above can be adapted for curves with an equation in which the coefficient of x^2 is not $+1$, for example, $y = 6x^2 - 13x + 6$ or $y = 6 - x - 2x^2$, as shown in the next example.

Maximum and minimum values

> **Worked example**

For the curve with equation $y = 6 - x - 2x^2$:

a) Will the turning point of the curve be a maximum or a minimum? Give a reason for your answer.

b) Write down the coordinates of the turning point.

c) State the equation of the line of symmetry.

Solution

a) The coefficient of x^2 is negative so the curve will be $\cap$-shaped. This means that the turning point will be a maximum.

b) First multiply the constant term and the coefficient of x^2, i.e. $6 \times -2 = -12$. Then find two whole numbers that multiply together to give this product.

Possible pairs are: 6 and -2, -6 and 2, 3 and -4, -3 and 4, 1 and -12, -1 and 12.

3 and -4 are the only pair with a sum of -1, so use this pair to split up the x term.
$$6 - x - 2x^2 = 6 + 3x - 4x - 2x^2$$
$$= 3(2 + x) - 2x(2 + x)$$
$$= (2 + x)(3 - 2x)$$

The graph of $y = 6 - x - 2x^2$ crosses the x-axis when $(2 + x)(3 - 2x) = 0$, i.e. when $x = -2$ and when $x = 1.5$.

$$x = \frac{-2 + 1.5}{2}$$
$$= -0.25$$

Substituting this value into the equation of the curve gives:
$$y = 6 - (-0.25) - 2(-0.25)^2$$
$$= 6.125$$

So the turning point is $(-0.25, 6.125)$

c) The equation of the line of symmetry is $x = -0.25$.

Continue to use any alternative methods of factorising that you are confident with.

Identify any of the pairs of numbers that can be added together to give the coefficient of x (-1).

Both expressions in the brackets must be the same. Notice the sign change is due to the sign in front of the 2.

The x-coordinate of the turning point is halfway between these two values.

Completing the square

The methods shown in the previous examples will always work for curves that cross the x-axis. For quadratic curves that do not cross the x-axis, you will need to use the method of **completing the square**, shown in the next example.

Another way of writing the quadratic expression $x^2 + 6x + 11$ is $(x + 3)^2 + 2$ and this is called **completed square form**. Written like this the expression consists of a squared term, $(x + 3)^2$, that includes the variable, x, and a constant term $+2$.

In the next example you see how to convert an ordinary quadratic expression into completed square form.

2 QUADRATIC FUNCTIONS

→ Worked example

a) Write $x^2 - 8x + 18$ in completed square form.

b) State whether the turning point is a maximum or minimum.

c) Sketch the curve $y = f(x)$.

Solution

a) Start by halving the coefficient of x and squaring the result.

$$-8 \div 2 = -4$$
$$(-4)^2 = 16$$

Now use this result to break up the constant term, +18, into two parts:

$$18 = 16 + 2$$

and use this to rewrite the original expression as:

$$f(x) = x^2 - 8x + 16 + 2$$
$$= (x - 4)^2 + 2$$
$$(x - 4)^2 \geq 0 \text{ (always)}$$
$$\Rightarrow f(x) \geq 2 \text{ for all values of } x$$

You will always have a perfect square in this expression.

In completed square form, $x^2 - 8x + 18 = (x - 4)^2 + 2$

b) $f(x) \geq 2$ for all values of x so the turning point is a minimum.

c) The function is a ∪-shaped curve because the coefficient of x^2 is positive. From the above, the minimum turning point is at $(4, 2)$ so the curve does not cross the x-axis. To sketch the graph, you will also need to know where it crosses the y-axis.

$f(x) = x^2 - 8x + 18$ crosses the y-axis when $x = 0$, i.e. at $(0, 18)$.

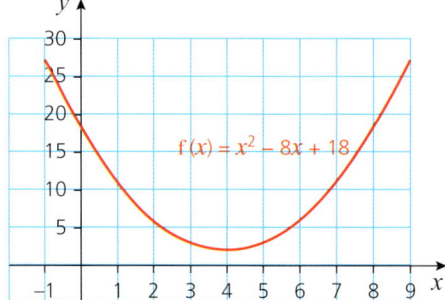

→ Worked example

Use the method of completing the square to work out the coordinates of the turning point of the quadratic function $f(x) = 2x^2 - 8x + 9$.

Solution

$$f(x) = 2x^2 - 8x + 9$$
$$= 2(x^2 - 4x) + 9$$
$$= 2((x - 2)^2 - 4) + 9$$
$$= 2(x - 2)^2 + 1$$

$(x - 2)^2 \geq 0$ (always), so the minimum value of f(x) is 1.

When f(x) = 1, x = 2.

Therefore the coordinates of the turning point (minimum value) of the function $f(x) = 2x^2 - 8x + 9$ are (2, 1).

Sometimes you will be asked to sketch the graph of a function f(x) for certain values of x. This set of values of x is called the **domain** of the function. The corresponding set of y-values is called the **range**.

→ Worked example

The domain of the function $y = 6x^2 + x - 2$ is $-3 \leq x \leq 3$.
Sketch the graph and find the range of the function.

Solution

The coefficient of x^2 is positive, so the curve is ∪-shaped and the turning point is a minimum.

The curve crosses the x-axis when $6x^2 + x - 2 = 0$.
$$6x^2 + x - 2 = (3x + 2)(2x - 1)$$
$$\Rightarrow (3x + 2)(2x - 1) = 0$$
$$\Rightarrow (3x + 2) = 0 \text{ or } (2x - 1) = 0$$

So the graph crosses the x-axis at $(-\frac{2}{3}, 0)$ and $(\frac{1}{2}, 0)$.

The curve crosses the y-axis when x = 0, i.e. at (0, -2).

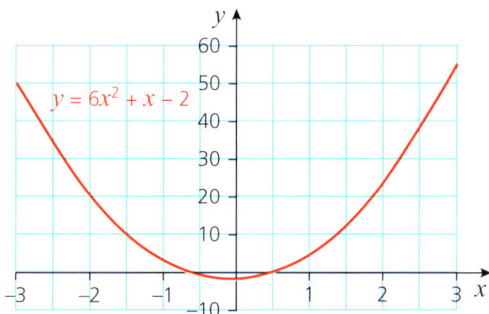

The curve has a vertical line of symmetry passing halfway between the two points where the curve intersects the x-axis. Therefore the equation of this line of symmetry is $x = \dfrac{-\frac{2}{3} + \frac{1}{2}}{2}$ or $x = -\dfrac{1}{12}$.

When $x = -\dfrac{1}{12}$, $y = 6\left(-\dfrac{1}{12}\right)^2 + \left(-\dfrac{1}{12}\right) - 2 = -2\dfrac{1}{24}$, the minimum value of the function.

To find the range, work out the values of y for x = -3 and x = +3.
When $x = -3$, $y = 6(-3)^2 + (-3) - 2 = 49$.
When $x = 3$, $y = 6(3)^2 + 3 - 2 = 55$.

The larger of these gives the maximum value.

The range of the function corresponding to the domain $-3 \leq x \leq 3$ is therefore $-2\dfrac{1}{24} \leq y \leq 55$.

2 QUADRATIC FUNCTIONS

Exercise 2.1

1. Solve each equation by factorising:
   a) $x^2 + x - 20 = 0$
   b) $x^2 - 5x + 6 = 0$
   c) $x^2 - 3x - 28 = 0$
   d) $x^2 + 13x + 42 = 0$

2. Solve each equation by factorising:
   a) $2x^2 - 3x + 1 = 0$
   b) $9x^2 + 3x - 2 = 0$
   c) $2x^2 - 5x - 7 = 0$
   d) $3x^2 + 17x + 10 = 0$

3. Solve each equation by factorising:
   a) $x^2 - 169 = 0$
   b) $4x^2 - 121 = 0$
   c) $100 - 64x^2 = 0$
   d) $12x^2 - 27 = 0$

4. For each of the following curves:
   i) Factorise the function.
   ii) Work out the coordinates of the turning point.
   iii) State whether the turning point is a maximum or minimum.
   iv) Sketch the graph, labelling the coordinates of the turning point and any points of intersection with the axes.
   a) $y = x^2 + 7x + 10$
   b) $f(x) = 16 - 6x - x^2$
   c) $y = 5 - 9x - 2x^2$
   d) $f(x) = 2x^2 + 11x + 12$

5. Write each quadratic expressions in the form $(x + a)^2 + b$:
   a) $x^2 + 4x + 9$
   b) $x^2 - 10x - 4$
   c) $x^2 + 5x - 7$
   d) $x^2 - 9x - 2$

6. Write each quadratic expression in the form $c(x + a)^2 + b$.
   a) $2x^2 - 12x + 5$
   b) $3x^2 + 12x + 20$
   c) $4x^2 - 8x + 5$
   d) $2x^2 + 9x + 6$

7. Solve the following quadratic equations. Leave your answers in the form $x = p \pm \sqrt{q}$.
   a) $x^2 + 4x - 9 = 0$
   b) $x^2 - 7x - 2 = 0$
   c) $2x^2 + 6x - 9 = 0$
   d) $3x^2 + 9x - 15 = 0$

8. For each of the following functions:
   i) Use the method of completing the square to find the coordinates of the turning point of the graph.
   ii) State whether the turning point is a maximum or a minimum.
   iii) Sketch the graph.
   a) $f(x) = x^2 + 6x + 15$
   b) $y = 8 + 2x - x^2$
   c) $y = 2x^2 + 2x - 9$
   d) $f : x \rightarrow x^2 - 8x + 20$

9. Sketch the graph and find the corresponding range for each function and domain.
   a) $y = x^2 - 7x + 10$ for the domain $1 \leqslant x \leqslant 6$
   b) $f(x) = 2x^2 - x - 6$ for the domain $-2 \leqslant x \leqslant 2$

Real-world activity

1. Draw a sketch of a bridge modelled on the equation $25y = 100 - x^2$ for $-10 \leqslant x \leqslant 10$. Label the origin O, point $A(-10, 0)$, point $B(10, 0)$ and point $C(0, 4)$.
2. 1 unit on your graph represents 1 metre. State the maximum height of the bridge, OC, and the span, AB.
3. Work out the equation of a similar bridge with a maximum height of 5 m and a span of 40 m.

The quadratic formula

The **roots** of a quadratic equation f(x) are those values of x for which $y = 0$ for the curve $y = f(x)$. In other words, they are the x-coordinates of the points where the curve either crosses or touches the x-axis. There are three possible outcomes.

1. The curve crosses the x-axis at two distinct points. In this case, the corresponding equation is said to have *two real distinct roots*.

2. The curve touches the x-axis, in which case the equation has *two equal (repeating) roots*.

3. The curve lies completely above or completely below the x-axis so it neither crosses nor touches the axis. In this case, the equation has *no real roots*.

The method of completing the square can be generalised to give a formula for solving quadratic equations. The next example uses this method in a particular case on the left-hand side and shows the same steps for the general case on the right-hand side, using algebra to derive the formula for solving quadratic equations.

➡ Worked example

Solve $2x^2 + x - 4 = 0$.

Solution	Generalisation
$2x^2 + x - 4 = 0$	$ax^2 + bx + c = 0$
$\Rightarrow \quad x^2 + \frac{1}{2}x - 2 = 0$	$\Rightarrow \quad x^2 + \frac{b}{a}x + \frac{c}{a} = 0$
$\Rightarrow \quad x^2 + \frac{1}{2}x = 2$	$\Rightarrow \quad x^2 + \frac{b}{a}x = -\frac{c}{a}$
$\Rightarrow \quad x^2 + \frac{1}{2}x + \left(\frac{1}{4}\right)^2 = 2 + \left(\frac{1}{4}\right)^2$	$\Rightarrow \quad x^2 + \left(\frac{b}{a}\right)x + \left(\frac{b}{2a}\right)^2 = -\frac{c}{a} + \left(\frac{b}{2a}\right)^2$
$\Rightarrow \quad \left(x + \frac{1}{4}\right)^2 = \frac{33}{16}$	$\Rightarrow \quad \left(x + \frac{b}{2a}\right)^2 = \frac{b^2}{4a^2} - \frac{c}{a}$
	$= \frac{b^2 - 4ac}{4a^2}$
$\Rightarrow \quad \left(x + \frac{1}{4}\right) = \pm\frac{\sqrt{33}}{4}$	$\Rightarrow \quad x + \frac{b}{2a} = \pm\sqrt{\frac{b^2 - 4ac}{4a^2}}$
	$= \pm\frac{\sqrt{b^2 - 4ac}}{2a}$
$\Rightarrow \quad x = -\frac{1}{4} \pm \frac{\sqrt{33}}{4}$	$\Rightarrow \quad x = -\frac{b}{2a} \pm \frac{\sqrt{b^2 - 4ac}}{2a}$
$= \frac{-1 \pm \sqrt{33}}{4}$	$= \frac{-b \pm \sqrt{b^2 - 4ac}}{2a}$

2 QUADRATIC FUNCTIONS

The result $x = \frac{-b \pm \sqrt{b^2 - 4ac}}{2a}$ is known as the **quadratic formula**. You can use it to solve any quadratic equation. One root is found by taking the + sign, and the other by taking the − sign. When the value of $b^2 - 4ac$ is negative, the square root cannot be found and so there is no real solution to that quadratic equation. This occurs when the curve does not cross the x-axis.

> In an equation of the form $(px + q)^2 = 0$, where p and q can represent either positive or negative numbers, $px + q = 0$ gives the only solution.

Note

The part $b^2 - 4ac$ is called the **discriminant** because it discriminates between quadratic equations with no roots, quadratic equations with one repeated root and quadratic equations with two real roots.
- If $b^2 - 4ac > 0$ there are 2 real roots.
- If $b^2 - 4ac = 0$ there is 1 repeated root.
- If $b^2 - 4ac < 0$ there are no real roots.

> **Note**
> You know that the square root of a positive number has two values, one positive and the other negative; so the square root of 9 is +3 or −3 and can be written as ±3. However, the square root symbol, $\sqrt{\;}$, means 'the positive square root of' so $\sqrt{9} = +3$. That is why you see $\pm\sqrt{\;}$ in the quadratic formula.

Worked example

a) Show that the equation $4x^2 - 12x + 9 = 0$ has a repeated root by:
   i) factorising
   ii) using the discriminant.

b) State with reasons how many real roots the following equations have:
   i) $4x^2 - 12x + 8 = 0$
   ii) $4x^2 - 12x + 10 = 0$

Solution

a) i) $\quad 4x^2 - 12x + 9 = 0$
   $\Rightarrow (2x - 3)(2x - 3) = 0$
   $\quad 2x - 3 = 0$
   $\Rightarrow \quad x = 1.5$

   ii) The equation has a repeated root because the discriminant $b^2 - 4ac = (-12)^2 - 4(4)(9) = 0$.

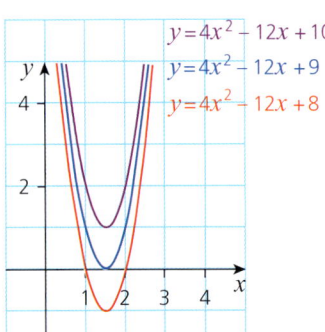

b) i) The curve $y = 4x^2 - 12x + 8$ is 1 unit below $y = 4x^2 - 12x + 9$ and crosses the x-axis at two points. So the equation has two real roots.

   ii) The curve $y = 4x^2 - 12x + 10$ is 1 unit above $y = 4x^2 - 12x + 9$ and does not cross the x-axis. So the equation $4x^2 - 12x + 10 = 0$ has no real roots.

The intersection of a line and a curve

In some cases, such as in the previous example, the factorisation is not straightforward. In such cases, evaluating the discriminant is a reliable method to obtain an accurate result.

Worked example

Show that the equation $3x^2 - 2x + 4 = 0$ has no real solution.

Solution

The most straightforward method is to look at the discriminant. If the discriminant is negative, there is no real solution.

For $3x^2 - 2x + 4 = 0$, $a = 3$, $b = -2$ and $c = 4$.
$$b^2 - 4ac = (-2)^2 - 4(3)(4)$$
$$= -44$$

Substituting these values into the discriminant

Since the discriminant is negative, there is no real solution.

The intersection of a line and a curve

The examples so far have considered whether or not a curve intersects, touches, or lies completely above or below the x-axis ($y = 0$). The next example considers the more general case of whether or not a curve intersects, touches or lies completely above or below a particular straight line.

The general equation of a straight line is $y = mx + c$. This has alternate forms, e.g. $ax + by + c = 0$.

Worked example

a) Find the coordinates of the points of intersection of the line $y = 4 - 2x$ and the curve $y = x^2 + x$.

b) Sketch the line and the curve on the same axes.

Solution

a) To find where the curve and the line insterpect, solve $y = x^2 + x$ simultaneously with $y = 4 - 2x$.
$$x^2 + x = 4 - 2x$$
$$\Rightarrow \quad x^2 + 3x - 4 = 0$$
$$\Rightarrow \quad (x + 4)(x - 1) = 0$$
$$\Rightarrow \quad x = -4 \text{ or } x = 1$$

The y-values of both equations are the same at the point(s) of intersection.

To find the y-coordinate, substitute into one of the equations.
When $x = -4$, $y = 4 - 2(-4) = 12$.
When $x = 1$, $y = 4 - 2(1) = 2$.
The line $y = 4 - 2x$ intersects the curve $y = x^2 + x$ at $(-4, 12)$ and $(1, 2)$.

It is more straightforward to substitute into the linear equation.

2 QUADRATIC FUNCTIONS

b) The curve has a positive coefficient of x^2 so is ∪-shaped.

It crosses the x-axis when $x^2 + x = 0$.

$\Rightarrow x(x + 1) = 0$

$\Rightarrow x = 0$ or $x = -1$

So the curve crosses the x-axis at $x = 0$ and $x = -1$.

It crosses the y-axis when $x = 0$.

Substituting $x = 0$ into $y = x^2 + x$ gives $y = 0$.

So the curve passes through the origin.

The line $2x + y = 4$ crosses the x-axis when $y = 0$. When $y = 0$, $x = 2$.

The line $2x + y = 4$ crosses the y-axis when $x = 0$. When $x = 0$, $y = 4$.

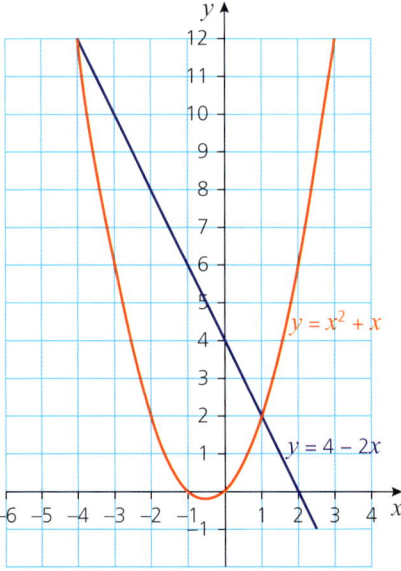

It is possible for a quadratic curve to touch a general line, either sloping or parallel to the x-axis. You can see this when you solve the equations of the line and the curve simultaneously. If you get a repeated root, it means that they touch at only one point. The line is a tangent to the curve. This is shown in the next example.

➜ Worked example

a) Use algebra to show that the line $y = 6x - 19$ touches the curve $y = x^2 - 2x - 3$ and find the coordinates of the point of contact.

b) Sketch the line and curve on the same axes.

The intersection of a line and a curve

Solution

a) Solving the equations simultaneously
$$x^2 - 2x - 3 = 6x - 19$$
$$\Rightarrow x^2 - 8x + 16 = 0$$
$$\Rightarrow (x - 4)^2 = 0$$
$$\Rightarrow x = 4$$

The repeated root $x = 4$ shows that the line and the curve touch.

Substitute $x = 4$ into either equation to find the value of the y-coordinate.
$$y = 6(4) - 19$$
$$= 5$$

It is more straightforward to substitute into the line equation.

Therefore the point of contact is (4, 5).

b) The coefficient of x^2 is positive so the curve is $\cup$-shaped.

Substituting $x = 0$ into $y = x^2 - 2x - 3$ shows that the curve intersects the y-axis at $(0, -3)$.

Substituting $y = 0$ into $y = x^2 - 2x - 3$ gives $x^2 - 2x - 3 = 0$.
$$\Rightarrow (x - 3)(x + 1) = 0$$
$$\Rightarrow x = -1 \text{ or } x = 3$$

So the curve intersects the x-axis at $(-1, 0)$ and $(3, 0)$.

You need two points to draw a line. It is best to choose points with whole numbers that are not too large, such as (3, −1) and (5, 11).

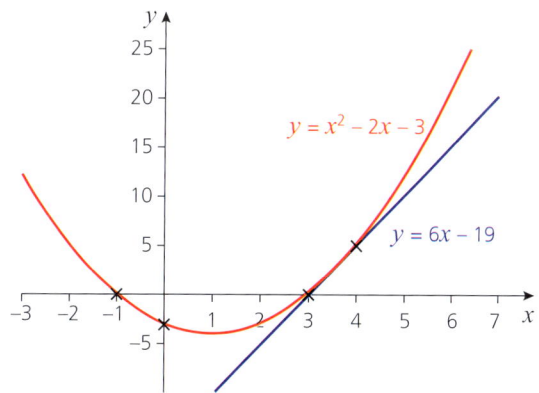

Discussion point

Why is it not possible for a quadratic curve to touch a line parallel to the y-axis?

There are many situations when a line and a curve do not intersect or touch each other. A straightforward example of this occurs when the graph of a quadratic function is a $\cup$-shaped curve completely above the x-axis, e.g. $y = x^2 + 3$, and the line is the x-axis.

2 QUADRATIC FUNCTIONS

You have seen how solving the equations of a curve and a line simultaneously gives a quadratic equation with *two* roots when the line crosses the curve, and a quadratic equation with a *repeated* root when it touches the curve. If solving the two equations simultaneously results in *no* real roots, i.e. the discriminant is negative, then they do not cross or touch.

→ Worked example

a) Sketch the graphs of the line $y = x - 3$ and the curve $y = x^2 - 2x$ on the same axes.

b) Use algebra to prove that the line and the curve don't meet.

Solution

a)

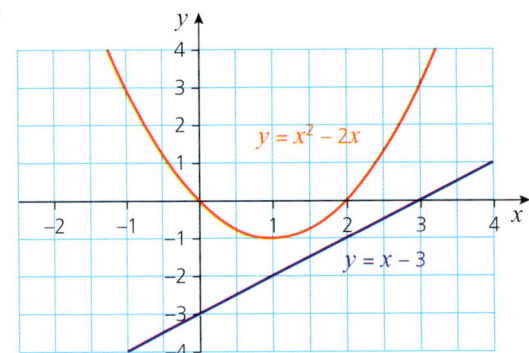

You do not actually need to write down the solution. Once you see that the value of the discriminant is negative, as in this case where it is -3, you know that the equation has no real roots, so the line and the curve don't meet.

b) $x^2 - 2x = x - 3$ ← Solving the two equations simultaneously
$\Rightarrow x^2 - 3x + 3 = 0$

This does not factorise, so solve using the quadratic formula
$$x = \frac{-b \pm \sqrt{b^2 - 4ac}}{2a}$$
$a = 1$, $b = -3$ and $c = 3$
$$x = \frac{-(-3) \pm \sqrt{(-3)^2 - 4(1)(3)}}{2(1)}$$
$$= \frac{3 \pm \sqrt{-3}}{2}$$

Since there is a negative value under the square root, there is no *real* solution. This implies that the line and the curve do not meet.

> **Note**
>
> It would have been sufficient to consider only the discriminant $b^2 - 4ac$. Solving a quadratic equation is equivalent to finding the point(s) where the curve crosses the horizontal axis (the roots).

34

Using quadratic equations to solve problems

→ Worked example

A triangle has a base of $(2x + 1)$ cm, a height of x cm and an area of $68\,\text{cm}^2$.

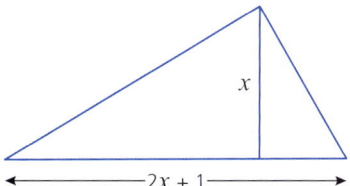

a) Show that x satisfies the equation $2x^2 + x - 136 = 0$.

b) Solve the equation and work out the base length of the triangle.

Solution

a) Using the formula for the area of a triangle, area $= \frac{1}{2}$base × height:

$$\text{Area} = \frac{1}{2} \times (2x + 1) \times x$$

$$= \frac{1}{2}(2x^2 + x)$$

The area is $68\,\text{cm}^2$, so:

$$\frac{1}{2}(2x^2 + x) = 68$$

$$\Rightarrow \quad 2x^2 + x = 136$$

$$\Rightarrow \quad 2x^2 + x - 136 = 0$$

b) It is not easy to factorise this equation – it is not even obvious that there are factors – so use the quadratic formula.

Alternatively, you can use a calculator to solve the equation.

$$x = \frac{-b \pm \sqrt{b^2 - 4ac}}{2a}$$

$a = 2$, $b = 1$ and $c = -136$

$$x = \frac{-1 \pm \sqrt{1^2 - 4(2)(-136)}}{2(2)}$$

$$\Rightarrow x = \frac{-1 \pm \sqrt{1089}}{4}$$

$$\Rightarrow x = \frac{-1 \pm 33}{4}$$

$$\Rightarrow x = 8 \text{ or } x = -8.5$$

Since x is a length, reject the negative solution.

Substitute $x = 8$ into the expression for the base of the triangle, $2x + 1$, and work out the length of the base of the triangle, $17\,\text{cm}$.

Check that this works with the information given in the original question.

$$\frac{1}{2} \times 17\,\text{cm} \times 8\,\text{cm} = 68\,\text{cm}^2$$

2 QUADRATIC FUNCTIONS

Solving quadratic inequalities

The quadratic inequalities in this section all involve quadratic expressions that factorise. This means that you can find a solution either by sketching the appropriate graph or by using line segments to reduce the quadratic inequality to two simultaneous linear inequalities. The example below shows two valid methods for solving quadratic inequalities. You should use whichever method you prefer. Your choice may depend on how easily you sketch graphs or if you have a graphic calculator that you can use to plot these graphs.

➡ Worked example

Solve these quadratic inequalities.

a) $x^2 - 2x - 3 < 0$

b) $x^2 - 2x - 3 \geq 0$

Solution

Method 1

$x^2 - 2x - 3 = (x + 1)(x - 3)$

So the graph of $y = x^2 - 2x - 3$ crosses the x-axis when $x = -1$ and $x = 3$.

Look at the two graphs below.

Here the end points are not included in the solution, so you draw open circles: ○

Here the end points are included in the solutions, so you draw solid circles: ●

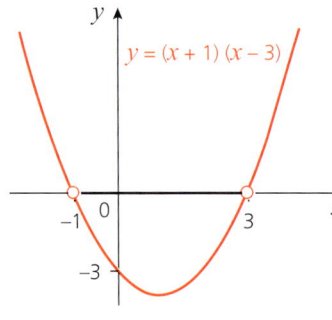
The solution is $-1 < x < 3$.

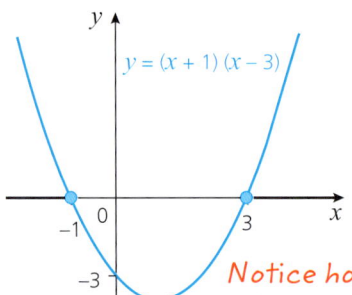
The solution is $x \leq -1$ or $x \geq 3$.

Notice how the solution is in two parts when there are two line segments.

a) The answer is the values of x for which $y < 0$, i.e. where the curve is below the x-axis.

b) The answer is the values of x for which $y \geq 0$, i.e. where the curve crosses or is above the x-axis.

Solving quadratic inequalities

Method 2

This method identifies the values of x for which each of the factors is 0 and considers the sign of each factor in the intervals between these critical values.

	$x < -1$	$x = -1$	$-1 < x < 3$	$x = 3$	$x > 3$
Sign of $(x + 1)$	$-$	0	$+$	$+$	$+$
Sign of $(x - 3)$	$-$	$-$	$-$	0	$+$
Sign of $(x + 1)(x - 3)$	$(-) \times (-) = +$	$(0) \times (-) = 0$	$(+) \times (-) = -$	$(+) \times (0) = 0$	$(+) \times (+) = +$

From the table, the solution to:

a) $(x + 1)(x - 3) < 0$ is $-1 < x < 3$

b) $(x + 1)(x - 3) \geq 0$ is $x \leq -1$ or $x \geq 3$

If the inequality to be solved contains $>$ or $<$, then the solution is described using $>$ and $<$. If the original inequality contains $\geq$ or $\leq$, then the solution is described using $\geq$ and $\leq$.

If the quadratic inequality has the variable on both sides, collect the terms involving the variable on one side first in the same way as you would before solving a quadratic equation.

→ Worked example

Solve $2x + x^2 > 3$.

Solution

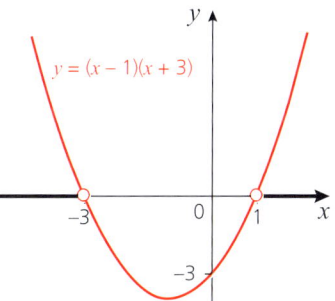

$2x + x^2 > 3 \Rightarrow x^2 + 2x - 3 > 0$
$\Rightarrow (x - 1)(x + 3) > 0$

From the graph, the solution is
$x < -3$ or $x > 1$.

Exercise 2.2

1 For each of the following equations, decide if there are two real and different roots, two equal roots or no real roots. Solve the equations with real roots.
  a) $x^2 + 3x + 2 = 0$ b) $t^2 - 9 = 0$ c) $x^2 + 16 = 0$
  d) $2x^2 - 5x = 0$ e) $p^2 + 3p - 18 = 0$ f) $x^2 + 10x + 25 = 0$
  g) $15a^2 + 2a - 1 = 0$ h) $3r^2 + 8r = 3$

2 Solve the following equations by:
  i) completing the square   ii) using the quadratic formula.
  Give your answers correct to two decimal places.
  a) $x^2 - 2x - 10 = 0$ b) $x^2 + x = 0$
  c) $2x^2 + 2x - 9 = 0$ d) $2x^2 + x - 8 = 0$

2 QUADRATIC FUNCTIONS

3 Try to solve each of the following equations. Where there is a solution, give your answers correct to two decimal places.
   a) $4x^2 + 6x - 9 = 0$ b) $9x^2 + 6x + 4 = 0$
   c) $(2x + 3)^2 = 7$ d) $x(2x - 1) = 9$

4 Use the discriminant to decide whether each of the following equations has two equal roots, two distinct roots or no real roots:
   a) $9x^2 - 12x + 4 = 0$ b) $6x^2 - 13x + 6 = 0$ c) $2x^2 + 7x + 9 = 0$
   d) $2x^2 + 9x + 10 = 0$ e) $3x^2 - 4x + 5 = 0$ f) $4x^2 + 28x + 49 = 0$

5 For each pair of equations determine if the line intersects the curve, is a tangent to the curve or does not meet the curve. Give the coordinates of any points where the line and curve touch or intersect.
   a) $y = x^2 + 12x$; $y = -9 + 6x$ b) $y = 2x^2 + 3x - 4$; $y = 2x - 6$
   c) $y = 6x^2 - 12x + 6$; $y = x$ d) $y = x^2 - 8x + 18$; $y = 2x + 3$
   e) $y = x^2 + x$; $2x + y = 4$ f) $y = 4x^2 + 9$; $y = 12x$
   g) $y = 3 - 2x - x^2$; $y = 9 + 2x$ h) $y = (3 - 2x)^2$; $y = 2 - 3x$

6 Solve the following inequalities:
   a) $x^2 - 6x + 5 > 0$ b) $a^2 + 3a - 4 \leq 0$ c) $4 - y^2 > 0$
   d) $x^2 - 4x + 4 > 0$ e) $8 - 2a > a^2$ f) $3y^2 + 2y - 1 > 0$

Real-world activity

Anna would like to design a pendant for her mother and decides that it should resemble an eye. She starts by making the scale drawing, shown below.

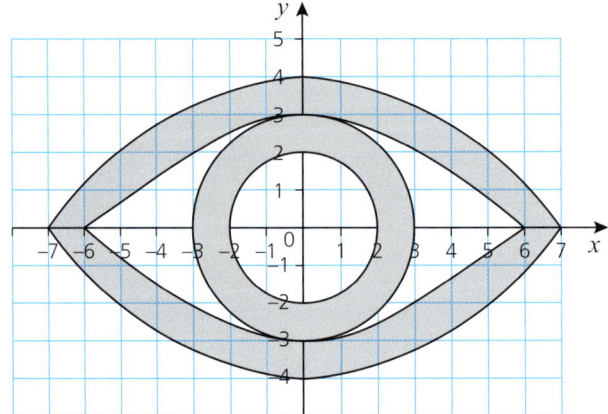

The pendant is made up of the shaded area.
The equations of the two circles are $x^2 + y^2 = 4$ and $x^2 + y^2 = 9$.
The rest of the pendant is formed by quadratic curves.
The scale is 2 units represents 1 cm.

1 Find the equations of the four quadratic curves.
2 Anna decides to make some earrings using a smaller version of the pendant design. She reduces the size by a factor of 2. Find the equations of the four quadratic curves for the earrings.

Solving quadratic inequalities

Past-paper questions

1. (i) Express $2x^2 - x + 6$ in the form $p(x - q)^2 + r$, where p, q and r are constants to be found. [3]
   (ii) Hence state the least value of $2x^2 - x + 6$ and the value of x at which this occurs. [2]

   Cambridge O Level Additional Mathematics (4037)
   Paper 21 Q5, June 2014
   Cambridge IGCSE Additional Mathematics (0606)
   Paper 21 Q5, June 2014

2. Find the set of values of k for which the curve $y = 2x^2 + kx + 2k - 6$ lies above the x-axis for all values of x. [4]

   Cambridge O Level Additional Mathematics (4037)
   Paper 12 Q4, June 2013
   Cambridge IGCSE Additional Mathematics (0606)
   Paper 12 Q4, June 2013

3. The line $y = mx + 2$ is a tangent to the curve $y = x^2 + 12x + 18$. Find the possible values of m. [4]

   Cambridge O Level Additional Mathematics (4037)
   Paper 13 Q3, November 2010
   Cambridge IGCSE Additional Mathematics (0606)
   Paper 13 Q3, November 2010

Now you should be able to:

★ find the maximum or minimum value of the quadratic function $f: x \mapsto ax^2 + bx + c$ by completing the square
★ use the maximum or minimum value of $f(x)$ to sketch the graph of $y = f(x)$ or determine the range for a given domain
★ know the conditions for $f(x) = 0$ to have:
  ● two real roots
  ● two equal roots
  ● no real roots
  and the related conditions for a given line to:
  ● intersect a given curve
  ● be a tangent to a given curve
  ● not intersect a given curve
★ solve quadratic equations for real roots
★ find the solution set for quadratic inequalities either graphically or algebraically.

2 QUADRATIC FUNCTIONS

Key points

- A **quadratic function** has the form $f(x) = ax^2 + bx + c$, where a, b and c can be any number (positive, negative or zero) provided that $a \neq 0$. The set of possible values of x is called the **domain** of the function and the set of y values is called the **range**.
- To plot the graph of a quadratic function, first calculate the value of y for each value of x in the given range.
- The graph of a quadratic function is symmetrical about a vertical line. It is ∪-shaped if the coefficient of x^2 is positive and ∩-shaped if the coefficient of x^2 is negative.
- To sketch the graph of a quadratic function:
  - look at the coefficient of x^2 to determine the shape
  - substitute $x = 0$ to determine where the curve crosses the vertical axis
  - solve $f(x) = 0$ to determine any values of x where the curve touches or crosses the horizontal axis.
- If there are no real values for x for which $f(x) = 0$, then the curve will be either completely above or completely below the x-axis.
- A **quadratic equation** is of the form $ax^2 + bx + c = 0$ with $a \neq 0$.
- To **factorise** a quadratic equation of the form $x^2 + bx + c = 0$, look for two numbers, p and q, with the sum b and the product c. The factorised form is then $(x - p)(x - q) = 0$. To factorise an equation of the form $ax^2 + bx + c = 0$, look for two numbers with the sum b and the product ac.
- The **discriminant** of a quadratic equation ($ax^2 + bx + c = 0$) is $b^2 - 4ac$. If $b^2 - 4ac > 0$, a quadratic equation will have two distinct solutions (or roots). If $b^2 - 4ac = 0$, the two roots are equal so there is one repeating root. If $b^2 - 4ac < 0$, the roots have no real values.
- An expression of the form $(px + q)^2$ is called a **perfect square**.
- $x^2 + bx + c$ can be written as $\left(x + \frac{b}{2}\right)^2 - \left(\frac{b}{2}\right)^2 + c$ using the method of **completing the square**. For expressions of the form $ax^2 + bx + c$, first take a out as a factor.
- The **quadratic formula** for solving an equation of the form $ax^2 + bx + c = 0$ is $x = \frac{-b \pm \sqrt{b^2 - 4ac}}{2a}$
- To find the point(s) where a line and a curve touch or intersect, substitute the expression for y from one equation into the other to give a quadratic equation in x.
- When solving a **quadratic inequality**, it is advisable to start by sketching the associated quadratic graph.

3 Factors of polynomials

*A **quadratic expression** is any expression of the form $ax^2 + bx + c$, where x is a variable and a, b and c are constants with $a \neq 0$.*
*An expression of the form $ax^3 + bx^2 + cx + d$ that includes a term in x^3 is called a **cubic expression**.*
*A **quartic** has a term in x^4 as its highest power, a **quintic** one with x^5 and so on.*
*All of these are **polynomials** and the highest power of the variable is called the **order** of the polynomial.*

There are things of an unknown number which when divided by 3 leave 2, by 5 leave 3, and by 7 leave 2. What is the smallest number?

Sun-Tzi (544–496BC)

Discussion point

Sun Tzi posed his problem in the Chinese Han dynasty and it is seen as the forerunner of the remainder theorem, which you will meet in this chapter. What is the answer? What is the next possible answer to Sun-Tzi's problem? How do you find further answers?

It is believed that the way that numbers were written during the Han dynasty laid the foundation for the abacus, an early form of hand calculator.

In Chapter 2 you met quadratic expressions like $x^2 - 4x - 12$ and solved quadratic equations such as $x^2 - 4x - 12 = 0$.

3 FACTORS OF POLYNOMIALS

Multiplication and division of polynomials

There are a number of methods for multiplying and dividing polynomials. One method is shown in the worked example, but if you already know and prefer an alternative method, continue to use it.

Multiplication

> **Worked example**

Multiply $(x^2 - 5x + 2)$ by $(2x^2 - x + 1)$.

Solution

This is an extension of the method you used to multiply two brackets that each contain two terms. If you are familiar with a different method, then use that.

$(x^2 - 5x + 2) \times (2x^2 - x + 1) = x^2(2x^2 - x + 1) - 5x(2x^2 - x + 1) + 2(2x^2 - x + 1)$
$= 2x^4 - x^3 + x^2 - 10x^3 + 5x^2 - 5x + 4x^2 - 2x + 2$
$= 2x^4 + x^3(-1 - 10) + x^2(1 + 5 + 4) + x(-5 - 2) + 2$
$= 2x^4 - 11x^3 + 10x^2 - 7x + 2$

Division

> **Worked example**

Divide $(x^3 - x^2 - 2x + 8)$ by $(x + 2)$.

Solution

Let $(x^3 - x^2 - 2x + 8) = (x + 2)(ax^2 + bx + c)$ ← *This bracket must be a quadratic expression.*

Multiplying each term in the second bracket by x and then by 2

$= x(ax^2 + bx + c) + 2(ax^2 + bx + c)$
$= ax^3 + bx^2 + cx + 2ax^2 + 2bx + 2c$

Collecting like terms

$= ax^3 + (b + 2a)x^2 + (c + 2b)x + 2c$

Comparing coefficients:

$a = 1$

Since $a = 1$ → $b + 2a = -1 \Rightarrow b = -3$

Since $b = -3$ → $c + 2b = -2 \Rightarrow c = 4$

Checking the constant term, $2c = 8$ which is correct.

This gives $(x^3 - x^2 - 2x + 8) \div (x + 2) = x^2 - 3x + 4$

Exercise 3.1

1. Multiply $(x^3 + 2x^2 - 3x - 4)$ by $(x + 1)$.
2. Multiply $(x^3 - 2x^2 + 3x + 2)$ by $(x - 1)$.
3. Multiply $(2x^3 - 3x^2 + 5)$ by $(2x - 1)$.
4. Multiply $(x^2 + 2x - 3)$ by $(x^2 - 2x + 3)$.
5. Multiply $(2x^2 - 3x + 4)$ by $(2x^2 - 3x - 4)$.
6. Simplify $(x^2 - 3x + 2)^2$.
7. Divide $(x^3 - 3x^2 + x + 1)$ by $(x - 1)$.
8. Divide $(x^3 - 3x^2 + x + 2)$ by $(x - 2)$.
9. Divide $(x^4 - 1)$ by $(x + 1)$.
10. Divide $(x^2 - 16)$ by $(x + 2)$.

Solving cubic equations

When a polynomial can be factorised, you can find the points where the corresponding curve crosses the x-axis either as whole numbers or simple fractions.

For example, $y = x^2 - 3x - 4$ factorises to give $y = (x + 1)(x - 4)$.

The graph of this equation is a curve that crosses the x-axis at the points where $y = 0$. These values, $x = -1$ and $x = 4$, are called the **roots** of the equation $x^2 - 3x - 4 = 0$.

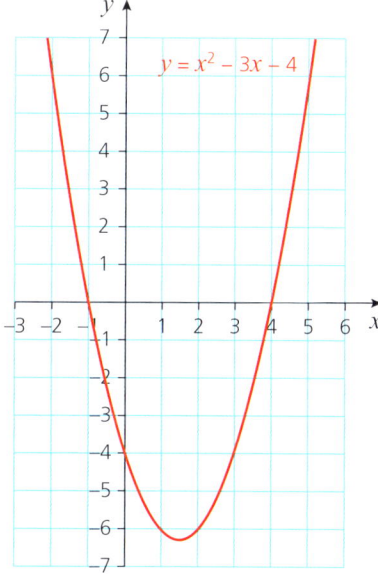

For a polynomial of the form $y = f(x)$, the roots are the solutions of $f(x) = 0$.

3 FACTORS OF POLYNOMIALS

> ## Worked example

a) Draw the graph of $y = x^3 - 5x^2 + 2x + 8$.

b) Hence solve the equation $x^3 - 5x^2 + 2x + 8 = 0$.

Solution

a) Start by setting up a table of values.

x	−2	−1	0	1	2	3	4	5
y	−24	0	8	6	0	−4	0	18

Then plot the curve.

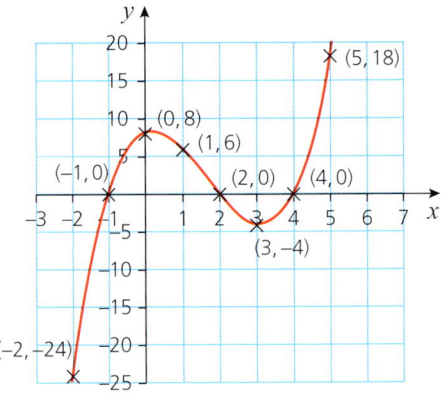

The solution is '$x = -1$ or $x = 2$ or $x = 4$' but the roots are '−1 and 2 and 4'.

b) The graph shows that the curve crosses the x-axis at the values −1, 2 and 4, giving the solution as $x = -1$, $x = 2$ or $x = 4$.

In some cases, a graph will not find all the roots but will allow you to find one or possibly two roots, or show you that there is only one root. The roots may not be whole numbers and may not even be rational as shown in the following examples.

> ## Worked example

Draw the graph of $y = 2x^3 - 7x^2 + 2x + 3$ and hence solve the equation $2x^3 - 7x^2 + 2x + 3 = 0$.

Solution

As before, start by setting up a table of values and then draw the curve.

x	−2	−1	0	1	2	3	4
y	−45	−8	3	0	−5	0	27

Finding factors and the factor theorem

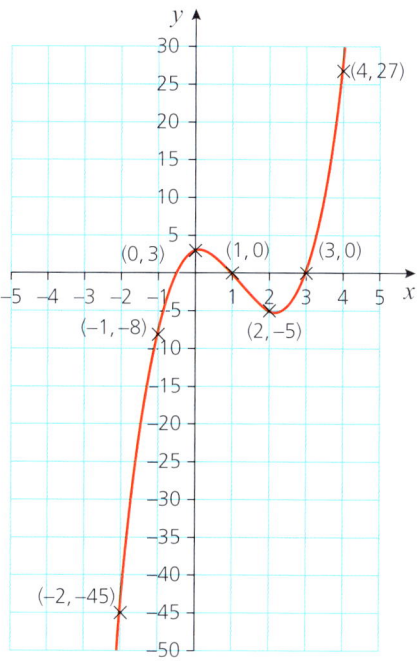

−0.5 is chosen as x since it is half way between −1 and 0.

The graph shows that the curve crosses the x-axis at 1, at 3 and again between −1 and 0. You can find this root using trial and improvement.

Let $x = -0.5$

$f(0.5) = 2(-0.5)^3 - 7(-0.5)^2 + 2(-0.5) + 3$
$= 0$

In this case, you were lucky and found the final root, −0.5, with only one iteration.

So the roots of the equation are −0.5, 1 and 3.

Finding factors and the factor theorem

The equation in the example above has roots that are whole numbers or exact fractions. This implies that it could have been factorised. Roots at $-\frac{1}{2}$, 1 and 3 suggest the factorised form:

$$\left(x + \frac{1}{2}\right)(x-1)(x-3)$$

However multiplying the x terms from all the brackets should give $2x^3$ so one of the brackets must be multiplied by 2.

$$2x^3 - 7x^2 + 2x + 3 = (2x+1)(x-1)(x-3)$$

It is not possible to factorise all polynomials. However, when a polynomial can be factorised, the solution to the corresponding equation follows immediately.

3 FACTORS OF POLYNOMIALS

$(2x+1)(x-1)(x-3) = 0 \Rightarrow (2x+1) = 0$ or $(x-1) = 0$ or $(x-3) = 0$
$\Rightarrow x = -0.5$ or $x = 1$ or $x = 3$

This leads to an important result known as the **factor theorem**.

If $(x - a)$ is a factor of f(x), then f(a) = 0 and $x = a$ is a root of the equation f(x) = 0.

Conversely, if f(a) = 0, then $(x - a)$ is a factor of f(x).

It is not necessary to try all integer values when you are looking for possible factors. For example, with $x^3 - 3x^2 - 2x + 6 = 0$ you need only try the factors of 6 as possible roots, i.e. ±1, ±2, ±3 and ±6.

➜ Worked example

a) Show that $x = 2$ is a root of the equation $x^3 - 3x^2 - 4x + 12 = 0$ and hence solve the equation.

b) Sketch the graph of $y = x^3 - 3x^2 - 4x + 12$.

Solution

a) $f(2) = 2^3 - 3(2^2) - 4(2) + 12 = 0$

This implies that $x = 2$ is a root of the equation and hence $(x - 2)$ is a factor of f(x).

Alternatively, you could factorise by long division (see p49).

Taking $(x - 2)$ as a factor gives

$x^3 - 3x^2 - 4x + 12 = (x - 2)(x^2 - x - 6)$
$= (x - 2)(x - 3)(x + 2)$

The solution to the equation is therefore $x = 2, x = 3$, or $x = -2$.

b) The graph crosses the x-axis at $x = -2$, $x = 2$ and $x = 3$ and the y-axis at $y = 12$.

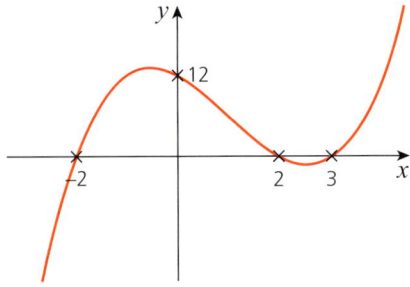

You will not be able to factorise the expression completely in all cases, but you may be able to find one factor by inspection as in the following example.

Finding factors and the factor theorem

> **Worked example**

Given that $f(x) = x^3 - x^2 - 10x + 12$

a) Show that $(x - 3)$ is a factor of $f(x)$.

b) Solve the equation $f(x) = 0$.

Solution

a) $f(3) = 3^3 - 3^2 - 10(3) + 12$
$= 27 - 9 - 30 + 12 = 0$

This shows that $(x - 3)$ is a factor of $f(x)$.

To show that $(x - 3)$ is a factor, you need to show that $f(3) = 0$.

b) Once you have found one linear factor of a cubic expression, the remaining factor is quadratic. With practice you will be able to do this step by inspection.

$x^3 - x^2 - 10x + 12 = (x - 3)(x^2 + 2x - 4)$

Check by multiplying out that you agree with this answer.

$x^2 + 2x - 4$ cannot be factorised so use the quadratic formula for the next step.

How do you know it can't be factorised?

$$ax^2 + bx + c = 0 \Rightarrow x = \frac{-b \pm \sqrt{b^2 - 4ac}}{2a}$$

In this example, $a = 1$, $b = 2$, $c = -4 \Rightarrow x = \dfrac{-2 \pm \sqrt{2^2 - 4(1)(-4)}}{2(1)}$

$\Rightarrow x = \dfrac{-2 \pm \sqrt{20}}{2}$

$\Rightarrow x = -1 \pm \sqrt{5}$

The solution to the equation is therefore $x = 3$ or $x = -1 \pm \sqrt{5}$.

Using the factor theorem to solve a cubic equation

This is very similar to earlier work except that the first step is to find a linear factor by inspection.

> **Worked example**

a) Work systematically to find a linear factor of $x^3 - 5x^2 - 2x + 24$.

b) Solve the equation $x^3 - 5x^2 - 2x + 24 = 0$.

c) Sketch the graph of $y = x^3 - 5x^2 - 2x + 24$.

d) Sketch $y = |x^3 - 5x^2 - 2x + 24|$ on a separate set of axes.

3 FACTORS OF POLYNOMIALS

Solution

a) Let $f(x) = x^3 - 5x^2 - 2x + 24$.
$f(1) = 1 - 5 - 2 + 24 = 18$
$f(-1) = -1 - 5 + 2 + 24 = 20$
$f(2) = 8 - 20 - 4 + 24 = 8$
$f(-2) = -8 - 20 + 4 + 24 = 0$

Start by working systematically through all factors of 24 until you find one giving $f(x) = 0$.

This shows that $(x+2)$ is a linear factor of $x^3 - 5x^2 - 2x + 24$.

b) Factorising by inspection:
$x^3 - 5x^2 - 2x + 24 = (x+2)(x^2 + ax + 12)$

The second bracket starts with x^2 to get the x^3 term and finishes with 12 since $2 \times 12 = 24$.

Looking at the x^2 term on both sides: $-5x^2 = 2x^2 + ax^2$
$\Rightarrow a = -7$

$x^3 - 5x^2 - 2x + 24 = 0 \Rightarrow (x+2)(x^2 - 7x + 12) = 0$
$\Rightarrow (x+2)(x-3)(x-4) = 0$
$\Rightarrow x = -2, x = 3 \text{ or } x = 4$

c) The graph is a cubic curve with a positive x^3 term that crosses the x-axis at -2, 3 and 4 and crosses the y-axis when $y = 24$.

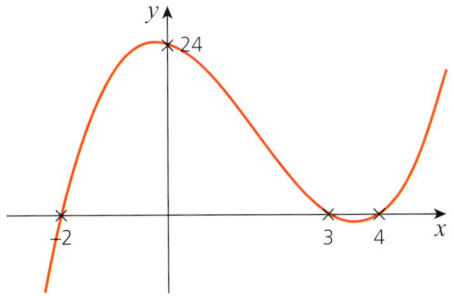

d) To sketch the curve $y = |x^3 - 5x^2 - 2x + 24|$, first sketch the curve as above, then reflect in the x-axis any part of the curve which is below it.

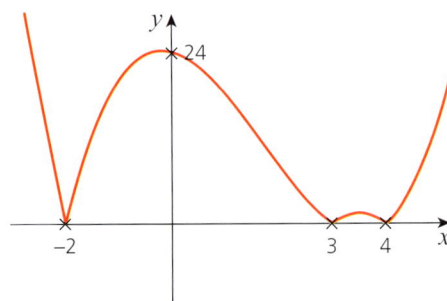

Exercise 3.2

1. Determine whether the following linear functions are factors of the given polynomials:
   a) $x^3 - 8x + 7$; $(x - 1)$
   b) $x^3 + 8x + 7$; $(x + 1)$
   c) $2x^3 + 3x^2 - 4x - 1$; $(x - 1)$
   d) $2x^3 - 3x^2 + 4x + 1$; $(x + 1)$
2. Use the factor theorem to find a linear factor of each of the following functions. Then factorise each function as a product of three linear factors and sketch its graph.
   a) $x^3 - 7x - 6$
   b) $x^3 - 7x + 6$
   c) $x^3 + 5x^2 - x - 5$
   d) $x^3 - 5x^2 - x + 5$
3. Factorise each of the following functions completely:
   a) $x^3 + x^2 + x + 1$
   b) $x^3 - x^2 + x - 1$
   c) $x^3 + 3x^2 + 3x + 2$
   d) $x^3 - 3x^2 + 3x - 2$
4. For what value of a is $(x - 2)$ a factor of $x^3 - 2ax + 4$?
5. For what value of c is $(2x + 3)$ a factor of $2x^3 + cx^2 - 4x - 6$?
6. The expression $x^3 - 6x^2 + ax + b$ is exactly divisible by $(x - 1)$ and $(x - 3)$.
   a) Find two simultaneous equations for a and b.
   b) Hence find the values of a and b.

The remainder theorem

> ### Worked example
>
> Is there an integer root to the equation $x^3 - 2x^2 + x + 1 = 0$?
>
> ### Solution
>
> Since the first term is x^3 and the last term is $+1$, the only possible factors are $(x + 1)$ and $(x - 1)$.
>
> $f(1) = 1$ and $f(-1) = -3$ so there is no integer root.

*This leads us to the **remainder theorem**.*

Any polynomial can be divided by another polynomial of lesser order using either long division or inspection. However, there will sometimes be a remainder. The steps for algebraic long division are very similar to those for numerical long division as shown below.

Look at $(x^3 - 2x^2 + x + 1) \div (x + 1)$.

Taking the first term from each (the dividend and the divisor) gives $x^3 \div x = x^2$, the first term on the top in the quotient.

$-3x^2 \div x$ gives $-3x$.
Similarly $4x \div x = 4$

$$\begin{array}{r} x^2 - 3x + 4 \\ x+1 \overline{\smash{)}x^3 - 2x^2 + x + 1} \\ \underline{-x^3 + x^2} \\ -3x^2 + x \\ \underline{--3x^2 - 3x} \\ 4x + 1 \\ \underline{--4x + 4} \\ -3 \end{array}$$

This result can be written as:
$$x^3 - 2x^2 + x + 1 = (x+1)(x^2 - 3x + 4) - 3.$$

3 FACTORS OF POLYNOMIALS

Substituting $x = -1$ into both sides gives a remainder of -3.

This means that $f(-1)$ will always be the remainder when a function $f(x)$ is divided by $(x + 1)$.

Generalising this gives the **remainder theorem**.

For any polynomial $f(x)$, $f(a)$ is the remainder when $f(x)$ is divided by $(x - a)$.

$$f(x) = (x - a)g(x) + f(a)$$

Worked example

Find the remainder when $f(x) = 2x^3 + 3x - 5$ is divided by $(x - 2)$.

Solution
Using the remainder theorem, the remainder is $f(2)$.

$$f(2) = 2(2)^3 + 3(2) - 5 = 17$$

Worked example

When $2x^3 - 3x^2 + ax - 5$ is divided by $x - 2$, the remainder is 7. Find the value of a.

Solution
To find the remainder, substitute $x = 2$ into $2x^3 - 3x^2 + ax - 5$.

$$2(2)^3 - 3(2)^2 + a(2) - 5 = 7$$
$$\Rightarrow 16 - 12 + 2a - 5 = 7$$
$$\Rightarrow -1 + 2a = 7$$
$$a = 4$$

Exercise 3.3

1. For each function, find the remainder when it is divided by the linear factor shown in brackets:
   a) $x^3 + 2x^2 - 3x - 4$; $(x - 2)$
   b) $2x^3 + x^2 - 3x - 4$; $(x + 2)$
   c) $3x^3 - 3x^2 - x - 4$; $(x - 4)$
   d) $3x^3 + 3x^2 + x + 4$; $(x + 4)$

2. When $f(x) = x^3 + ax^2 + bx + 10$ is divided by $(x + 1)$, there is no remainder. When it is divided by $(x - 1)$, the remainder is 4. Find the values of a and b.

3. The equation $f(x) = x^3 + 4x^2 + x - 6$ has three integer roots. Solve $f(x) = 0$.

4. $(x - 2)$ is a factor of $x^3 + ax^2 + a^2x - 14$. Find all possible values of a.

5. When $x^3 + ax + b$ is divided by $(x - 1)$, the remainder is -12. When it is divided by $(x - 2)$, the remainder is also -12. Find the values of a and b and hence solve the equation $x^3 + ax + b = 0$.

6. Sketch each curve by first finding its points of intersection with the axes:
   a) $y = x^3 + 2x^2 - x - 2$
   b) $y = x^3 - 4x^2 + x + 6$
   c) $y = 4x - x^3$
   d) $y = 2 + 5x + x^2 - 2x^3$

The remainder theorem

Past-paper questions

1. The polynomial $f(x) = ax^3 - 15x^2 + bx - 2$ has a factor of $2x - 1$ and a remainder of 5 when divided by $x - 1$.
   - (i) Show that $b = 8$ and find the value of a. [4]
   - (ii) Using the values of a and b from part (i), express $f(x)$ in the form $(2x - 1) g(x)$, where $g(x)$ is a quadratic factor to be found. [2]
   - (iii) Show that the equation $f(x) = 0$ has only one real root. [2]

   Cambridge O Level Additional Mathematics (4037)
   Paper 11 Q6, June 2015
   Cambridge IGCSE Additional Mathematics (0606)
   Paper 11 Q6, June 2015

2. A function f is such that $f(x) = 4x^3 + 4x^2 + ax + b$. It is given that $2x - 1$ is a factor of both $f(x)$ and $f'(x)$.
   - (i) Show that $b = 2$ and find the value of a. [5]

   Using the values of a and b from part (i),
   - (ii) find the remainder when $f(x)$ is divided by $x + 3$, [2]
   - (iii) express $f(x)$ in the form $f(x) = (2x - 1)(px^2 + qx + r)$, where p, q and r are integers to be found, [2]
   - (iv) find the values of x for which $f(x) = 0$. [2]

   Cambridge O Level Additional Mathematics (4037)
   Paper 12 Q10, November 2012
   Cambridge IGCSE Additional Mathematics (0606)
   Paper 12 Q10, November 2012

3. It is given that $f(x) = 6x^3 - 5x^2 + ax + b$ has a factor of $x + 2$ and leaves a remainder of 27 when divided by $x - 1$.
   - (i) Show that $b = 40$ and find the value of a. [4]
   - (ii) Show that $f(x) = (x + 2)(px^2 + qx + r)$, where p, q and r are integers to be found. [2]
   - (iii) Hence solve $f(x) = 0$. [2]

   Cambridge O Level Additional Mathematics (4037)
   Paper 12 Q7, June 2013
   Cambridge IGCSE Additional Mathematics (0606)
   Paper 12 Q7, June 2013

Now you should be able to:
- ★ know and use the remainder and factor theorems
- ★ find factors of polynomials
- ★ solve cubic equations.

3 FACTORS OF POLYNOMIALS

Key points

- An expression of the form $ax^3 + bx^2 + cx + d$ where $a \neq 0$ is called a **cubic expression**.
- The graph of a cubic expression can be plotted by first calculating the value of y for each value of x in the range.
- The solution to a cubic equation is the set of values for which the corresponding graph crosses the x-axis.
- The **factor theorem** states: if $(x - a)$ is a factor of f(x), then f$(a) = 0$ and $x = a$ is a root of the equation f$(x) = 0$.
- The **remainder theorem** states: for any polynomial f(x), f(a) is the remainder when f(x) is divided by $(x - a)$. This can be generalised to $f(x) = (x - a)g(x) + f(a)$.

Review exercise 1

Ch 1 **1** The functions f and g are defined by:
$f(x) = \sqrt{1-x}$ for $x \leq 1$
$g(x) = \dfrac{x+2}{3x}$ for $x > 0$

  a Find fg(3). [2]
  b Find an expression for $g^2(x)$, giving your answer in the form $\dfrac{ax+b}{cx+d}$, where a, b, c and d are constants to be found. [3]
  c Find an expression for $f^{-1}(x)$. [2]
  d On axes like the ones here, sketch the graphs of $y = f(x)$ and $y = f^{-1}(x)$, stating the domain and range of $f^{-1}(x)$. [4]

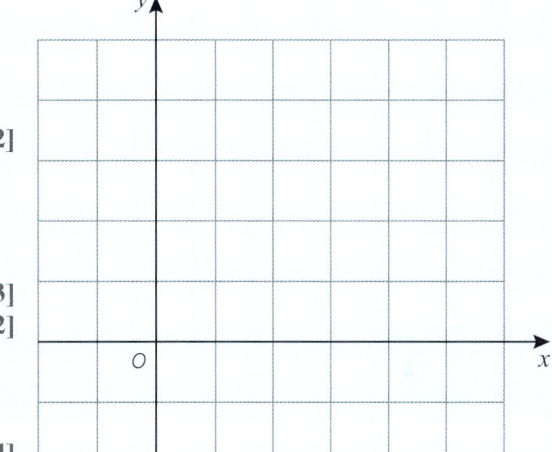

2 The function f is defined by $f(x) = x^2 - 1$ for $x \geq 0$.
  a Find an expression for $f^{-1}(x)$, stating its domain and range. [3]
  b On axes like the ones below, sketch the graphs of $y = f(x)$ and $y = f^{-1}(x)$, indicating the geometrical relationship between the two. [3]

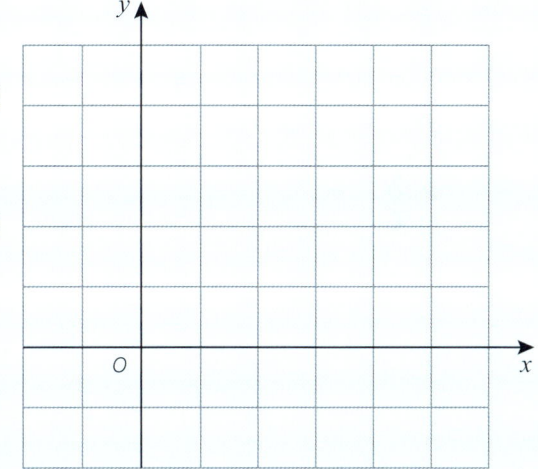

3 a The function f is defined by $f(x) = \sqrt{1+x^2}$, for all real values of x. The graph of $y = f(x)$ is given below.
  i Explain, with reference to the graph, why f does not have an inverse. [1]
  ii Find $f^2(x)$. [2]
  b The function g is defined, for $x > k$, by $g(x) = \sqrt{1+x^2}$ and g has an inverse.
  i Write down a possible value for k. [1]
  ii Find $g^{-1}(x)$. [2]

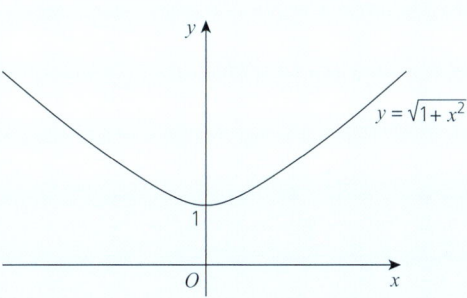

Cambridge O Level Additional Mathematics (4037)
Paper 22 Q10 a & b, February/March 2018
Cambridge IGCSE Additional Mathematics (0606)
Paper 22 Q10 a & b, February/March 2018

REVIEW EXERCISE 1

Ch 2

4 a Express $2x^2 + 5x - 3$ in the form $p(x-q)^2 + r$, where p, q and r are constants to be found. [3]

b Hence state the least value of $2x^2 + 5x - 3$ and the value of x at which this occurs. [2]

5 a Find the values of x for which $(2x+1)^2 \leq 3x + 4$. [3]

b Show that, whatever the value of k, the equation $\dfrac{x^2}{4} + kx + k^2 + 1 = 0$ has no real roots. [3]

Cambridge O Level Additional Mathematics (4037)
Paper 22 Q4, February/March 2019
Cambridge IGCSE Additional Mathematics (0606)
Paper 22 Q4, February/March 2019

6 Find the values of k for which the line $y = kx + 8$ is a tangent to the curve $y = 2x^2 + 8x + k$. [5]

7 Find the set of values of k for which the curve $y = x^2 - 2kx + k + 2$ lies above the x-axis for all values of x. [4]

Ch 2, 3

8 The polynomial p(x) is $x^4 - 2x^3 - 3x^2 + 8x - 4$.
 i Show that p(x) can be written as $(x-1)(x^3 - x^2 - 4x + 4)$. [1]
 ii Hence write p(x) as a product of its linear factors, showing all your working. [4]

Cambridge O Level Additional Mathematics (4037)
Paper 22 Q3, February/March 2017
Cambridge IGCSE Additional Mathematics (0606)
Paper 22 Q3, February/March 2017

9 It is given that f(x) = $3x^3 + 8x^2 + ax + b$ has a factor of $x - 1$ and leaves a remainder of -50 when divided by $x + 4$.
 a Show that $b = -6$ and find the value of a. [4]
 b Show that f(x) = $(x-1)(px^2 + qx + r)$ where p, q and r are integers to be found. [2]
 c Find the values of x for which f(x) = 0. [2]

10 It is given that p(x) = $2x^3 + x^2 + ax + 8$ has a factor of $x + 2$. When p(x) is divided by $x - 1$ the remainder is b.
 a Show that $a = -2$ and find the value of b. [4]
 b Express p(x) in the form $(x+2)$q(x) where q(x) is a quadratic to be found. [2]
 c Show that the equation p(x) = 0 has only one real root. [2]

4 Equations, inequalities and graphs

It is India that gave us the ingenious method of expressing all numbers by means of ten symbols, each symbol receiving a value of position as well as an absolute value; a profound and important idea which appears so simple to us now that that we ignore its true merit.

Pierre-Simon, Marquis de Laplace (1749–1827)

The picture shows a quadrat. This is a tool used by biologists to select a random sample of ground; once it is in place they will make a record of all the plants and creatures living there. Then they will throw the quadrat so that it lands somewhere else.

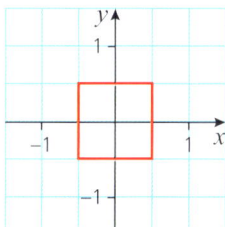

This diagram illustrates a 1 metre square quadrat. The centre point is taken to be the origin and the sides to be parallel to the x- and y-axes.

Discussion point

What is the easiest way to describe the region it covers?

4 EQUATIONS, INEQUALITIES AND GRAPHS

Many practical situations involve the use of inequalities.

How economical is a Formula 1 car?

The Monaco Grand Prix, consisting of 78 laps and a total distance of approximately 260 km, is a well-known Formula 1 race. In 2017 it was won by Sebastian Vettel in 1 hour 44 minutes. Restrictions on the amount and use of fuel mean that drivers need to manage the performance of their car very carefully throughout the race.

A restriction in 2017 was that the total amount of fuel used during the race was limited to 105 kg which is approximately 140 litres.

Using: f to denote the total amount of fuel used in litres

d to represent the distance travelled in kilometres

E to represent the fuel economy in litres per kilometre $\left(E = \frac{f}{d}\right)$

the restriction can be represented as $E \leqslant \frac{140}{260}$

$$\Rightarrow E \leqslant 0.538$$

This shows that, at worst, the fuel economy of Vettel's Ferrari Formula 1 car is 0.538 litres per kilometre.

> **Discussion point**
>
> How does this compare with an average road car?

Modulus functions and graphs

For any real number, the **absolute value**, or **modulus**, is its positive size whether that number is positive or negative. It is denoted by a vertical line on each side of the quantity. For example, $|5| = 5$ and $|-5| = 5$ also. The absolute value of a number will always be positive or zero. It can be thought of as the distance between that point on the x-axis and the origin.

You have already met graphs of the form $y = 3x + 4$ and $y = x^2 - 3x - 4$. However, you might not be as familiar with graphs of the form $y = |3x + 4|$ and $y = |x^2 - 3x - 4|$.

For any real number x, the modulus of x is denoted by $|x|$ and is defined as:
$|x| = x$ if $x \geqslant 0$
$|x| = -x$ if $x < 0$.

Solving modulus equations

 Worked example

Set up a table for the graphs $y = x + 2$ and $y = |x + 2|$ for $-6 \leqslant x \leqslant 2$. Draw both graphs on the same axes.

Solution

y	−6	−5	−4	−3	−2	−1	0	1	2
$x + 2$	−4	−3	−2	−1	0	1	2	3	4
$|x + 2|$	4	3	2	1	0	1	2	3	4

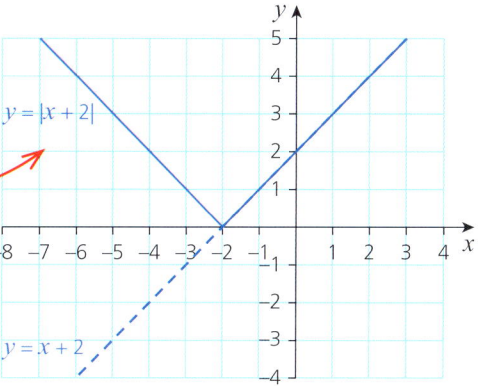

The equation of this part of the graph is $y = -(x + 2)$.

Notice that the effect of taking the modulus is a transformation in which the positive part of the original graph (above the x-axis) remains the same and the negative part of the original graph (below the x-axis) is reflected in the x-axis.

Solving modulus equations

 Worked example

Solve the equation $|2x + 3| = 5$

a) graphically

b) algebraically.

4 EQUATIONS, INEQUALITIES AND GRAPHS

Solution

a) First draw the graph of $y = 2x + 3$.

Start by choosing three values of x and calculating the corresponding values of y, for example, $(-2, -1)$, $(0, 3)$ and $(2, 7)$.

Then reflect in the x-axis any part of the graph that is below the x-axis to give the graph of $y = |2x + 3|$.

Next draw the line $y = 5$.

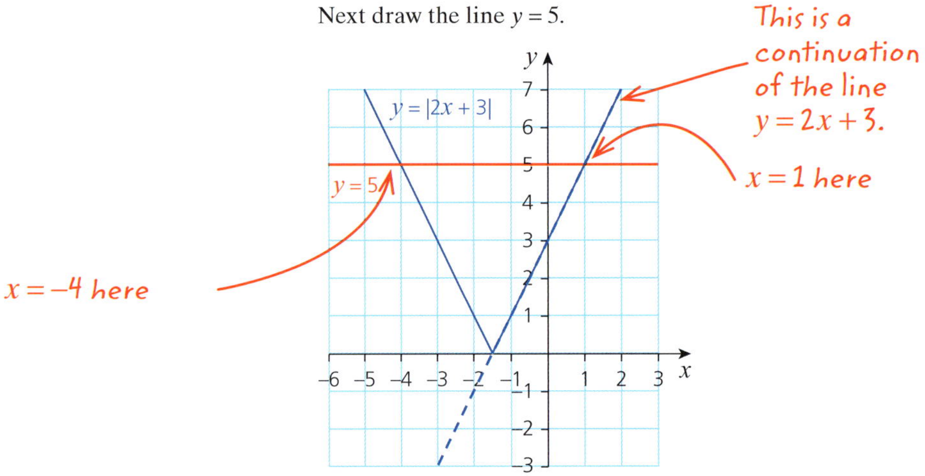

This is a continuation of the line $y = 2x + 3$.

$x = 1$ here

$x = -4$ here

The solution is given by the values of x where the V-shaped graph meets the line $y = 5$

$\Rightarrow x = 1$ or $x = -4$.

b) $|2x + 3| = 5 \Rightarrow 2x + 3 = 5$ or $2x + 3 = -5$

$\qquad\qquad \Rightarrow 2x = 2 \quad$ or $\quad 2x = -8$

$\qquad\qquad \Rightarrow x = 1 \quad$ or $\quad x = -4$

Discussion point

Notice that in the solution three points are used to draw the straight line when only two are necessary. Why is this good practice?

Solving modulus equations

➤ Worked example

Solve the equation $|3x - 1| = x + 5$

a) graphically

b) algebraically.

Solution

a) Start by drawing the graphs of $y = |3x - 1|$ and $y = x + 5$ on the same axes.

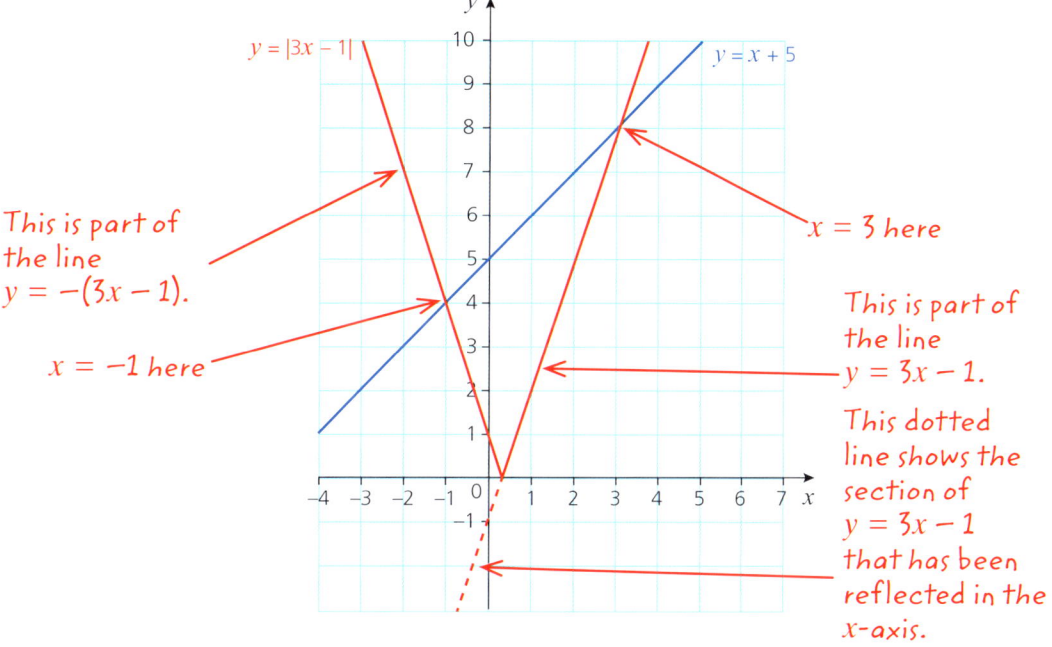

The solution is given by the values of x where the V-shaped graph meets the line $y = x + 5$.

$\Rightarrow x = -1$ or $x = 3$

b) $|3x - 1| = x + 5$ $\quad\Rightarrow 3x - 1 = x + 5 \quad$ or $-(3x - 1) = x + 5$

$\quad\quad\quad\quad\quad\quad\quad\quad\Rightarrow 2x = 6 \quad\quad\quad$ or $\quad -4x = 4$

$\quad\quad\quad\quad\quad\quad\quad\quad\Rightarrow x = 3 \quad\quad\quad$ or $\quad x = -1$

Either of these methods can be extended to find the points where two V-shaped graphs intersect. However, the graphical method will not always give an accurate solution.

4 EQUATIONS, INEQUALITIES AND GRAPHS

> ## Worked example
>
> Solve the equation $|2x + 5| = |x - 4|$.
>
> ### Solution
> Start by drawing the graphs of $y = |2x + 5|$ and $y = |x - 4|$ on the same axes.

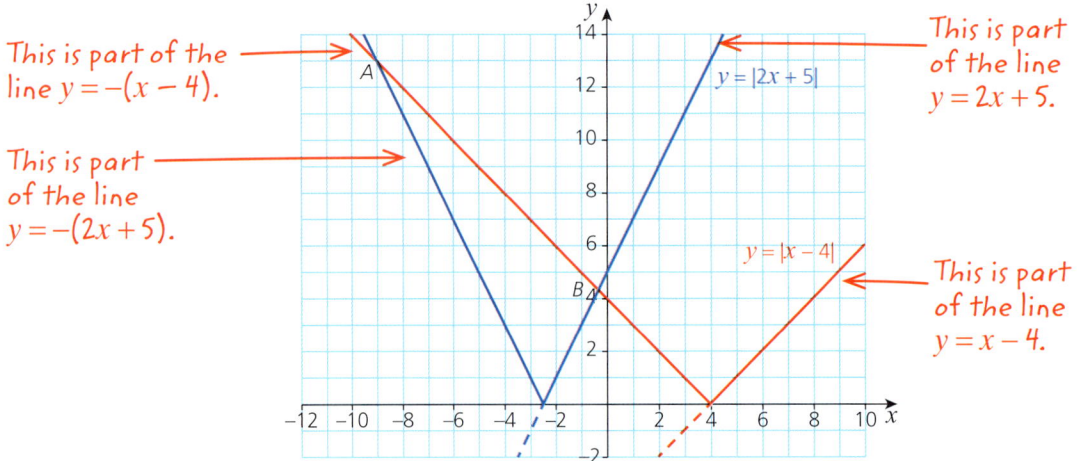

This is part of the line $y = -(x - 4)$.

This is part of the line $y = -(2x + 5)$.

This is part of the line $y = 2x + 5$.

This is part of the line $y = x - 4$.

The graph shows that the point A is $(-9, 13)$, but the coordinates of B are not clear.

This shows a failing of the graphical method. However, the graph is useful in determining the equation of the line required for an algebraic solution.

The graph shows that both points of intersection occur where the reflected part of the line $y = x - 4$, i.e. the line $y = -(x - 4)$ intersects the graph of $y = |2x + 5|$.

At A, $y = 4 - x$ meets $y = -2x - 5$

$\Rightarrow 4 - x = -2x - 5$

$\Rightarrow 2x - x = -5 - 4$

$\Rightarrow x = -9$

When $x = -9$, $y = 4 - (-9) = 13$, i.e. A is the point $(-9, 13)$.

At B, $y = 4 - x$ meets $y = 2x + 5$

$\Rightarrow 2x + 5 = 4 - x$

$\Rightarrow 3x = -1$

$\Rightarrow x = -\frac{1}{3}$

When $x = -\frac{1}{3}$, $y = 4 - \left(-\frac{1}{3}\right) = 4\frac{1}{3}$, i.e. B is the point $\left(-\frac{1}{3}, 4\frac{1}{3}\right)$.

Solving modulus equations

> **Note**
> Modulus equations, like the one in the previous example, can also be solved by using the **method of squaring.**

➜ Worked example

Solve the equation $|3x - 2| = |x + 4|$.

Solution

Square both sides of the equation. This ensures that both sides are positive so you do not need to consider individual cases.

$|3x - 2| = |x + 4|$

$\Rightarrow (3x - 2)^2 = (x + 4)^2$

$\Rightarrow 9x^2 - 12x + 4 = x^2 + 8x + 16$

$\Rightarrow 8x^2 - 20x - 12 = 0$ ← *Rearrange to form a quadratic equation, then solve for x.*

$\Rightarrow 2x^2 - 5x - 3 = 0$

$\Rightarrow (2x + 1)(x - 3) = 0$

$2x + 1 = 0 \Rightarrow x = -\frac{1}{2}$

$x - 3 = 0 \Rightarrow x = 3$

So, $x = -\frac{1}{2}$ or $x = 3$

➜ Worked example

a) Draw the graph of $y = |x^2 - 8x + 14|$.

b) Use the graph to solve $|x^2 - 8x + 14| = 2$.

c) Use algebra to verify your answer to part **b**.

Solution

a)

These are parts of the curve $y = x^2 - 8x + 14$.

This is part of the curve $y = -(x^2 - 8x + 14)$.

This dotted line shows the section of $y = x^2 - 8x + 14$ that has been reflected in the x-axis.

$4 - \sqrt{2}$ $4 + \sqrt{2}$

61

4 EQUATIONS, INEQUALITIES AND GRAPHS

b)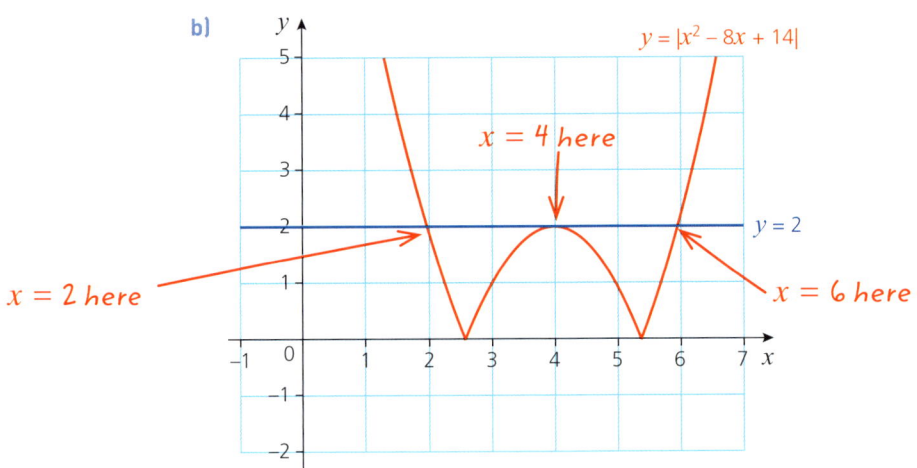

So, from the graph $x = 2$, $x = 4$ or $x = 6$.

c) $|x^2 - 8x + 14| = 2 \Rightarrow x^2 - 8x + 14 = 2$ or $-(x^2 - 8x + 14) = 2$
$\Rightarrow x^2 - 8x + 12 = 0$ $\Rightarrow -x^2 + 8x - 16 = 0$
$\Rightarrow (x - 2)(x - 6) = 0$ $\Rightarrow x^2 - 8x + 16 = 0$
$\Rightarrow x = 2$ or $x = 6$ $\Rightarrow (x - 4)^2 = 0$
                                                                                     $\Rightarrow x = 4$

So $x = 2$, $x = 4$ or $x = 6$.

Exercise 4.1

For questions 1–3, sketch each pair of graphs on the same axes.

1. a) $y = x$ and $y = |x|$
   b) $y = x - 1$ and $y = |x - 1|$
   c) $y = x - 2$ and $y = |x - 2|$
2. a) $y = 2x$ and $y = |2x|$
   b) $y = 2x - 1$ and $y = |2x - 1|$
   c) $y = 2x - 2$ and $y = |2x - 2|$
3. a) $y = 2 - x$ and $y = |2 - x|$
   b) $y = 3 - x$ and $y = |3 - x|$
   c) $y = 4 - x$ and $y = |4 - x|$
4. a) Draw the graph of $y = |x + 1|$.
   b) Use the graph to solve the equation $|x + 1| = 5$.
   c) Use algebra to verify your answer to part **b**.
5. a) Draw the graph of $y = |x - 1|$.
   b) Use the graph to solve the equation $|x - 1| = 5$.
   c) Use algebra to verify your answer to part **b**.
6. a) Draw the graph of $y = |2x + 3|$.
   b) Use the graph to solve the equation $|2x + 3| = 7$.
   c) Use algebra to verify your answer to part **b**.

Exercise 4.1 (cont)

7. a) Draw the graph of $y = |2x - 3|$.
   b) Use the graph to solve the equation $|2x - 3| = 7$.
   c) Use algebra to verify your answer to part **b**.
8. a) On the same set of axes, draw the graphs of $y = 2x - 3$, $y = |2x - 3|$ and $y = x + 3$ for $-1 \leq x \leq 7$.
   b) Use your graphs to solve $|2x - 3| = x + 3$.
   c) Use algebra to verify your answer to part **b**.
9. a) On the same set of axes, draw the graphs of $y = |3x - 3|$ and $y = 5x + 11$ for $-3 \leq x \leq 3$.
   b) Use your graphs to solve $|3x - 3| = 5x + 11$.
   c) Use algebra to verify your answer to part **b**.
10. Solve the equation $|x + 1| = |x - 1|$ both graphically and algebraically.
11. Solve the equation $|x + 5| = |x - 5|$ both graphically and algebraically.
12. Solve the equation $|2x + 4| = |2x - 4|$ both graphically and algebraically.
13.

   a) Use the diagram above to solve $|x^2 + 10x + 17| = 8$.
   b) Use algebra to verify your answer to part **a**.
14. Solve algebraically the equation $|x^2 - 16x + 58| = 5$.
15. Solve algebraically the equation $|2x^2 - 3x - 2| = 7$.

4 EQUATIONS, INEQUALITIES AND GRAPHS

Solving modulus inequalities

When illustrating an inequality in one variable:
- An open circle at the end of a line shows that the end point is excluded.
- A solid circle at the end of a line shows that the end point is included.
- The line is drawn either in colour or as a solid line.

For example, the inequality $-2 < x \leq 3$ is shown as:

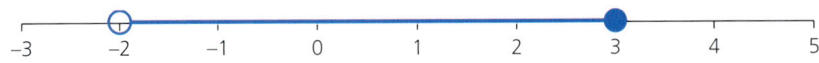

Worked example

a) Solve algebraically the inequality $|x - 3| > 2$.
b) Illustrate the solution on a number line.

An open circle is used to show that the value there is not part of the solution.

Solution

a) $|x - 3| > 2 \Rightarrow x - 3 > 2$ or $x - 3 < -2$

$\Rightarrow x > 5$ or $x < 1$

The blue lines show the required parts of the number line.

b)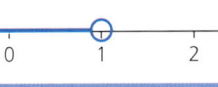

Worked example

Write the inequality $-3 \leq x \leq 9$ in the form $|x - a| \leq b$ and show a and b on a number line.

Solution

$|x - a| \leq b \Rightarrow -b \leq x - a \leq b$

$\Rightarrow a - b \leq x \leq a + b$ ← *You are finding the values of x within $a \pm b$.*

Solve $a + b = 9$ and $a - b = -3$ simultaneously.

Adding: $2a = 6$, so $a = 3$

Subtracting: $2b = 12$, so $b = 6$

Substituting in $a - b \leq x \leq a + b$ gives $-3 \leq x \leq 9$.
Substituting in $|x - a| \leq b$ gives $|x - 3| \leq 6$.

Solving modulus inequalities

→ Worked example

Solve the inequality $|3x + 2| \leq |2x - 3|$.

Solution

Draw the graphs of $y = |3x + 2|$ and $y = |2x - 3|$. The inequality is true for values of x where the unbroken blue line is below or crosses the unbroken red line, i.e. between (and including) the points A and B.

Draw the line $y = 3x + 2$ as a straight line through $(0, 2)$ with a gradient of $+3$. Reflect in the x-axis the part of the line that is below this axis.

Draw the line $y = 2x - 3$ as a straight line through $(0, -3)$ with a gradient of $+2$. Reflect in the x-axis the part of the line that is below this axis.

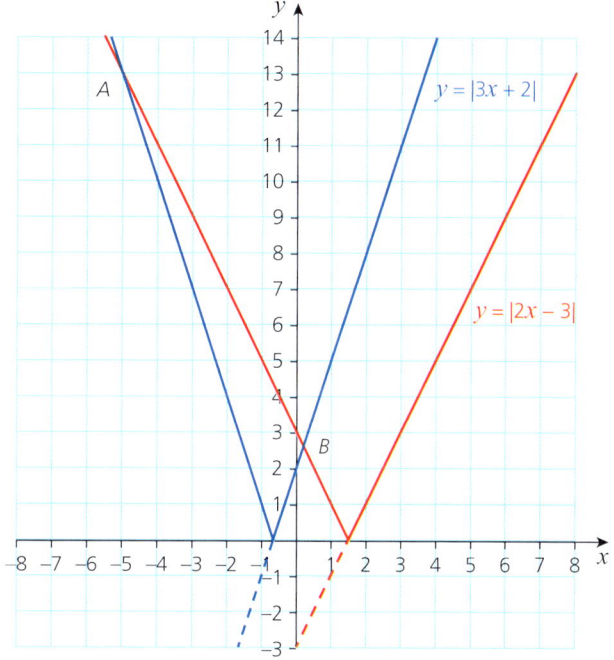

The graph shows that $x = -5$ at A, but the exact value for x at B is not clear. The algebraic solution gives a more precise value.

At A, $-(3x + 2) = -(2x - 3) \Rightarrow 3x + 2 = 2x - 3$

$$\Rightarrow x = -5$$

Substituting in either of the equations gives $y = 13$, so A is the point $(-5, 13)$.

At B, $3x + 2 = -(2x - 3) \Rightarrow 3x + 2 = -2x + 3$

$$\Rightarrow 5x = 1$$

$$\Rightarrow x = 0.2$$

Substituting in either equation gives $y = 2.6$, so B is the point $(0.2, 2.6)$.

The inequality is satisfied for values of x between A and B, i.e. for $-5 \leq x \leq 0.2$.

4 EQUATIONS, INEQUALITIES AND GRAPHS

➡ Worked example

Solve the inequality $|x + 7| < |4x|$.

Solution

The question does not stipulate a particular method, so start with a sketch graph.

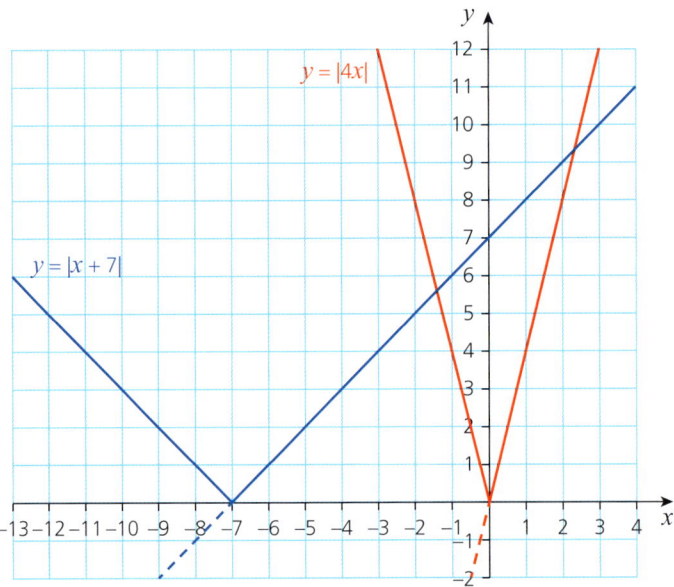

The sketch graph shows that the points where the two graphs intersect do not have integer coordinates. This means that a graphical method is unlikely to give an accurate solution.

Use algebra to find the points when $|x + 7| = |4x|$, i.e. when $x + 7 = 4x$ and when $x + 7 = -4x$.

Discussion point

Why is it sufficient to consider only these two cases? Why do you not need to consider when $-(x + 7) = 4x$?

$x + 7 = 4x$ when $x = \frac{7}{3}$.

$x + 7 = -4x$ when $x = -\frac{7}{5}$.

Think about a point to the left of $x = -\frac{7}{5}$, such as $x = -2$.

When $x = -2$, $|x + 7| < |4x|$ gives $5 < 8$. This is true so the inequality is satisfied.

This tells you that part of the solution is to the left of $x = -\frac{7}{5}$, i.e. $x < -\frac{7}{5}$.

Next think about a value of x in the interval $\left(-\frac{7}{5}, \frac{7}{3}\right)$, for example, $x = 0$.

When $x = 0$, $|x + 7| < |4x|$ gives $7 < 0$, which is false.

Any other value in this interval will also give a false result.

Finally consider a value greater than $\frac{7}{3}$, for example, $x = 3$.

When $x = 3$, $|x + 7| < |4x|$ gives $10 < 12$. This is true so the inequality is satisfied.

Therefore, the solution is $x < -\frac{7}{5}$ or $x > \frac{7}{3}$.

Solving modulus inequalities

Inequalities in two dimensions are illustrated by regions. For example, $x > 2$ is shown by the part of the x–y plane to the right of the line $x = 2$ and $x < -1$ by the part to the left of the line $x = -1$.

If you are asked to illustrate the region $x \geq 2$, then the line $x = 2$ must be included as well as the region $x > 2$.

> **Note**
> - When the boundary line is included, it is drawn as a **solid line**; when it is excluded, it is drawn as a **dotted line**.
> - The answer to an inequality of this type is a region of the x–y plane, not simply a set of points. It is common practice to specify the region that you want (called the **feasible region**) by shading out the unwanted region. This keeps the feasible region clear so that you can see clearly what you are working with.

➡ Worked example

Illustrate the inequality $3y - 2x \geq 0$ on a graph.

Solution

Draw the line $3y - 2x = 0$ as a solid line through $(0, 0)$, $(3, 2)$ and $(6, 4)$.

> **Discussion point**
>
> Why are these points more suitable than, for example, $\left(1, \frac{2}{3}\right)$?

Choose a point which is not on the line as a test point, for example, $(1, 0)$.

Using these values, $3y - 2x = -2$. This is less than 0, so this point is not in the feasible region. Therefore shade out the region containing the point $(1, 0)$.

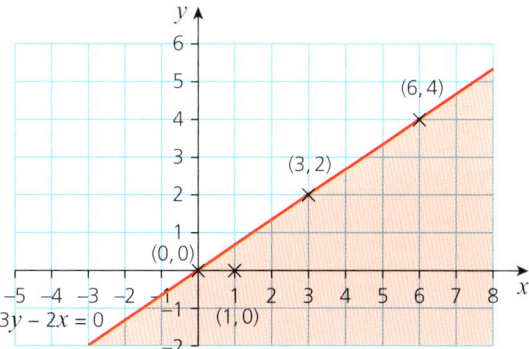

4 EQUATIONS, INEQUALITIES AND GRAPHS

Exercise 4.2

1. Write each of the following inequalities in the form $|x - a| \leq b$:
   a) $-3 \leq x \leq 15$
   b) $-4 \leq x \leq 16$
   c) $-5 \leq x \leq 17$

2. Write each of the following expressions in the form $a \leq x \leq b$:
   a) $|x - 1| \leq 2$
   b) $|x - 2| \leq 3$
   c) $|x - 3| \leq 4$

3. Solve the following inequalities:
   a) $|x - 1| < 4$
   b) $|x - 1| > 4$
   c) $|2x + 3| < 5$
   d) $|2x + 3| > 5$

4. Illustrate each of the following inequalities graphically by shading the unwanted region:
   a) $y - 2x > 0$
   b) $y - 2x \leq 0$
   c) $2y - 3x > 0$
   d) $2y - 3x \leq 0$

5. Solve the following inequalities: **i)** graphically **ii)** algebraically.
   a) $|x - 1| < |x + 1|$
   b) $|x - 1| > |x + 1|$
   c) $|2x - 1| \leq |2x + 1|$
   d) $|2x - 1| \geq |2x + 1|$

6. Each of the following graphs represents an inequality. Name the inequality.

   a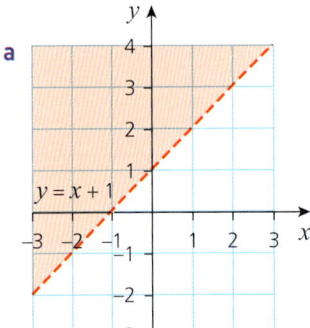
   b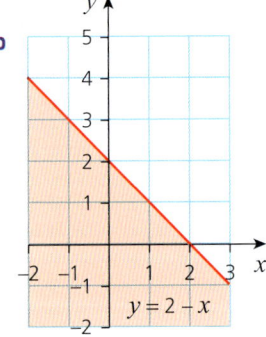

   c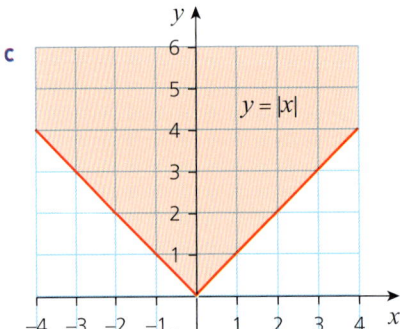

7. a) On the same set of axes, draw the graphs of $y = |5x - 4|$ and $y = x + 4$ for $-1 \leq x \leq 4$.
   b) Use your graphs to solve $|5x - 4| \leq x + 4$.
   c) Use algebra to verify your answer to part **b**.

8. Solve the following inequalities:
   a) $2|5x + 6| > 8$
   b) $5|2x - 7| \leq 2$
   c) $|4x - 3| \leq 2x + 6$
   d) $|2x + 5| \geq |x - 1|$
   e) $2|7x - 1| \leq |2x + 8|$

9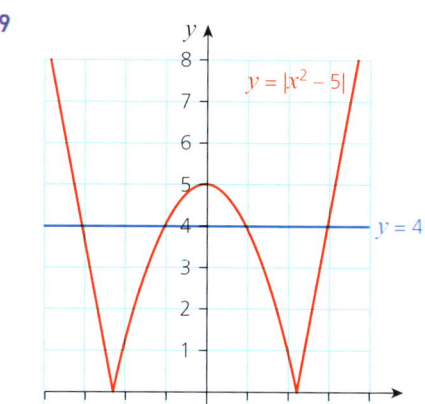

a) Use the diagram above to solve $|x^2 - 5| \leq 4$.
b) Use algebra to verify your answer to part b.

10 Solve algebraically the inequality $|2x^2 + 3x - 5| > 3$.

Using substitution to solve quadratic equations

Sometimes you will meet an equation which includes a square root. Although this is not initially a quadratic equation, you can use a substitution to solve it in this way, as shown in the following example.

➡ Worked example

Use the substitution $x = u^2$ to solve the equation $x - 3\sqrt{x} = -2$.

Solution

Substituting $x = u^2$ in the equation $x - 3\sqrt{x} = -2$ gives $u^2 - 3u = -2$

$\Rightarrow u^2 - 3u + 2 = 0$

Factorising $\Rightarrow (u - 1)(u - 2) = 0$

$\Rightarrow u = 1$ or $u = 2$

Checking these values Since $x = u^2$, $x = 1$ or $x = 4$.

When $x = 1$, $1 - 3\sqrt{1} = -2$, so $x = 1$ is a valid solution.
When $x = 4$, $4 - 3\sqrt{4} = -2$, so $x = 4$ is also a valid solution.

It is always advisable to check possible solutions, since in some cases not all values of u will give a valid solution to the equation, as shown in the following example.

4 EQUATIONS, INEQUALITIES AND GRAPHS

 Worked example

Solve the equation $x - \sqrt{x} = 6$.

Solution

Substituting $x = u^2$ in the equation $x - \sqrt{x} = 6$ gives $u^2 - u = 6$

$\Rightarrow u^2 - u - 6 = 0$

Factorising $\longrightarrow$ $\Rightarrow (u-3)(u+2) = 0$

$\Rightarrow u = 3$ or $u = -2$

Checking these values $\longrightarrow$ Since $x = u^2$, $x = 9$ or $x = 4$.

When $x = 9$, $9 - \sqrt{9} = 6$, so $x = 9$ is a possible solution.

When $x = 4$, $4 - \sqrt{4} = 2$, so reject $x = 4$ as a possible solution.

The only solution to this equation is $x = 9$.

Using graphs to solve cubic inequalities

Cubic graphs have distinctive shapes determined by the coefficient of x^3.

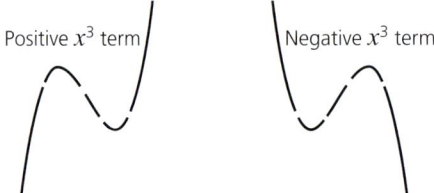

Positive x^3 term   Negative x^3 term

The centre part of each of these curves may not have two distinct turning points like those shown above, but may instead 'flatten out' to give a **point of inflexion.** When the modulus of a cubic function is required, any part of the curve below the x-axis is reflected in that axis.

 Worked example

a) Sketch the graph of $y = 3(x+2)(x-1)(x-7)$. Identify the points where the curve cuts the axes.

b) Sketch the graph of $y = |3(x+2)(x-1)(x-7)|$.

You are asked for a sketch graph, so although it must show the main features, it does not need to be absolutely accurate. You may find it easier to draw the curve first, with the positive x^3 term determining the shape of the curve, and then position the x-axis so that the distance between the first and second intersections is about half that between the second and third, since these are 3 and 6 units respectively.

Using graphs to solve cubic inequalities

Solution

a) The curve crosses the x-axis at −2, 1 and 7. Notice that the distance between consecutive points is 3 and 6 units respectively, so the y-axis is between the points −2 and 1 on the x-axis, but closer to the 1.

The curve crosses the y-axis when $x = 0$, i.e. when $y = 3(2)(-1)(-7) = 42$.

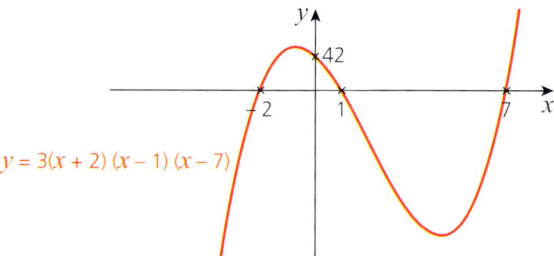

b) To obtain a sketch of the modulus curve, reflect any part of the curve which is below the x-axis in the x-axis.

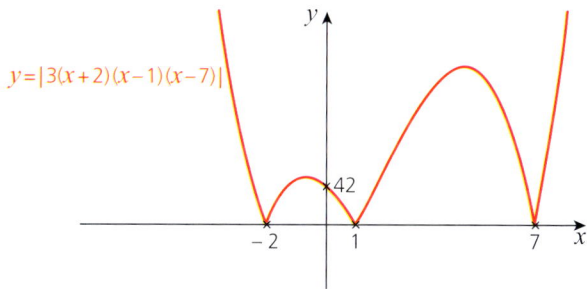

→ Worked example

Solve the inequality $3(x + 2)(x - 1)(x - 7) \leq -100$ graphically.

Solution

Because you are solving the inequality graphically, you will need to draw the curve as accurately as possible on graph paper, so start by drawing up a table of values.

$y = 3(x + 2)(x - 1)(x - 7)$

x	−3	−2	−1	0	1	2	3	4	5	6	7	8
$(x + 2)$	−1	0	1	2	3	4	5	6	7	8	9	10
$(x - 1)$	−4	−3	−2	−1	0	1	2	3	4	5	6	7
$(x - 7)$	−10	−9	−8	−7	−6	−5	−4	−3	−2	−1	0	1
y	−120	0	48	42	0	−60	−120	−162	−168	−120	0	210

4 EQUATIONS, INEQUALITIES AND GRAPHS

The solution is given by the values of x that correspond to the parts of the curve on or below the line $y = -100$.

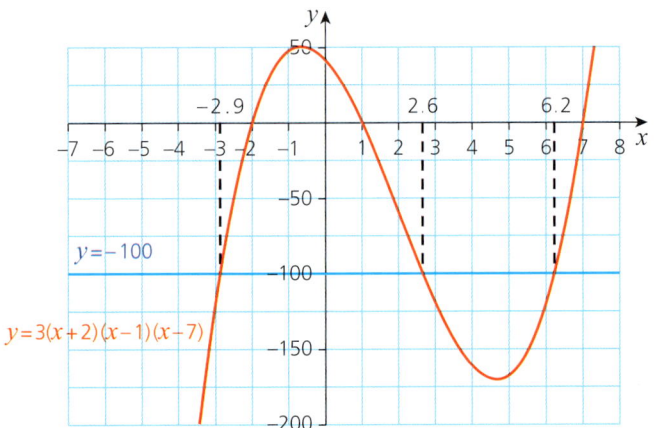

From the graph, the solution is $x \leq -2.9$ or $2.6 \leq x \leq 6.2$.

Discussion point

The example above used the graph to solve $3(x+2)(x-1)(x-7) \leq -100$. How would the answer change if the inequality sign was $\geq$ instead of $\leq$? What about $>$? What about $<$?

Exercise 4.3

Remember, $\sqrt{x}$ means the positive square root of x.

1 Where possible, use the substitution $x = u^2$ to solve the following equations:
   a) $x - 4\sqrt{x} = -4$
   b) $x + 2\sqrt{x} = 8$
   c) $x - 2\sqrt{x} = 15$
   d) $x + 6\sqrt{x} = -5$

2 Use the substitution $x = u^3$ to solve the equation $x^{\frac{2}{3}} + 3x^{\frac{1}{3}} = 4$.

3 Use the substitution $x = u^{\frac{3}{2}}$ to solve the equation $x^{\frac{4}{3}} - 10x^{\frac{2}{3}} = -9$.

4 Using a suitable substitution, solve the following equations:
   a) $x - 7\sqrt{x} = -12$
   b) $x - 2\sqrt{x} + 1 = 0$
   c) $x^{\frac{2}{3}} + 3x^{\frac{1}{3}} = 10$

5 a) Use the substitution $x = u^{\frac{1}{2}}$ to solve the equation $x^4 - 5x^2 + 4 = 0$.
   b) Using the same substitution, show that the equation $x^4 + 5x^2 + 4 = 0$ has no solution.
   c) Solve where possible:
      i) $x^{\frac{4}{3}} - 5x^{\frac{2}{3}} + 4 = 0$
      ii) $x^{\frac{4}{3}} + 5x^{\frac{2}{3}} + 4 = 0$

6 Sketch the following graphs, indicating the points where they cross the x-axis:
   a) $y = x(x-2)(x+2)$
   b) $y = |x(x-2)(x+2)|$
   c) $y = 3(2x-1)(x+1)(x+3)$
   d) $y = |3(2x-1)(x+1)(x+3)|$

7 Solve the following equations graphically. You will need to use graph paper.
   a) $x(x+2)(x+3) \geq 1$
   b) $x(x+2)(x+3) \leq -1$
   c) $(x+2)(x-1)(x-3) > 2$
   d) $(x+2)(x-1)(x-3) < -2$

Using graphs to solve cubic inequalities

8 Identify the following cubic graphs:

a)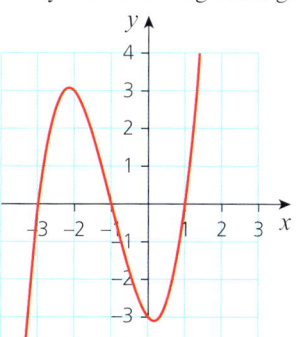

b)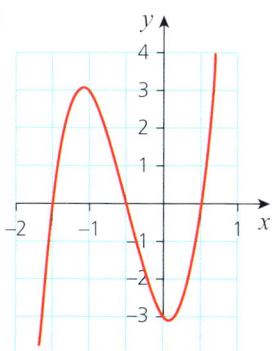

c)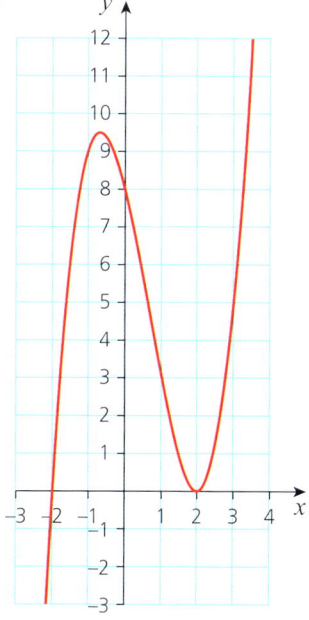

9 Identify these graphs. (They are the moduli of cubic graphs.)

a)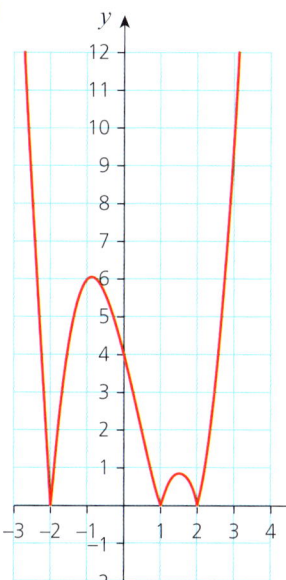

b)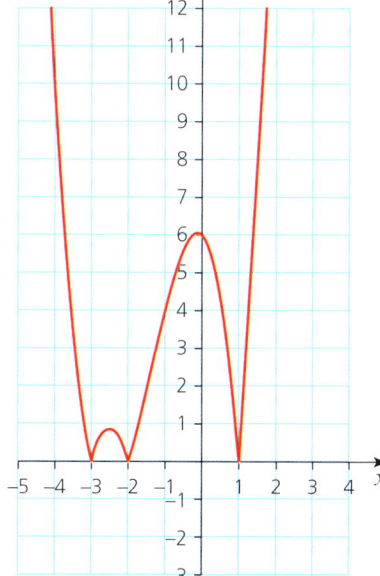

c)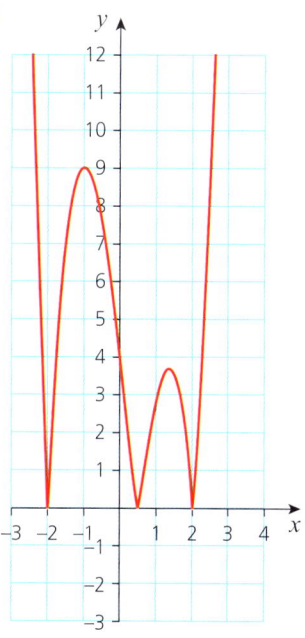

4 EQUATIONS, INEQUALITIES AND GRAPHS

Exercise 4.3 (cont)

10 Why is it not possible to identify the following graph without further information?

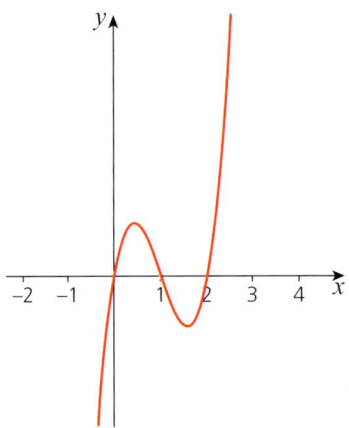

Past-paper questions

1 (i) Sketch the graph of $y = |(2x + 3)(2x - 7)|$. [4]

 (ii) How many values of x satisfy the equation
 $|(2x + 3)(2x - 7)| = 2x$? [2]

Cambridge O Level Additional Mathematics (4037)
Paper 23 Q6, November 2011
Cambridge IGCSE Additional Mathematics (0606)
Paper 23 Q6, November 2011

2 (i) On a grid like the one below, sketch the graph of
 $y = |(x - 2)(x + 3)|$ for $-5 \leq x \leq 4$, and state the coordinates of the points where the curve meets the coordinate axes. [4]

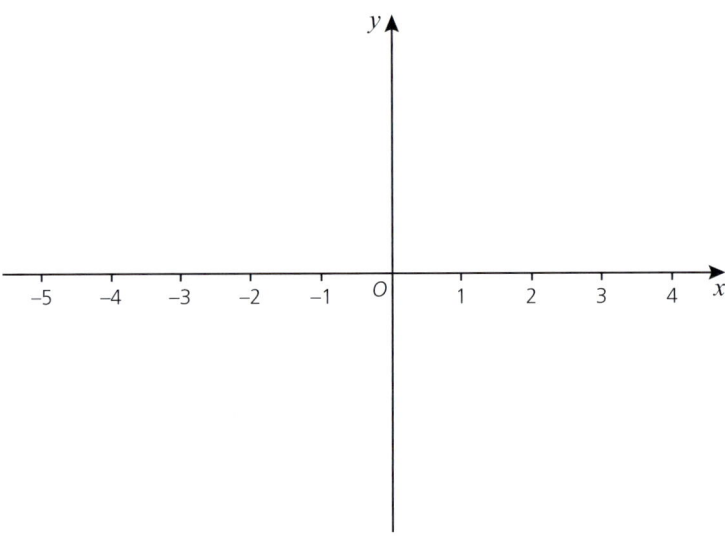

Using graphs to solve cubic inequalities

(ii) Find the coordinates of the stationary point on the curve
$y = |(x - 2)(x + 3)|$. [2]

(iii) Given that k is a positive constant, state the set of values of k for which $|(x - 2)(x + 3)| = k$ has 2 solutions only. [1]

Cambridge O Level Additional Mathematics (4037)
Paper 12 Q8, November 2013
Cambridge IGCSE Additional Mathematics (0606)
Paper 12 Q8, November 2013

3 Solve the inequality $9x^2 + 2x - 1 < (x + 1)^2$. [3]

Cambridge O Level Additional Mathematics (4037)
Paper 22 Q2, November 2014
Cambridge IGCSE Additional Mathematics (0606)
Paper 22 Q2, November 2014

Now you should be able to:
★ solve equations of the type
  ● $|ax + b| = c \ (c \geq 0)$
  ● $|ax + b| = cx + d$
  ● $|ax + b| = |cx + d|$
  ● $|ax^2 + bx + c| = d$
  using algebraic or graphical methods
★ solve graphically or algebraically inequalities of the type
  ● $k|ax + b| > c \ (c \geq 0)$
  ● $k|ax + b| \leq c \ (c \geq 0)$
  ● $k|ax + b| \leq |cx + d|$
  where $k > 0$
  ● $|ax + b| \leq cx + d$
  ● $|ax^2 + bx + c| > d$
  ● $|ax^2 + bx + c| \leq d$
★ use substitution to form and solve a quadratic equation in order to solve a related equation
★ sketch the graphs of cubic polynomials and their moduli, when given as a product of three linear factors
★ solve graphically cubic inequalities of the form
  ● $f(x) \geq d$
  ● $f(x) > d$
  ● $f(x) \leq d$
  ● $f(x) < d$
  where $f(x)$ is a product of three linear factors and d is a constant.

4 EQUATIONS, INEQUALITIES AND GRAPHS

Key points

- For any real number x, the **modulus** of x is denoted by $|x|$ and is defined as:
  $$|x| = x \text{ if } x \geq 0$$
  $$|x| = -x \text{ if } x < 0.$$
- A modulus equation of the form $|ax + b| = b$ can be solved either graphically or algebraically.
- A modulus equation of the form $|ax + b| = |cx + d|$ can be solved graphically by first drawing both graphs on the same axes and then, if necessary, identifying the solution algebraically.
- A modulus inequality of the form $|x - a| < b$ is equivalent to the inequality $a - b < x < a + b$ and can be illustrated on a number line with an open circle marking the ends of the interval to show that these points are not included. For $|x - a| \leq b$, the interval is the same but the end points are marked with solid circles.
- A modulus inequality of the form $|x - a| > b$ or $|x - a| \geq b$ is represented by the parts of the line outside the intervals above.
- A modulus inequality in two dimensions is identified as a region on a graph, called the **feasible region.** It is common practice to shade out the region not required to keep the feasible region clear.
- It is sometimes possible to solve an equation involving both x and $\sqrt{x}$ by making a substitution of the form $x = u^2$. You must check all answers in the original equation.
- The graph of a cubic function has a distinctive shape determined by the coefficient of x^3.

Positive x^3 term   Negative x^3 term

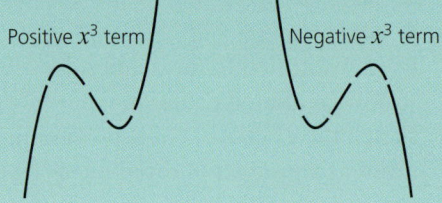

5 Simultaneous equations

The beauty of mathematics only shows itself to more patient followers.

Maryam Mirzakhani (1977–2017)

Sometimes, in mathematics, it is necessary to use two or more variables to describe a situation. In such cases, you need more than one equation to find the individual values of the variables: two equations for two variables, three for three variables and so on. Although the focus of this chapter is on solving two equations to find the values of two variables, the techniques introduced can be extended to cover situations involving more than two variables.

The following problem can be solved using simultaneous equations.

> A school has two netball teams and there is great rivalry between them. Last season, each team played 16 games and their positions in the league were tied with 26 points each. A team is awarded points for a win or a draw only. Team A won 6 games and drew 8; team B won 8 games and drew 2. How many points are awarded for:
>
> **a)** a win   **b)** a draw?

Let w denote the number of points for a win and d denote the number of points for a draw. Each team's results can be expressed as an equation, but unlike equations you have met previously, each one contains two variables.

$6w + 8d = 26$

$8w + 2d = 26$

5 SIMULTANEOUS EQUATIONS

In this case, since the values of w and d will be small integers, you could find the answer by trial and improvement, but in this chapter we will look at some structured methods for solving simultaneous equations.

Simultaneous equations can be solved graphically or algebraically using elimination or substitution. Graphical methods often give only approximate answers and so the main focus of this chapter is algebraic methods.

Solving linear simultaneous equations

You will have already used graphs to solve simultaneous equations. The example below shows that this method does not always give an accurate solution. This is why it is not the focus of this chapter.

➡ Worked example

Solve the following pairs of simultaneous equations graphically and comment on your answers.

a) $x + y = 4$
$y = 2x + 1$

b) $x + y = 4$
$y = 4x + 2$

Solution

a)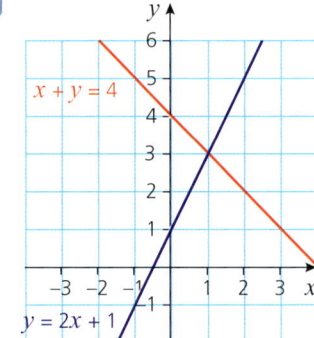

b)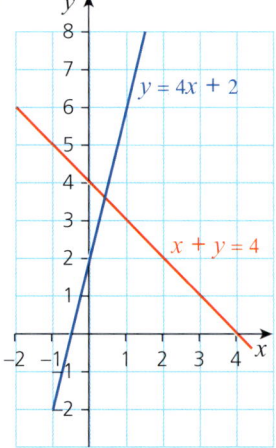

Here you can see that the two lines intersect at the point (1, 3) so the solution is $x = 1$, $y = 3$.

It is not clear from the graph exactly where these two lines intersect.

Solving linear simultaneous equations

The elimination method

 Worked example

Solve the following simultaneous equations by elimination.

$2x + y = 10$

$x - y = 2$

Solution

Since one equation contains $+y$ and one contains $-y$ adding them will eliminate y.

$$2x + y = 10$$
$$x - y = 2$$
$$\overline{3x \quad\quad = 12}$$
$$\Rightarrow x = 4$$

The rule Same Signs, Subtract (and Opposite Signs, Add) is useful here.

Substitute this value into one of the equations.

$4 - y = 2$
$\Rightarrow y = 2$

You would generally choose the simpler one, in this case $x - y = 2$.

Check these values in the other equation.

$2x + y = 8 + 2$
$\quad\quad\quad = 10$

The solution to the simultaneous equations is $x = 4$, $y = 2$.

Sometimes you need to multiply one of the equations before you can eliminate a variable, as in the example below.

 Worked example

Solve the following simultaneous equations by elimination.

$3x + y = 13$

$x + 2y = 11$

Solution

Start by multiplying the first equation by 2 so that the coefficient of y is the same in both equations.

$$6x + 2y = 26$$
$$x + 2y = 11$$
$$\overline{5x \quad\quad = 15}$$
$$\Rightarrow x = 3$$

Subtract to eliminate y

79

5 SIMULTANEOUS EQUATIONS

An alternative starting point is to multiply the second equation by 3 so that the coefficient of x is the same in both equations. Try it and show that it gives the same answer.

Substitute this into the second equation (since it is the simpler one).

$3 + 2y = 11$
$\Rightarrow 2y = 8$
$\Rightarrow y = 4$

The solution is therefore $x = 3$, $y = 4$.

Sometimes it is necessary to manipulate both of the original equations in order to eliminate one of the variables easily, as in the following example.

Worked example

Solve the following simultaneous equations by elimination.

$3x + 2y = 1$
$2x + 3y = 4$

Solution

Here you need to multiply each equation by a suitable number so that either the coefficients of x or the coefficients of y are the same.

$3x + 2y = 1 \Rightarrow 9x + 6y = 3$ — *3× the first equation*

$2x + 3y = 4 \Rightarrow 4x + 6y = 8$ — *2× the second equation*

$9x + 6y = 3$
$4x + 6y = 8$
Subtracting
$\Rightarrow 5x = -5$
$\Rightarrow x = -1$

Substitute $x = -1$ into the first of the two original equations.

$-3 + 2y = 1$
$\Rightarrow 2y = 4$
$\Rightarrow y = 2$ — *You could choose either one. Remember to check these values in the second of the original equations.*

The solution is therefore $x = -1$, $y = 2$.

The substitution method

Worked example

Solve the following simultaneous equations by substitution:

$3x - y = -10$
$x = 2 - y$

Solution

Substitute the expression for x from the second equation into the first.

$$3(2-y) - y = -10$$
$$\Rightarrow 6 - 3y - y = -10$$
$$\Rightarrow 16 = 4y$$
$$\Rightarrow y = 4$$

Since it is the simpler one

Substitute $y = 4$ into the second equation.

$$x = 2 - 4$$
$$\Rightarrow x = -2$$

It is a good idea to check these values using the equation you did not substitute into.

The solution is therefore $x = -2$, $y = 4$.

Since each of these equations can be represented by a straight line, solving them simultaneously gives the coordinates of their point of intersection.

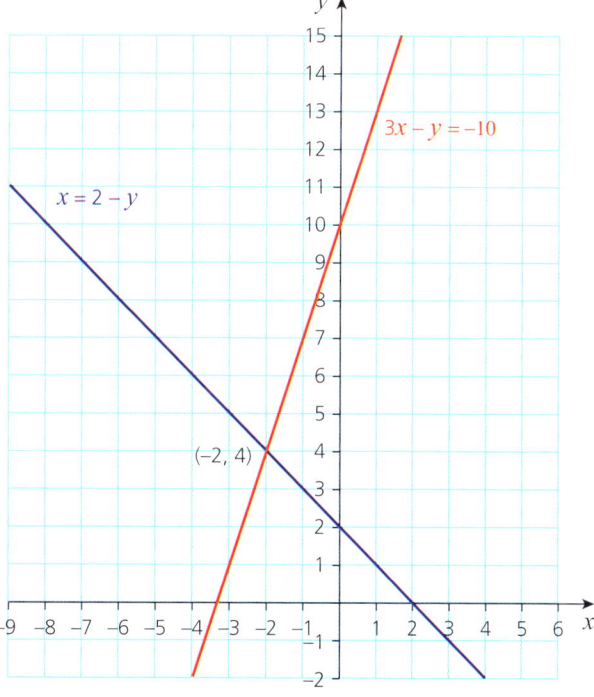

Sometimes simultaneous equations may arise in everyday problems as in the following example.

→ Worked example

A taxi firm charges a fixed amount plus so much per kilometre. A journey of three kilometres costs $4.60 and a journey of seven kilometres costs $9.40. How much does a journey of six kilometres cost?

5 SIMULTANEOUS EQUATIONS

Solution

Let f be the fixed amount and m be the cost per kilometre. Writing this information as a pair of simultaneous equations:

$$f + 3m = 4.6$$
$$f + 7m = 9.4$$

Subtracting the first equation from the second:

$$4m = 4.8$$
$$\Rightarrow m = 1.2$$

Substituting into the first equation:

$$f + 3 \times 1.2 = 4.6$$
$$\Rightarrow f = 1.0$$

A journey of six kilometres will cost $1.0 + 6(1.2) = \$8.20$.

At this stage you must remember to answer the question as it is set.

Exercise 5.1

1. Solve the following pairs of simultaneous equations graphically:
   a) $y = x + 2$
      $y = 2x - 3$
   b) $x + 2y = 3$
      $2x - y = -4$

   Use the substitution method to solve the simultaneous equations in questions 2–5.

2. a) $2x + y = 13$
      $y = 2x + 1$
   b) $x + 2y = 13$
      $x = 2y + 1$

3. a) $3x + 4y = 2$
      $y = 4x + 10$
   b) $4x + 3y = 2$
      $x = 4y + 10$

4. $x - 3y = -2$
   $y = 3x - 2$

5. $x + 4y = -13$
   $x = 3y + 1$

   Use the elimination method to solve the simultaneous equations in questions 6–9.

6. a) $x + y = 4$
      $x - y = 2$
   b) $x + 2y = 4$
      $x - 2y = 2$

7. a) $3x + y = 9$
      $2x - y = 1$
   b) $3x + 2y = 9$
      $x - y = 0.5$

8. $2x + 3y = -4$
   $4x + 2y = 0$

9. $5x - 2y = -23$
   $3x + y = -5$

10. 3 pencils and 4 rulers cost $5.20. 5 pencils and 2 rulers cost $4. Find the cost of 6 pencils and a ruler.

11. At the cinema, 3 packets of popcorn and 2 packets of nuts cost $16 and 2 packets of popcorn and 1 packet of nuts cost $9. What is the cost of one packet of each?

12 Two adults and one child paid $180 to go to the theatre and one adult and three children paid $190. What is the cost for two adults and five children?

13 A shop is trying to reduce their stock of books by holding a sale. $20 will buy either 8 paperback and 4 hardback books or 4 paperbacks and 7 hardbacks. How much change would I get from $40 if I bought 10 paperbacks and 10 hardbacks?

Solving non-linear simultaneous equations

The substitution method is particularly useful when one of the equations represents a curve, as in the following example.

Worked example

a) Sketch the graphs of $y = x^2 + 3x + 2$ and $2x = y - 8$ on the same axes.

b) Use the method of substitution to solve these equations simultaneously

Solution

a)

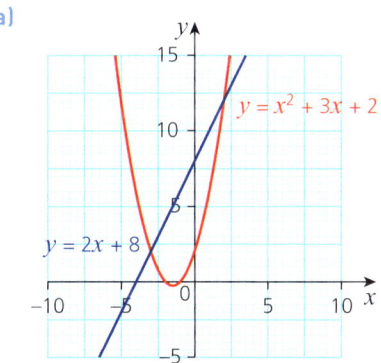

b) $y = x^2 + 3x + 2$
$2x = y - 8$
$2x = (x^2 + 3x + 2) - 8$
$\Rightarrow 0 = x^2 + x - 6$
$\Rightarrow (x + 3)(x - 2) = 0$
$\Rightarrow x = -3$ or $x = 2$

Substituting $x^2 + 3x + 2$ for y in the second equation

Collecting all the terms on one side

Take each of these values in turn, and substitute into the **linear** equation to find the corresponding values of y.

When $x = -3$, $-6 = y - 8 \Rightarrow y = 2$.

This means that one possible solution is $x = -3$, $y = 2$.

When $x = 2$, $4 = y - 8 \Rightarrow y = 12$.

This gives the other solution as $x = 2$, $y = 12$.

The full solution is therefore $x = -3$, $y = 2$ or $x = 2$, $y = 12$.

5 SIMULTANEOUS EQUATIONS

> **Note**
>
> 1. It is equally acceptable to start by substituting for x from the second equation into the first. This gives $2x + 8 = x^2 + 3x + 2$ and leads to the same result.
> 2. Once you have found values of one variable, you must substitute into the **linear** equation. If you substitute into the non-linear equation, you **could** find other 'rogue' values appearing erroneously as solutions (but not in this case).

The original equations represent a curve and a line, so the two solutions give the coordinates of their points of intersection as in the graph above.

In the previous example, it would also have been possible to solve the two equations by simply drawing graph because the solution had integer values. However, this is often not the case so this method will not always give an accurate answer.

➜ Worked example

a) Use an algebraic method to find the points of intersection of the curve $x^2 + y^2 = 5$ and the line $y = x + 1$.

b) Given that the curve is the circle with centre the origin and radius $\sqrt{5}$, illustrate your answer on a graph.

Solution

a) In this case you should use the substitution method.

$$x^2 + (x+1)^2 = 5 \quad \leftarrow \text{Substituting from the linear equation into the equation for the curve}$$
$$\Rightarrow x^2 + (x^2 + 2x + 1) = 5$$
$$\Rightarrow 2x^2 + 2x - 4 = 0$$
$$\Rightarrow x^2 + x - 2 = 0 \quad \leftarrow \text{Dividing by 2 to simplify}$$
$$\Rightarrow (x+2)(x-1) = 0$$
$$\Rightarrow x = -2 \text{ or } x = 1$$

When $x = -2$, $y = -2 + 1 = -1$. $\leftarrow$ Substituting each of these values in turn into the **linear** equation

When $x = 1$, $y = 1 + 1 = 2$.

The points of intersection are therefore $(-2, -1)$ and $(1, 2)$.

b)

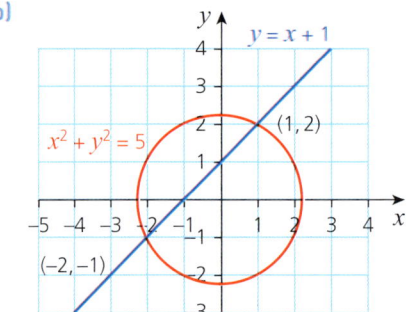

Solving non-linear simultaneous equations

→ Worked example

Solve this pair of simultaneous equations:

$xy^2 - 5 = 0$

$xy = 4$

Solution

As xy^2 can be written xyy, substituting $xy = 4$ into the first equation gives

Solve to find y → $4y - 5 = 0$

$\Rightarrow 4y = 5$

$\Rightarrow y = \dfrac{5}{4} = 1.25$

Substituting $y = 1.25$ into $xy = 4$ gives

$1.25x = 4$

$\Rightarrow x = \dfrac{4}{1.25} = 3.2$

So you get the solution $x = 3.2$, $y = 1.25$

→ Worked example

Solve this pair of simultaneous equations:

$\dfrac{x}{y} + \dfrac{3y}{x} = 4$

$y = 2x - 1$

Solution

Substituting $y = 2x - 1$ into the first equation gives

$$\dfrac{x}{2x-1} + \dfrac{3(2x-1)}{x} = 4$$

Write with a common denominator in order to combine as a single fraction →

$\Rightarrow \dfrac{x^2}{x(2x-1)} + \dfrac{3(2x-1)^2}{x(2x-1)} = 4$

$\Rightarrow \dfrac{x^2 + 3(2x-1)^2}{x(2x-1)} = 4$

$\Rightarrow x^2 + 3(2x-1)^2 = 4x(2x-1)$

$\Rightarrow x^2 + 3(4x^2 - 4x + 1) = 4x(2x-1)$

$\Rightarrow x^2 + 12x^2 - 12x + 3 = 8x^2 - 4x$

$\Rightarrow 5x^2 - 8x + 3 = 0$

5 SIMULTANEOUS EQUATIONS

$\Rightarrow (5x-3)(x-1) = 0$

$\Rightarrow 5x-3 = 0$ or $x-1 = 0$

$\Rightarrow x = 0.6$ or $x = 1$

If $x = 0.6$, $y = 2 \times 0.6 - 1 = 0.2$

If $x = 1$, $y = 2 \times 1 - 1 = 1$

So, the solution has two pairs of values. Either $x = 0.6$, $y = 0.2$ or $x = 1$, $y = 1$.

Exercise 5.2

1 Solve this pair of simultaneous equations algebraically:
   $x^2 + y^2 = 13$
   $x = 2$

2 Solve this pair of simultaneous equations algebraically and sketch a graph to illustrate your solution:
   $y = x^2 - x + 8$
   $y = 5x$

3 Solve this pair of simultaneous equations:
   $xy = 4$
   $y = x - 3$

4 Solve this pair of simultaneous equations:
   $y = 8x^2 - 2x - 10$
   $4x + y = 5$

5 Solve this pair of simultaneous equations:
   $xy^2 = 6$
   $xy - 2 = 0$

6 The diagram shows the circle $x^2 + y^2 = 25$ and the line $x - 7y + 25 = 0$. Find the coordinates of A and B.

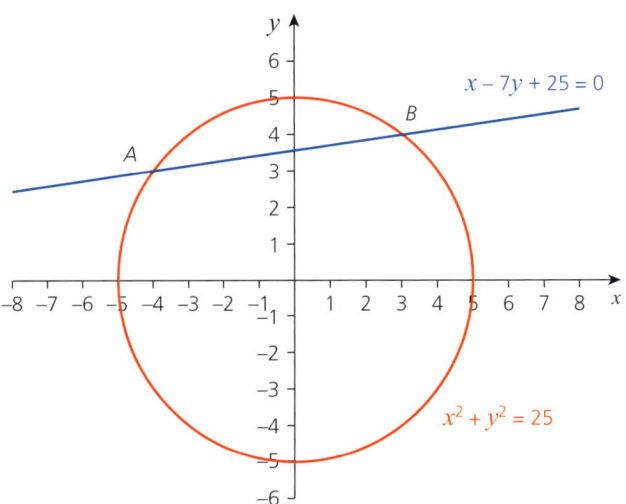

Solving non-linear simultaneous equations

7 The diagram shows a circular piece of card of radius r cm, from which a smaller circle of radius x cm has been removed. The area of the remaining card is 209π cm². The circumferences of the two circles add up to 38π. Write this information as a pair of simultaneous equations and hence find the values of r and x.

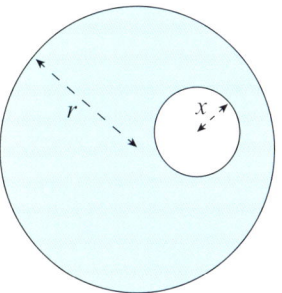

8 a) Solve this pair of simultaneous equations:
$y = x^2 + 1$
$y = 2x$

b) Why is there only one solution? Illustrate this using a sketch.

9 a) Explain what happens when you try to solve this pair of simultaneous equations:
$y = 2x^2 - 3x + 4$
$y = x - 1$

b) Illustrate your explanation with a sketch graph.

10 Solve this pair of simultaneous equations:
$\dfrac{x}{y} + \dfrac{2y}{x} = 3$
$y = 2x - 3$

Past-paper questions

1 The line $y = 2x + 10$ intersects the curve $2x^2 + 3xy - 5y + y^2 = 218$ at the points A and B.

Find the equation of the perpendicular bisector of AB. [9]

Cambridge O Level Additional Mathematics (4037)
Paper 23 Q10, November 2011
Cambridge IGCSE Additional Mathematics (0606)
Paper 23 Q10, November 2011

2 The curve $y = xy + x^2 - 4$ intersects the line $y = 3x - 1$ at the points A and B. Find the equation of the perpendicular bisector of the line AB. [8]

Cambridge O Level Additional Mathematics (4037)
Paper 11 Q5, June 2015
Cambridge IGCSE Additional Mathematics (0606)
Paper 11 Q5, June 2015

3 Find the set of values of k for which the line $y = 3x - k$ does not meet the curve $y = kx^2 + 11x - 6$. [6]

Cambridge O Level Additional Mathematics (4037)
Paper 23 Q3, November 2013
Cambridge IGCSE Additional Mathematics (0606)
Paper 23 Q3, November 2013

5 SIMULTANEOUS EQUATIONS

Now you should be able to:

★ solve simultaneous equations in two unknowns by elimination or substitution.

Key points

Simultaneous equations may be solved using these methods.

✔ **Graphically:** This method may be used for any two simultaneous equations. The advantage is that it is generally easy to draw graphs, although it can be time-consuming. The disadvantage is that it may not give an answer to the level of accuracy required.

✔ **Elimination:** This is the most useful method when solving two linear simultaneous equations.

✔ **Substitution:** This method is best for one linear and one non-linear equation. You start by isolating one variable in the linear equation and then substituting it into the non-linear equation.

6 Logarithmic and exponential functions

To forget one's ancestors is to be a brook without a source, a tree without a root.

Chinese proverb

Discussion point

You have two parents and each of them has (or had) two parents so you have four grandparents. Going back you had $2^3 = 8$ great grandparents, $2^4 = 16$ great great grandparents and so on going backwards in time. Assuming that there is one generation every 30 years, and that all your ancestors were different people, estimate how many ancestors you had living in the year 1700. What about the year 1000?

The graph below shows an estimate of the world population over the last 1000 years. Explain why your answers are not realistic. What assumption has caused the problem?

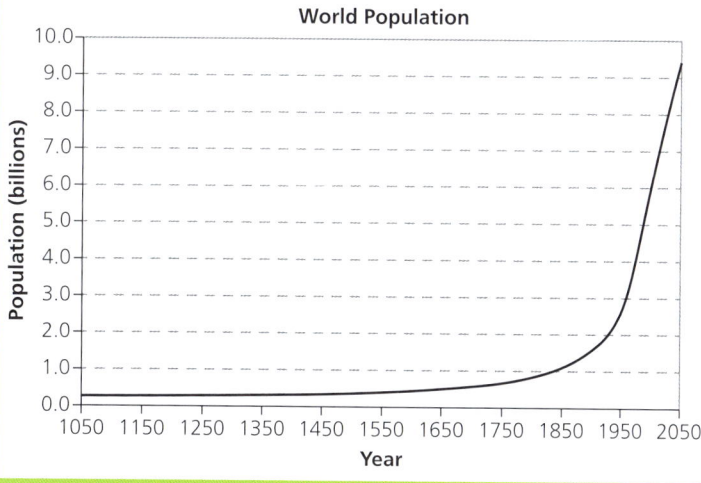

6 LOGARITHMIC AND EXPONENTIAL FUNCTIONS

Recent DNA analysis shows that almost everyone in Europe is descended from just seven women. Arriving at different times during the last 45 000 years, they survived wolves, bears and ice ages to form different clans that eventually became today's population. Another 26 maternal lineages have been uncovered on other continents.

Researching family history is a popular hobby and there are many internet sites devoted to helping people find out details of their ancestry. Most of us only know about parents, grandparents, and possibly great-grandparents, but it is possible to go back much further.

Worked example

Assuming that a new generation occurs, on average, every 30 years, how many direct ancestors will be on your family tree if you go back 120 years? What about if you were able to go back 300 years?

Solution

30 years ago, you would have information about your two parents.

Each of these would have had two parents, so going back a further 30 years there are also four grandparents, another 30 years gives eight great-grandparents and so on.

If you tabulate these results, you can see a sequence starting to form.

Number of years	Number of people
30	2
60	$4 = 2^2$
90	$8 = 2^3$
120	$16 = 2^4$

← *This is a **geometric sequence** of numbers. You will meet these sequences in Chapter 12.*

For each period of 30 years, the number of direct ancestors is double the number in the previous generation. After 120 years, the total number of ancestors is $2 + 4 + 8 + 16 = 30$.

300 years ago is ten periods of 30 years, so following the pattern, there are $2^{10} = 1024$ direct ancestors in this generation.

In practice, family trees are much more complicated, since most families have more than one child. It gets increasingly difficult the further back in time you research.

Discussion point

How many years would you expect to need to go back to find over 1 billion direct ancestors?

What date would that be?

Look at the graph on the previous page and say why this is not a reasonable answer.

Where has the argument gone wrong?

Logarithms

You may have answered the discussion point by continuing the pattern in the table on the previous page. Or you may have looked for the smallest value of n for which 2^n is greater than 1 billion. You can find this by trial and error but, as you will see, it is quicker to use logarithms to solve equations and inequalities like this.

Logarithms

Logarithm is another word for **index** or **power**.

For example, if you want to find the value of x such that $2^x = 8$, you can do this by checking powers of 2. However, if you have $2^x = 12$, for example, it is not as straightforward and you would probably need to resort to trial and improvement.

The equation $2^3 = 8$ can also be written as $\log_2 8 = 3$. The number 2 is referred to as the **base** of the logarithm.

Read this as 'log to base 2 of 8 equals 3'.

Similarly, $2^x = 12$ can be written as $\log_2 12 = x$.

In general,
$$a^x = y \Leftrightarrow x = \log_a y.$$

Most calculators have three buttons for logarithms.
- **log** which uses 10 as the base.
- **ln** which has as its base the number 2.718..., denoted by the letter e, which you will meet later in the chapter.
- **log ▫** which allows you to choose your own base.

➡ Worked example

Find the logarithm to base 2 of each of these numbers. Do not use a calculator.

a) 32   b) $\frac{1}{4}$ c) 1   d) $\sqrt{2}$

Solution

This is equivalent to being asked to find the power when the number is written as a power of 2.

a) $32 = 2^5$, so $\log_2 32 = 5$

b) $\frac{1}{4} = 2^{-2}$, so $\log_2 \frac{1}{4} = -2$

c) $1 = 2^0$, so $\log_2 1 = 0$ ← *$\log_n 1 = 0$ for all positive values of n.*

d) $\sqrt{2} = 2^{\frac{1}{2}}$, so $\log_2 \sqrt{2} = \frac{1}{2}$

6 LOGARITHMIC AND EXPONENTIAL FUNCTIONS

Graphs of logarithms

The graph of $y = \log_a x$ has the same general shape for all values of the base a where $a > 1$.

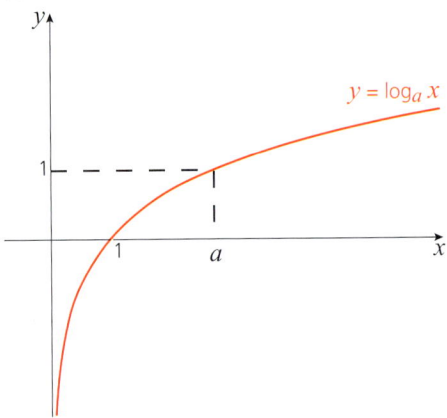

The graph has the following properties:
- The curve only exists for positive values of x.
- The gradient of the graph is always positive. As the value of x increases, the gradient of the curve decreases.
- It crosses the x-axis at $(1, 0)$.
- The line $x = 0$ is an asymptote, i.e. the curve approaches it ever more closely but never actually touches or crosses it.
- The graph passes through the point $(a, 1)$.
- $\log_a x$ is negative for $0 < x < 1$.

Graphs of other logarithmic functions are obtained from this basic graph by applying one or more transformations – translations, stretches or reflections – as shown in the following examples.

> **Note**
>
> - A **translation** moves the graph – horizontally, vertically or in both directions – to a different position. It does not change in shape. When $a > 0$:
>   - replacing x by $(x - a)$ moves the graph a units to the right (the positive direction)
>   - replacing x by $(x + a)$ moves the graph a units to the left (the negative direction)
>   - replacing y by $(y - a)$ moves the graph a units upwards (the positive direction)
>   - replacing y by $(y + a)$ moves the graph a units downwards (the negative direction).
> - A **reflection** gives a mirror image. In this book only reflections in the coordinate axes are considered.
>   - Replacing x by $(-x)$ reflects the graph in the y-axis.
>   - Replacing y by $(-y)$ reflects the graph in the x-axis.

Logarithms

> **Worked example**

Sketch each pair of graphs and describe the transformation shown. In each pair, join (2, log 2) and (3, log 3) to their images.

a) $y = \log x$ and $y = \log(x - 3)$

b) $y = \log x$ and $y = \log(x + 2)$

c) $y = \log x$ and $y = -\log x$

d) $y = \log x$ and $y = \log(-x)$

Solution

a) The graph of $y = \log(x - 3)$ is a translation of the graph of $y = \log x$ 3 units to the right.

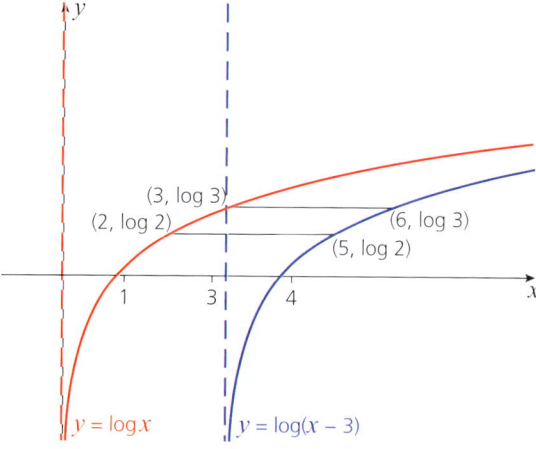

b) The graph of $y = \log(x + 2)$ is a translation of the graph of $y = \log x$ 2 units to the left.

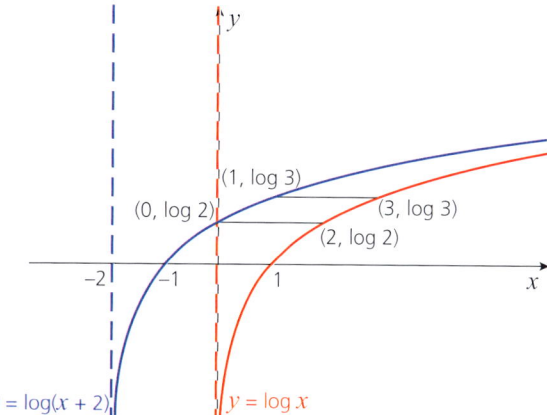

c) The graph of $y = -\log x$ (which is the same as $-y = \log x$) is a reflection of the graph of $y = \log x$ in the x-axis.

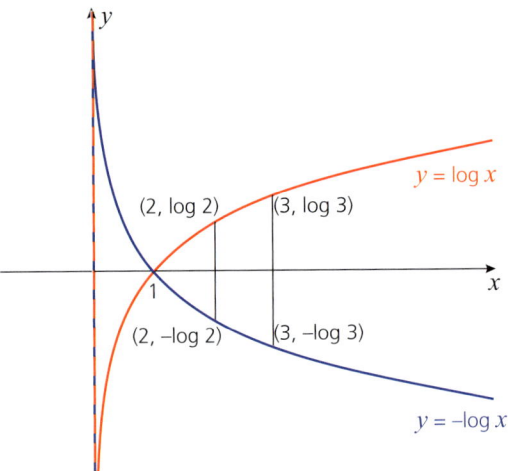

d) The graph of $y = \log(-x)$ is a reflection of the graph of $y = \log x$ in the y-axis.

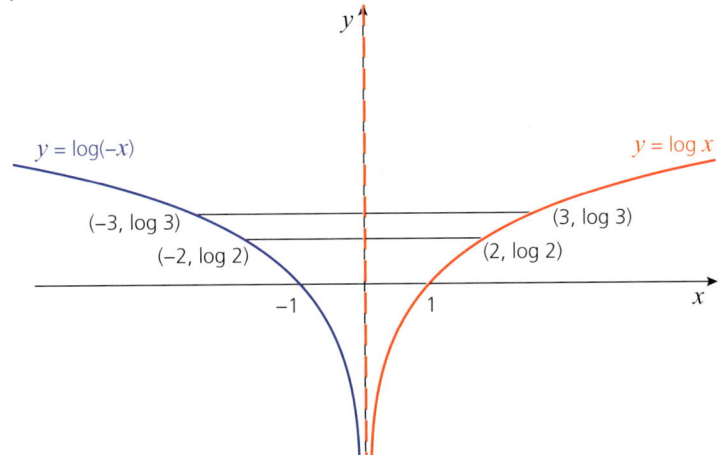

Worked example

You are given the curve of $y = \log x$ and told that $\log 3 = 0.48$ (2 d.p.).

a) Sketch the graph of $y = \log 3 + \log x$.

b) What is the relationship between the graphs of $y = \log x$ and $y = \log 3 + \log x$?

c) Sketch the graphs of $y = \log x$ and $y = \log 3x$ on the same axes.

d) What do you notice?

Solution

a)

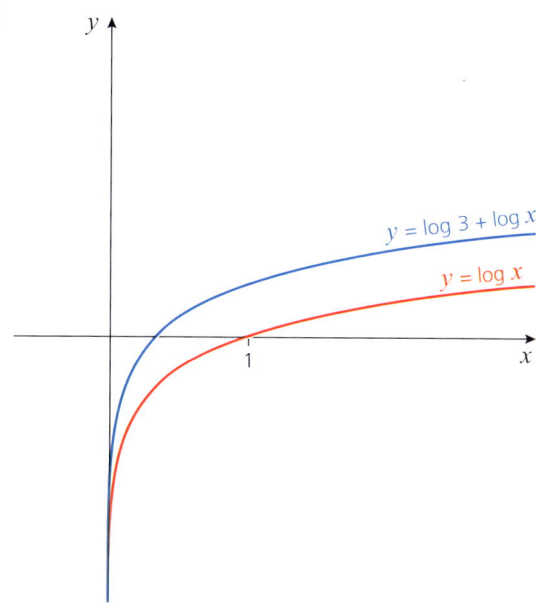

b) The graph of $y = \log 3 + \log x$ is a translation of the graph of $y = \log x$ upwards by a distance of $\log 3$.

c)

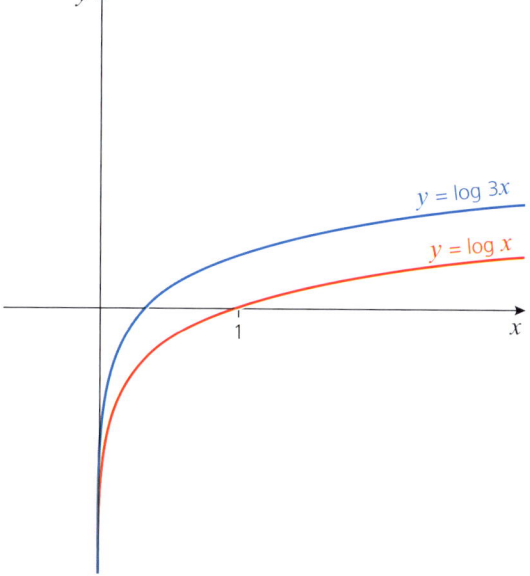

You can use graphing software to show that the graph of $y = \log 3x$ is the same as the graph of $y = \log 3 + \log x$. This confirms one of the 'laws of logarithms' introduced below.

d) The graph of $y = \log 3x$ looks the same as the graph of $y = \log 3 + \log x$.

6 LOGARITHMIC AND EXPONENTIAL FUNCTIONS

Laws of logarithms

If a logarithmic expression is true for any base, the base is often omitted.

There are a number of rules for manipulating logarithms. They are derived from the rules for manipulating indices. These laws are true for all logarithms to any positive base.

Operation	Law for indices	Law for logarithms
Multiplication	$a^x \times a^y = a^{x+y}$	$\log_a xy = \log_a x + \log_a y$
Division	$a^x \div a^y = a^{x-y}$	$\log_a \frac{x}{y} = \log_a x - \log_a y$
Powers	$(a^x)^n = a^{nx}$	$\log_a x^n = n \log_a x$
Roots	$(a^x)^{\frac{1}{n}} = a^{\frac{x}{n}}$	$\log_a \sqrt[n]{x} = \frac{1}{n} \log_a x$
Logarithm of 1	$a^0 = 1$	$\log_a 1 = 0$
Reciprocals	$\frac{1}{a^x} = a^{-x}$	$\log_a \frac{1}{x} = \log_a 1 - \log_a x = -\log_a x$
Log to its own base	$a^1 = a$	$\log_a a = 1$

You can use these laws, together with the earlier work on translations, to help you sketch the graphs of a range of logarithmic expressions by breaking them down into small steps as shown below.

➡ Worked example

Write $2 + 3\log 2 - \log 4$ as a single base 10 logarithm.

Solution

$2 + 3\log 2 - \log 4$

$\log_a a = 1$ so $\log 10 = 1$

$= 2\log 10 + 3\log 2 - \log 4$

$= \log 10^2 + \log 2^3 - \log 4$ ⟵ *using $n\log_a x = \log_a x^n$*

$= \log 100 + \log 8 - \log 4$

combine using $\log_a x + \log_a y = \log_a xy$

$= \log(100 \times 8) - \log 4$

$= \log 800 - \log 4$

$= \log \frac{800}{4}$ ⟵ *combine using $\log_a x - \log_a y = \log_a \frac{x}{y}$*

$= \log 200$

Logarithms

> **Worked example**

Sketch the graph of $y = 3\log(x - 2)$.

Solution

Transforming the graph of the curve $y = \log x$ into $y = 3\log(x - 2)$ involves two stages. Translating the graph of $y = \log x$ two units to the right gives the graph of $y = \log(x - 2)$.

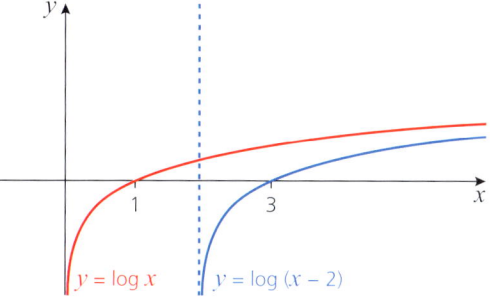

Multiplying $\log(x - 2)$ by 3 stretches the new graph in the y direction by a **scale factor** of 3.

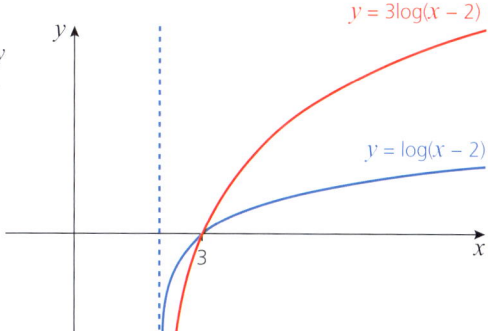

Logarithms to different bases

All the graphs you have met so far in the chapter could have been drawn to any base a greater than 1.
» When a logarithm is to the base 10 it can be written either as $\log_{10}$ or as lg. So, for example lg7 means $\log_{10} 7$.
» Base e is the other common base for logarithms.

The graphs of logarithms with a base number that is not 10 are very similar to the graphs of logarithms with base 10.

6 LOGARITHMIC AND EXPONENTIAL FUNCTIONS

Notice that when you use a different base for the logarithm, the graph has a similar shape and still passes through the point (1, 0).

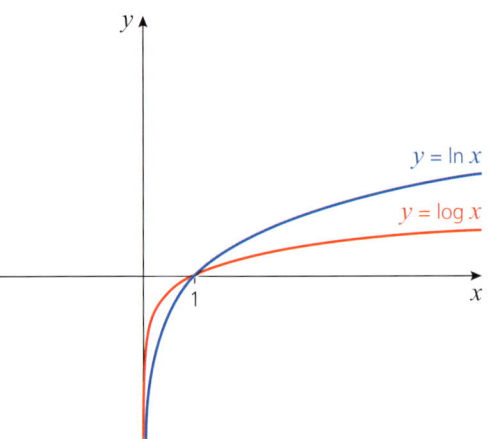

Change of base of logarithms

It is sometimes useful to change the base of a logarithm.

$$x = \log_a b \Leftrightarrow a^x = b$$
$$\Leftrightarrow \log_c a^x = \log_c b$$
$$\Leftrightarrow x \log_c a = \log_c b$$
$$\Leftrightarrow x = \frac{\log_c b}{\log_c a}$$

Some calculators can manipulate logarithms to any positive base. Check whether yours is one of them.

→ Worked example

Write $\log_8 12$ as a logarithm to base 2.

Solution

$x = \log_a b \Leftrightarrow x = \dfrac{\log_c b}{\log_c a}$

$$\log_8 12 = \frac{\log_2 12}{\log_2 8} = \frac{1}{3} \log_2 12$$

$\log_2 8 = 3$ as $2^3 = 8$

→ Worked example

Write $\dfrac{1}{\log_4 e}$ as a natural logarithm.

Solution

$$\log_4 e = \frac{\log_e e}{\log_e 4} = \frac{1}{\ln 4}$$

$x = \log_a b \Leftrightarrow x = \dfrac{\log_c b}{\log_c a}$

Hence

$$\frac{1}{\log_4 e} = \frac{1}{\frac{1}{\ln 4}} = \ln 4$$

98

Logarithms

Using logarithms to solve equations

Logarithms can be used to solve equations involving powers, to any level of accuracy.

→ Worked example

Solve the equation $3^x = 2000$.

Solution

Taking logarithms to the base 10 of both sides:

$\lg 3^x = \lg 2000 \quad \Rightarrow \quad x \lg 3 = \lg 2000$

$$\Rightarrow x = \frac{\lg 2000}{\lg 3} = 6.92 \text{ (3 s.f.)}$$

This question does not ask for any particular base. In this case base 10 is used but you could alternatively have used base e. These are the two bases for logarithms on nearly all calculators.

Logarithms can also be used to solve more complex equations.

→ Worked example

Solve the equation $4e^{3x} = 950$.

Solution

$$4e^{3x} = 950$$
$$\Rightarrow \quad e^{3x} = 237.5$$
$$\Rightarrow \quad 3x = \ln 237.5$$
$$\Rightarrow \quad 3x = 5.470\,167\ldots$$
$$\Rightarrow \quad x = 1.82 \text{ (3 s.f.)}$$

Taking the logarithms to base e of both sides

When there is a term of the form e^x, it is easier to use logarithms to base e, i.e. the ln button on your calculator.

→ Worked example

Solve the equation $5^{2x} - 5^x - 20 = 0$

Solution

$$5^{2x} - 5^x - 20 = 0$$
$$\Rightarrow (5^x)^2 - 5^x - 20 = 0$$

Substituting $5^x = u$

$$\Rightarrow u^2 - u - 20 = 0$$
$$\Rightarrow (u+4)(u-5) = 0$$
$$\Rightarrow u = -4 \text{ or } u = 5$$

Factorise

6 LOGARITHMIC AND EXPONENTIAL FUNCTIONS

> *Remember, the graph of $y = \log_a x$ only exists for positive values of x. The logarithm of a negative number is undefined.*

Since $5^x = u$

$\Rightarrow 5^x = -4$ or $5^x = 5$

$5^x = -4 \Rightarrow x = \log_5 -4$ which is not a valid solution.

$5^x = 5 \Rightarrow x = \log_5 5 = 1$ ⟵ $a^x = y \Leftrightarrow x = \log_a y$

Discussion point

It is always a good idea to check that your solution is correct.

For the example above, verify that the solution $x = 1$ satisfies the original equation $5^{2x} - 5^x - 20 = 0$.

➡ Worked example

Use logarithms to solve the equation $3^{5-x} = 2^{5+x}$. Give your answer correct to 3 s.f.

Solution

No base is mentioned, so you can use logarithms to any base. Using base 10:

$3^{5-x} = 2^{5+x}$

$\Rightarrow \lg 3^{5-x} = \lg 2^{5+x}$

$\Rightarrow (5-x)\lg 3 = (5+x)\lg 2$

$\Rightarrow 5\lg 3 - x\lg 3 = 5\lg 2 + x\lg 2$

$\Rightarrow 5\lg 3 - 5\lg 2 = x\lg 2 + x\lg 3$

$\Rightarrow 5(\lg 3 - \lg 2) = x(\lg 2 + \lg 3)$

$\Rightarrow x = \dfrac{5(\lg 3 - \lg 2)}{(\lg 2 + \lg 3)}$

$\Rightarrow x = 1.13$

Note that any base will yield the same answer. Using base 2:

$3^{5-x} = 2^{5+x}$

$\Rightarrow \log_2 3^{5-x} = \log_2 2^{5+x}$

> *Remember, $\log_2 2 = 1$.*

$\Rightarrow (5-x)\log_2 3 = (5+x)\log_2 2$

$\Rightarrow 5\log_2 3 - x\log_2 3 = 5 + x$

$\Rightarrow 5\log_2 3 - 5 = x + x\log_2 3$

$\Rightarrow 5(\log_2 3 - 1) = x(1 + \log_2 3)$

$\Rightarrow x = \dfrac{5(\log_2 3 - 1)}{1 + \log_2 3}$

$\Rightarrow x = 1.13$ (3 s.f.)

Logarithms

> **Discussion point**
>
> Did you find one of these methods easier than the other? If so, which one?

Using logarithms to solve inequalities

Logarithms are also useful to solve inequalities occurring, for example, in problems involving interest or depreciation.

When an inequality involves logs, it is often better to solve it as an equation first and then address the inequality. If you choose to solve it as an inequality, you may need to divide by a negative quantity and will therefore need to reverse the direction of the inequality sign. Both methods are shown in the example below.

Worked example

A second-hand car is bought for $20 000 and is expected to depreciate at a rate of 15% each year. After how many years will it first be worth less than $10 000?

Solution

The rate of depreciation is 15% so after one year the car will be worth 85% of the initial cost.

At the end of the second year, it will be worth 85% of its value at the end of Year 1, so $(0.85)^2 \times \$20\,000$.

Continuing in this way, its value after n years will be $(0.85)^n \times \$20\,000$.

Method 1: Solving as an equation

Solving the equation $(0.85)^n \times 20\,000 = 10\,000$

$$\Rightarrow (0.85)^n = 0.5$$
$$\Rightarrow \lg 0.85^n = \lg 0.5$$
$$\Rightarrow n \lg 0.85 = \lg 0.5$$
$$\Rightarrow n = \frac{\lg 0.5}{\lg 0.85}$$
$$\Rightarrow n = 4.265\ldots$$

The car will be worth $10 000 after 4.265 years, so it is **5 years** before it is first worth less than $10 000.

6 LOGARITHMIC AND EXPONENTIAL FUNCTIONS

Method 2: Solving as an inequality

Solving the inequality $(0.85)^n \times 20\,000 < 10\,000$

$\Rightarrow (0.85)^n < 0.5$

$\Rightarrow \lg 0.85^n < \lg 0.5$

$\Rightarrow n \lg 0.85 < \lg 0.5$

$\Rightarrow n > \dfrac{\lg 0.5}{\lg 0.85}$

$\Rightarrow n > 4.265\ldots$

The car will be worth less than $10\,000 after 5 years.

> *Remember that lg 0.85 is negative and when you divide an inequality by a negative number, you must change the direction of the inequality.*

Exercise 6.1

In some of the following questions you are instructed not to use your calculator for the working, but you may use it to check your answers.

1 By first writing each of the following equations using powers, find the value of y without using a calculator:
  a) $y = \log_2 8$ **b)** $y = \log_3 1$ **c)** $y = \log_5 25$ **d)** $y = \log_2 \dfrac{1}{4}$

2 $3^2 = 9$ can be written using logarithms as $\log_3 9 = 2$. Using your knowledge of indices, find the value of each of the following without using a calculator:
  a) $\log_2 16$ **b)** $\log_3 81$ **c)** $\log_5 125$ **d)** $\log_4 \dfrac{1}{64}$

3 Find the following without using a calculator:
  a) $\lg 100$ **b)** $\lg(\text{one million})$
  c) $\lg \dfrac{1}{1000}$ **d)** $\lg(0.000\,001)$

> *Remember that lg means $\log_{10}$.*

4 Using the rules for manipulating logarithms, rewrite each of the following as a single logarithm. For example, $\log 6 + \log 2 = \log(6 \times 2) = \log 12$.
  a) $\log 3 + \log 5$ **b)** $3 \log 4$
  c) $\log 12 - \log 3$ **d)** $\dfrac{1}{2} \log 25$
  e) $2 \log 3 + 3 \log 2$ **f)** $4 \log 3 - 3 \log 4$
  g) $\dfrac{1}{2} \log 4 + 4 \log \dfrac{1}{2}$

5 Write each of the following as a single base 10 logarithm:
  a) $\lg 3 + 2 \lg 6$ **b)** $2 \lg 9 - 3 \lg 3$ **c)** $3 + 2 \lg 6 - \lg 8$

6 Write each of the following as a logarithm to the given base:
  a) $\log_4 25$ to base 2   **b)** $\log_3 10$ to base 9   **c)** $\ln 1000$ to base 10

7 Write each of the following as a single natural logarithm:
  a) $\dfrac{1}{\log_6 e}$ **b)** $\dfrac{2}{\log_7 e}$ **c)** $\dfrac{1}{\log_3 e} + \dfrac{3}{\log_4 e}$

8 Express each of the following in terms of $\log x$:
  a) $\log x^5 - \log x^2$ **b)** $\log x^3 + 3 \log x$ **c)** $5 \log \sqrt{x} - 3 \log \sqrt[3]{x}$

Logarithms

9 This cube has a volume of 800 cm³.

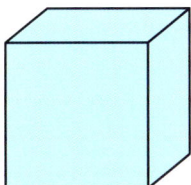

  a) Use logarithms to calculate the side length correct to the nearest millimetre.
  b) What is the surface area of the cube?

10 Starting with the graph of $y = \ln x$, list the transformations required, in order when more than one is needed, to sketch each of the graphs. Use the transformations you have listed to sketch each graph.
  a) $y = 3\ln x$
  b) $y = \ln(x + 3)$
  c) $y = 3\ln 2x$
  d) $y = 3\ln x + 2$
  e) $y = -3\ln(x + 1)$
  f) $y = \ln(2x + 4)$

11 Match each equation from **i** to **vi** with the correct graph **a** to **f**.
  i) $y = \log(x + 1)$
  ii) $y = \log(x - 1)$
  iii) $y = -\ln x$
  iv) $y = 3\ln x$
  v) $y = \log(2 - x)$
  vi) $y = \ln(x + 2)$

a)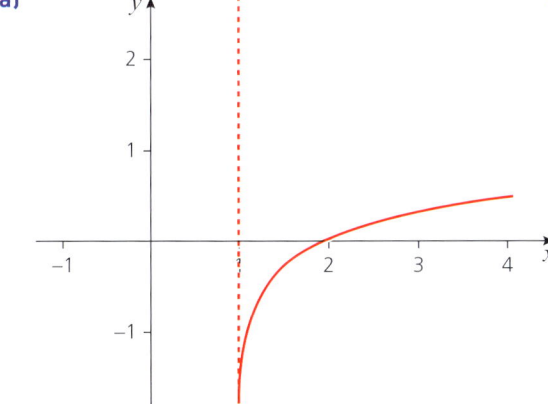

b)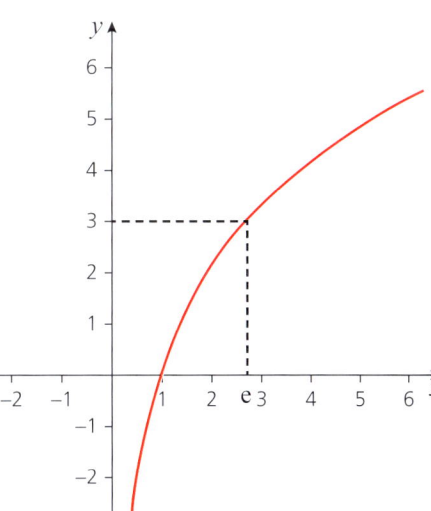

6 LOGARITHMIC AND EXPONENTIAL FUNCTIONS

Exercise 6.1 (cont)

c)
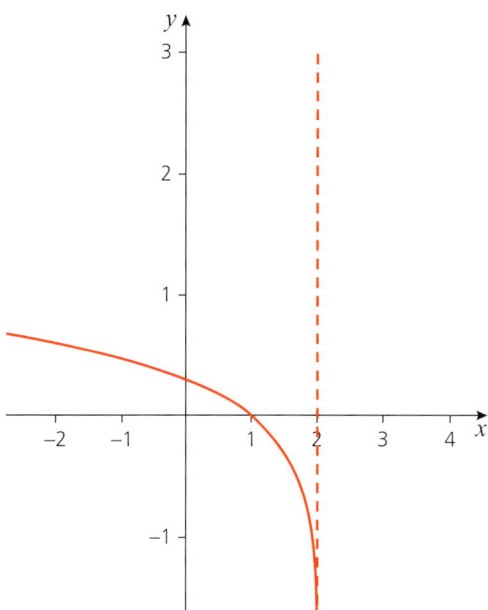

d)
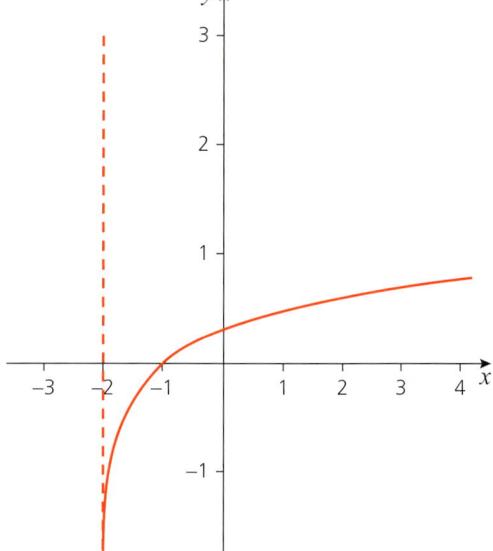

e)
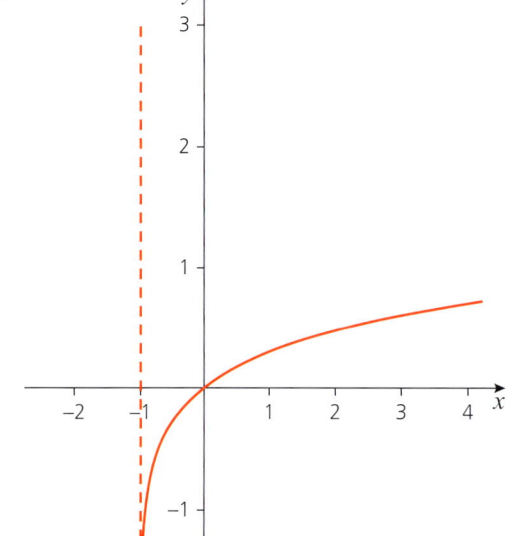

f)
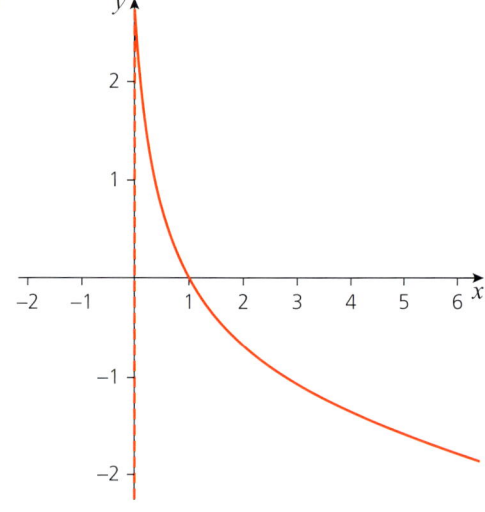

12 Solve the following equations for x, given that $\ln a = 3$:
   a) $a^{2x} = e^3$
   b) $a^{3x} = e^2$
   c) $a^{2x} - 3a^x + 2 = 0$

13 Before photocopiers were commonplace, school examination papers were duplicated using a process where each copy produced was only $c\%$ as clear as the previous copy. The copy was not acceptable if the writing was less than 50% as clear as the original. What is the value of c if the machine could produce only 100 acceptable copies from the original?

14 Use logarithms to solve the equation $5^{2x-1} = 4^{x+3}$. Give the value of x correct to 3 s.f.

15 a) $\$20\,000$ is invested in an account that pays interest at 2.4% per annum. The interest is added at the end of each year. After how many years will the value of the account first be greater than $\$25\,000$?
   b) What percentage interest should be added each month if interest is to be accrued monthly?
   c) How long would the account take to reach $\$25\,000$ if the interest was added:
      i) every month
      ii) every day?

16 Where possible, solve each of the following equations:
   a) $8^{2x} + 8^x - 6 = 0$
   b) $10^{2x} + 6 \times 10^x + 9 = 0$
   c) $(\lg x)^2 - 7\lg x + 12 = 0$
   d) $(\lg x)^2 + 2\lg x - 8 = 0$

Exponential functions

The expression $y = \log_a x$ can be written as $x = a^y$. Therefore, the graphs of these two expressions are identical.

For any point, interchanging the x- and y-coordinates has the effect of reflecting the original point in the line $y = x$, as shown below.

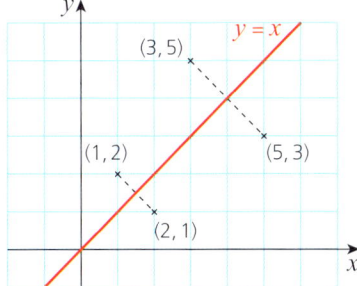

Interchanging x and y for the graph $y = \log_a x$ (shown in red) gives the curve $x = \log_a y$ (shown in blue).

6 LOGARITHMIC AND EXPONENTIAL FUNCTIONS

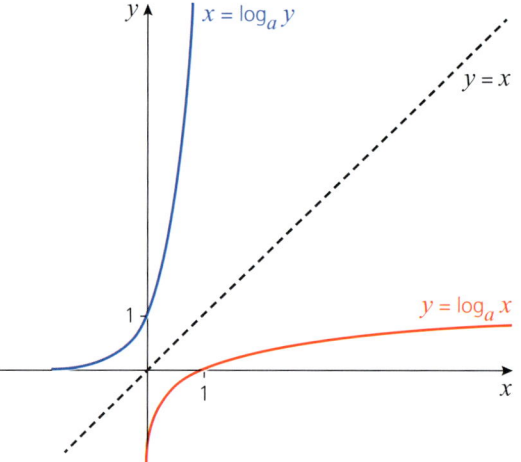

When rewritten with y as the subject of the equation, $x = \log_a y$ becomes $y = a^x$.

The function $y = a^x$ is called an **exponential function** and is the **inverse** of the logarithm function.

The most commonly used exponential function, known as **the exponential function**, is e^x, where e is the base of the logarithmic function $\ln x$ and is approximately equal to 2.718. You can manipulate exponential functions using the same rules as any other functions involving powers.

» $e^{a+b} = e^a \times e^b$

» $e^{a-b} = e^a \div e^b$

Graphs of e^x and associated exponential functions

The graph of $y = e^x$ has a similar shape to the graph of $y = a^x$ for positive value of a. The difference lies in the steepness of the curve.

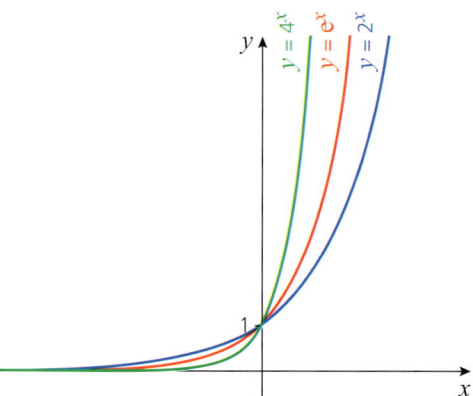

Exponential functions

As the base number increases (i.e. 2, e and 4 in the equations above), the curve becomes steeper for positive values of x. All the y-values are positive and all the curves pass through and 'cross over' at the point $(0, 1)$.

For positive integer values of n, curves of the form $y = e^{nx}$ are all related as shown below. Notice again, that the graphs all pass through the point $(0, 1)$ and, as the value of n increases, the curves become steeper.

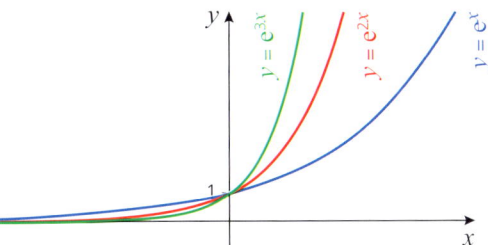

The graph of $y = e^{-x}$ is a reflection in the y-axis of the graph of $y = e^x$. The graphs of $y = e^{nx}$ and $y = e^{-nx}$ are related in a similar way for any integer value of n.

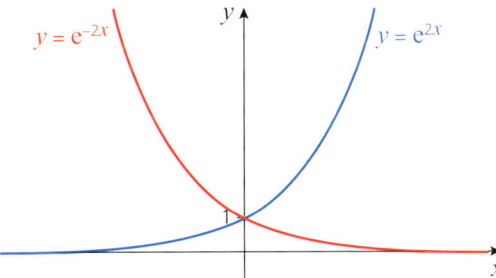

The family of curves $y = ke^x$, where k is a positive integer, is a set of different transformations of the curve $y = e^x$. These represent stretches of the curve $y = e^x$ in the y-direction.

Notice that the curve $y = ke^x$ crosses the y-axis at $(0, k)$.

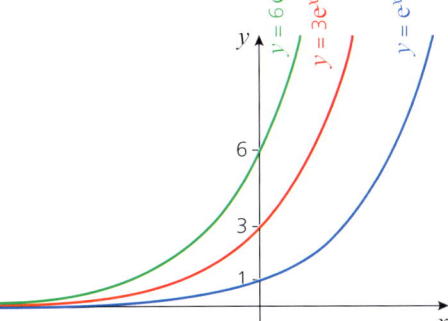

6 LOGARITHMIC AND EXPONENTIAL FUNCTIONS

Similarly, for a fixed value of n, graphs of the family $y = ke^{nx}$ are represented by stretches of the graph $y = e^{nx}$ by scale factor k in the y-direction.

One additional transformation gives graphs of the form $y = ke^{nx} + a$.

→ Worked example

Sketch the graph of $y = 3e^{2x} + 1$.

Solution

Start with $y = e^x$.

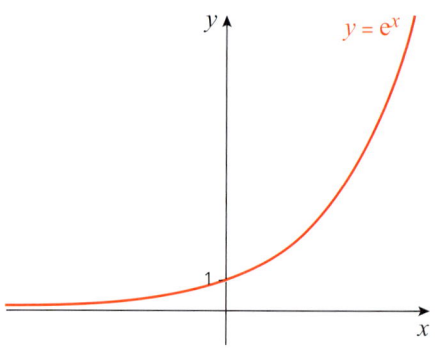

Transform to $y = e^{2x} = (e^x)^2$. The y values are squared, giving smaller values for $x < 0$ (where $y < 1$) and larger values for $x > 0$.

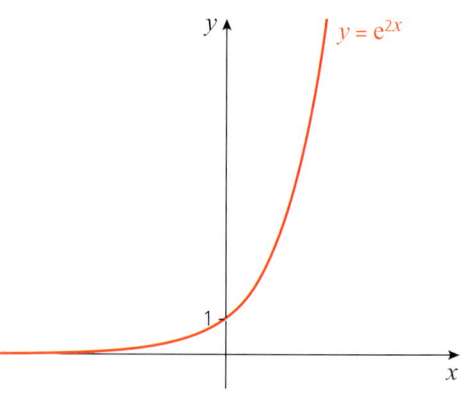

Stretch in the y-direction with a scale factor of 3 to give the graph of $y = 3e^{2x}$.

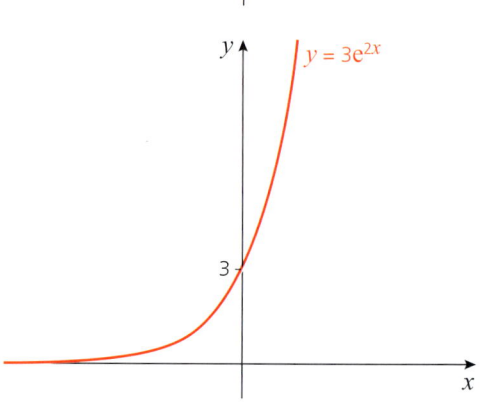

Translate 1 unit upwards to give $y = 3e^{2x} + 1$.

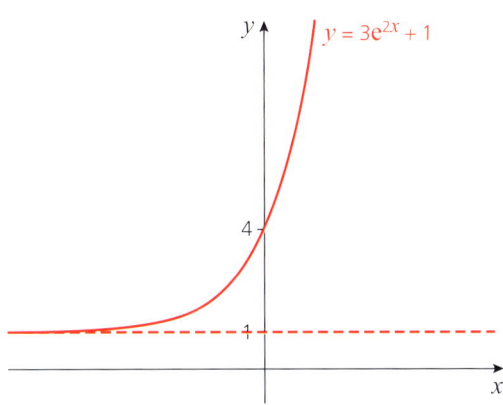

Worked example

Solve the equation $2e^x = 5 + 3e^{-x}$.

Solution

$2e^x = 5 + 3e^{-x}$

Multiply through by e^x

$\Rightarrow 2(e^x)^2 = 5e^x + 3e^{-x}e^x$

$2e^{2x} = 5e^x + 3$

Rearrange to form a quadratic which equals 0

$2e^{2x} - 5e^x - 3 = 0$

Substituting $e^x = u$

$\Rightarrow 2u^2 - 5u - 3 = 0$

Factorise

$\Rightarrow (2u + 1)(u - 3) = 0$

$\Rightarrow 2u + 1 = 0$ or $u - 3 = 0$

$\Rightarrow u = -\frac{1}{2}$ or $u = 3$

Since $e^x = u$

$\Rightarrow e^x = -\frac{1}{2}$ or $e^x = 3$

Remember, the graph of $y = \log_a x$ only exists for positive values of x. The logarithm of a negative number is undefined.

$e^x = -\frac{1}{2} \Rightarrow x = \ln\left(-\frac{1}{2}\right)$ which is not a valid solution.

$e^x = 3 \Rightarrow x = \ln 3 = 1.099$ (3 d.p.)

6 LOGARITHMIC AND EXPONENTIAL FUNCTIONS

> **Worked example**

Solve the equation $3(\ln 4x)^2 + 5(\ln 4x) - 2 = 0$.

Solution

Substituting $\ln 4x = u$ gives

$3u^2 + 5u - 2 = 0$

$\Rightarrow (3u - 1)(u + 2) = 0$

$\Rightarrow 3u - 1 = 0$ or $u + 2 = 0$

$\Rightarrow u = \frac{1}{3}$ or $u = -2$

Since $\ln 4x = u$

$\Rightarrow \ln 4x = \frac{1}{3}$ or $\ln 4x = -2$

$\ln 4x = \frac{1}{3} \Rightarrow 4x = e^{\frac{1}{3}} \Rightarrow x = \frac{e^{\frac{1}{3}}}{4} = 0.349$ (3 d.p.)

$\ln 4x = -2 \Rightarrow 4x = e^{-2} \Rightarrow x = \frac{e^{-2}}{4} = 0.034$ (3 d.p.)

Exponential growth and decay

The word 'exponential' is often used to refer to things that increase or decrease at a very rapid rate.

Any function of the form $y = a^x$ is referred to as an exponential function. When $x > 0$, the function $y = a^x$ is referred to as **exponential growth**; when $x < 0$ it is **exponential decay**.

*$y = e^x$ is called **the** exponential function.*

> **Worked example**

During the growth of an organism, a cell divides into two approximately every 6 hours. Assuming that the process starts with a single cell, and none of the cells die, how many cells will there be after 1 week?

Solution

It is possible to work this out without any special formulae:

2 cells after 6 hours

4 cells after 12 hours

8 cells after 18 hours…

However as the numbers get larger, the working becomes more tedious.

Notice the pattern here using 6 hours as 1 time unit.

2^1 cells after 1 time unit

2^2 cells after 2 time units

2^3 cells after 3 time units…

1 day of 24 hours is 4 time units, so 1 week of 7 days is 28 time units. So after 1 week there will be $2^{28} = 268\,435\,456$ cells.

Worked example

A brand of 'invisible' ink fades rapidly once it is applied to paper. After each minute the intensity is reduced by one quarter. It becomes unreadable to the naked eye when the intensity falls below 5% of the original value.

a) What is the intensity, as a percentage of the original value, after 3 minutes?

b) After how many minutes does it become unreadable to the naked eye? Give your answer to the nearest whole number.

Solution

a) After 1 minute it is $\frac{3}{4}$ of the original value.

After 2 minutes it is $\frac{3}{4}\left(\frac{3}{4}\right) = \left(\frac{3}{4}\right)^2$ of the original value.

After 3 minutes it is $\left(\frac{3}{4}\right)^3 = \frac{27}{64}$ or approximately 42% of the original value.

b) Using the pattern developed above:

It would be very tedious to continue the method in used above until the ink becomes unreadable.

After t minutes it is approximately $\left(\frac{3}{4}\right)^t$ of the original value.

The situation is represented by: $\left(\frac{3}{4}\right)^t < \frac{5}{100}$ ← $5\% = \frac{5}{100}$

Using logarithms to solve the inequality as an equation:

$\lg\left(\frac{3}{4}\right)^t = \lg\frac{5}{100} \Rightarrow t\lg\left(\frac{3}{4}\right) = \lg\left(\frac{5}{100}\right)$

$\Rightarrow t\lg 0.75 = \lg 0.05$

$\Rightarrow t = \frac{\lg 0.05}{\lg 0.75}$

$\Rightarrow t = 10.4$

Since the question asks for the time as a whole number of minutes, and the time is increasing, the answer is 11 minutes.

Exercise 6.2

It is a good idea to check the graphs you draw in questions 1–4 using any available graphing software.

1. For each set of graphs:
   i) Sketch the graphs on the same axes.
   ii) Give the coordinates of any points of intersection with the axes.
   a) $y = e^x$, $y = e^x + 1$ and $y = e^{x+1}$
   b) $y = e^x$, $y = 2e^x$ and $y = e^{2x}$
   c) $y = e^x$, $y = e^x - 3$ and $y = e^{x-3}$

2. Sketch the graphs of $y = e^{3x}$ and $y = e^{3x} - 2$.

3. Sketch the graphs of $y = e^{2x}$, $y = 3e^{2x}$ and $y = 3e^{2x} - 1$.

6 LOGARITHMIC AND EXPONENTIAL FUNCTIONS

Exercise 6.2 (cont)

4 Sketch each curve and give the coordinates of any points where it cuts the y-axis.
   a) $y = 2 + e^x$
   b) $y = 2 - e^x$
   c) $y = 2 + e^{-x}$
   d) $y = 2 - e^{-x}$

5 Solve the following equations:
   a) $5e^{0.3t} = 65$
   b) $13e^{0.5t} = 65$
   c) $e^{t+2} = 10$
   d) $e^{t-2} = 10$

6 The value, V, of an investment after t years is given by the formula $V = Ae^{0.03t}$, where A is the initial investment.
   a) How much, to the nearest dollar, will an investment of $4000 be worth after 3 years?
   b) To the nearest year, how long will I need to keep an investment for it to double in value?

7 The path of a projectile launched from an aircraft is given by the equation $h = 5000 - e^{0.2t}$, where h is the height in metres and t is the time in seconds.
   a) From what height was the projectile launched?

   The projectile is aimed at a target at ground level.
   b) How long does it take to reach the target?

8 Match each equation from **i** to **vi** to the correct graph **a** to **f**.
   i) $y = e^{2x}$
   ii) $y = e^x + 2$
   iii) $y = 2 - e^x$
   iv) $y = 2 - e^{-x}$
   v) $y = 3e^{-x} - 5$
   vi) $y = e^{-2x} - 1$

a)

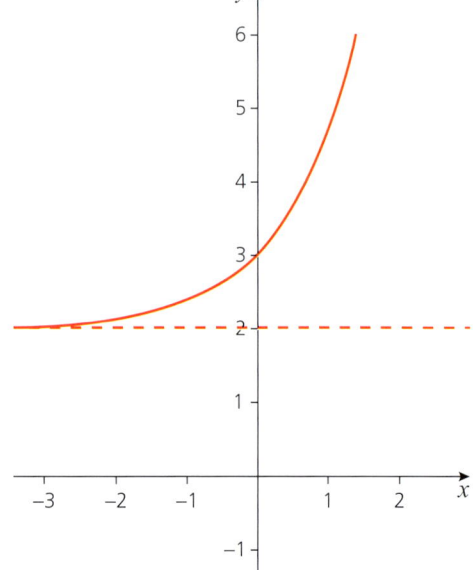

b)

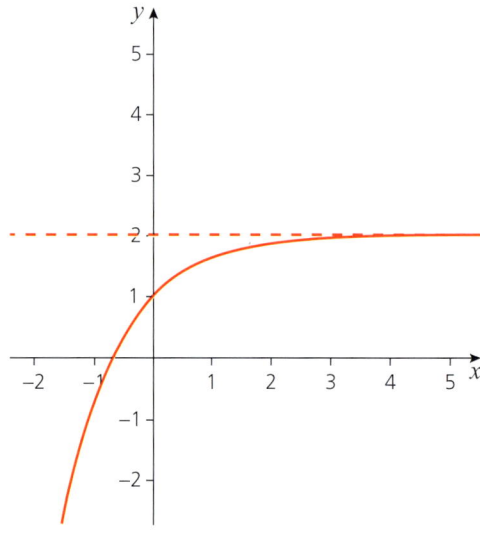

Exponential functions

c)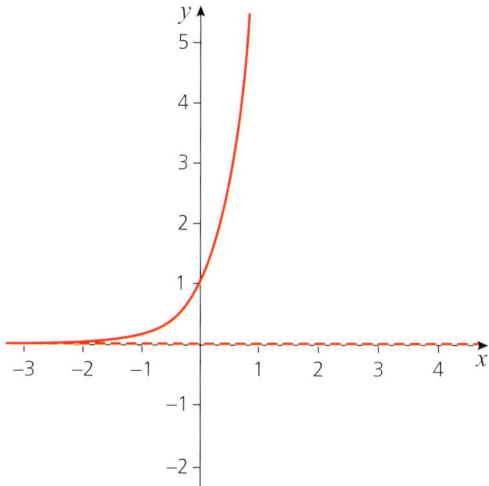

d)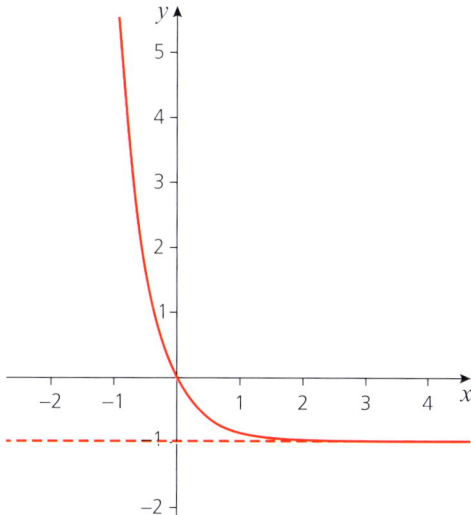

e)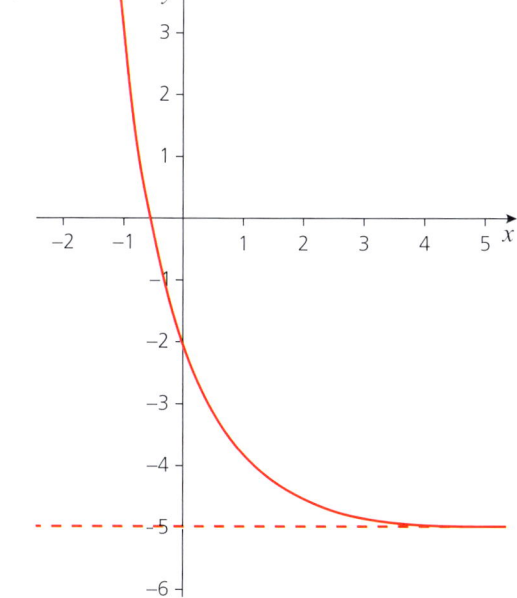

f)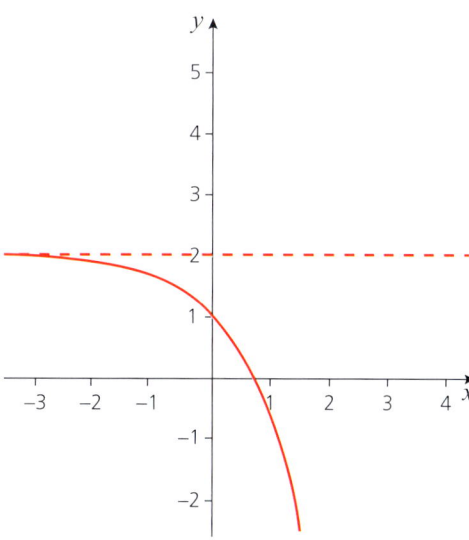

6 LOGARITHMIC AND EXPONENTIAL FUNCTIONS

Exercise 6.2 (cont)

9 A radioactive substance of mass 100g is decaying such that after t days the amount remaining, M, is given by the equation $M = 100e^{-0.002t}$.
   a) Sketch the graph of M against t.
   b) What is the half-life of the substance (i.e. the time taken to decay to half the initial mass)?

10 When David started his first job, he earned $15 per hour and was promised an annual increment (compounded) of 3.5%.
   a) What is his hourly rate in his 5th year?
   After 5 years he was promoted. His hourly wage increased to $26 per hour, with the same compounded annual increment.
   b) For how many more years will he need to work before his hourly rate reaches $30 per hour?

11 a) Solve the equation $e^{2x} + e^x - 12 = 0$.
   b) Hence solve the equation $e^{4x} + e^{2x} - 12 = 0$.

12 a) Solve $2(3^{2x}) - 5(3^x) + 2 = 0$
   b) Solve $e^x e^{x+1} = 10$
   c) Solve $2^{2x} - 5(2^x) + 4 = 0$
   d) Solve $2e^{2x} - 13e^x + 15 = 0$.
   e) Solve $3e^x = 11 - 10e^{-x}$.
   f) Solve $2(\ln 3x)^2 - 3(\ln 3x) - 14 = 0$.

Past-paper questions

1 Given that $\log_a pq = 9$ and $\log_a p^2 q = 15$, find the value of
   (i) $\log_a p$ and $\log_a q$, [4]
   (ii) $\log_p a + \log_q a$. [2]

   Cambridge O Level Additional Mathematics (4037)
   Paper 12 Q4, November 2012
   Cambridge IGCSE Additional Mathematics (0606)
   Paper 12 Q4, November 2012

2 Solve the simultaneous equations
   $\log_3 a = 2 \log_3 b$,
   $\log_3 (2a - b) = 1$. [5]

   Cambridge O Level Additional Mathematics (4037)
   Paper 13 Q5, November 2010
   Cambridge IGCSE Additional Mathematics (0606)
   Paper 13 Q5, November 2010

3 The number of bacteria B in a culture, t days after the first observation, is given by
   $B = 500 + 400e^{0.2t}$
   (i) Find the initial number present. [1]
   (ii) Find the number present after 10 days. [1]
   (iv) Find the value of t when $B = 10000$. [3]

   Cambridge O Level Additional Mathematics (4037)
   Paper 22 Q5 i, ii & iv, November 2014
   Cambridge IGCSE Additional Mathematics (0606)
   Paper 22 Q5 i, ii & iv, November 2014

Exponential functions

Now you should be able to:

★ know and use simple properties and graphs of the logarithmic and exponential functions, including ln x and e^x
★ know and use the laws of logarithms, including change of base of logarithms
★ solve equations of the form $a^x = b$.

Key points

✔ **Logarithm** is another word for **index** or **power**.
✔ The laws for logarithms are valid for all bases greater than 0 and are related to those for indices.

Operation	Law for indices	Law for logarithms
Multiplication	$a^x \times a^y = a^{x+y}$	$\log_a xy = \log_a x + \log_a y$
Division	$a^x \div a^y = a^{x-y}$	$\log_a \frac{x}{y} = \log_a x - \log_a y$
Powers	$(a^x)^n = a^{nx}$	$\log_a x^n = n \log_a x$
Roots	$(a^x)^{\frac{1}{n}} = a^{\frac{x}{n}}$	$\log_a \sqrt[n]{x} = \frac{1}{n} \log_a x$
Logarithm of 1	$a^0 = 1$	$\log_a 1 = 0$
Reciprocals	$\frac{1}{a^x} = a^{-x}$	$\log_a \frac{1}{x} = \log_a 1 - \log_a x = -\log_a x$
Log to its own base	$a^1 = a$	$\log_a a = 1$

✔ The graph of $y = \log x$:
is only defined for $x > 0$
has the y-axis as an asymptote
has a positive gradient
passes through (0, 1) for all bases.
✔ Notation.
The logarithm of x to the base a is written $\log_a x$.
The logarithm of x to the base 10 is written $\lg x$ or $\log x$.
The logarithm of x to the base e is written $\ln x$.
✔ An **exponential function** is of the form $y = a^x$.
✔ The exponential function is the inverse of the log function.
$y = \log_a x \Leftrightarrow a^y = x$
✔ For $a > 0$, the graph of $y = a^x$:
has the x-axis as an asymptote
has a positive gradient
passes through (0, 1).
✔ For $a > 0$, the graph of $y = a^{-x}$:
has the x-axis as an asymptote
has a negative gradient
passes through (0, 1).

Review exercise 2

1 a Solve the equation $|5 - 2x| = 12$. [3]
  b Solve the inequality $|x - 3| \leq |2x|$. [4]

2 a On a grid like the one below, sketch the graph of $y = |2x^2 + x - 10|$, stating the coordinates of any points where the curve meets the coordinate axes. [4]

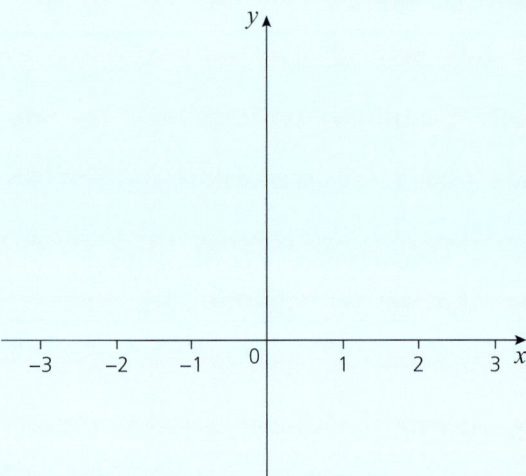

  b How many values of x satisfy the equation $|2x^2 + x - 10| = 3x$? [2]

3 Solve the inequality $3x^2 + x - 3 \leq (x - 2)^2$. [3]

4 It is given that the polynomial $p(x) = x^3 - x^2 - 4x + 4$ has a factor of $x - 2$.
  a Write $p(x)$ as a product of its linear factors. [3]
  b i On axes like the ones below, sketch the graph of $y = p(x)$.
     ii State the coordinates of any points where the curve meets the coordinate axes. [3]
  c Solve the inequality $x^3 + 4 \geq x^2 + 4x$. [2]

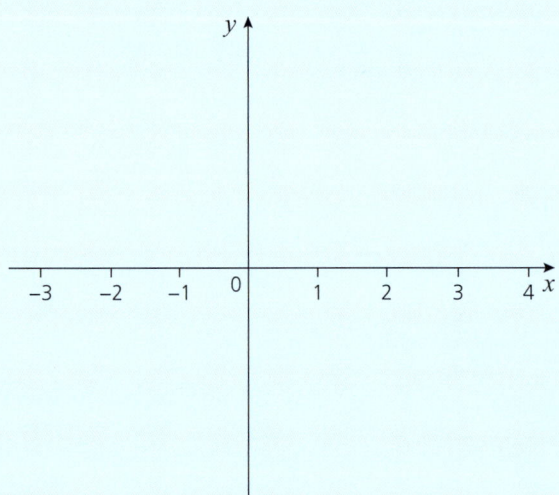

Review exercise 2

5 Solve the equations
$y - x = 4$,
$x^2 + y^2 - 8x - 4y - 16 = 0$. [5]

Cambridge O Level Additional Mathematics (4037)
Paper 11 Q1, May/June 2018
Cambridge IGCSE Additional Mathematics (0606)
Paper 11 Q1, May/June 2018

6 Find the set of values of k for which the line $y = 2x + k$ intersects the curve $y = kx^2 - 2x + 5$ at two distinct points. [6]

7 Solve the simultaneous equations
$\log_6 x = 2\log_6 y$
$\log_6(5y - x) = 1$ [5]

8 The value, V dollars, of a car aged t years is given by $V = 12\,000e^{-0.2t}$.
  i Write down the value of the car when it was new. [1]
  ii Find the time it takes for the value to decrease to $\frac{2}{3}$ of the value when it was new. [2]

Cambridge O Level Additional Mathematics (4037)
Paper 22 Q2, February/March 2017
Cambridge IGCSE Additional Mathematics (0606)
Paper 22 Q2, February/March 2017

9 On separate grids, like the one below, sketch the graphs of the following.
State the equations of any asymptotes and the coordinates of any point where the curves meet the coordinate axes.
  a $y = \ln(x - 1)$ [3]
  b $y = |\ln(x - 1)|$ [1]

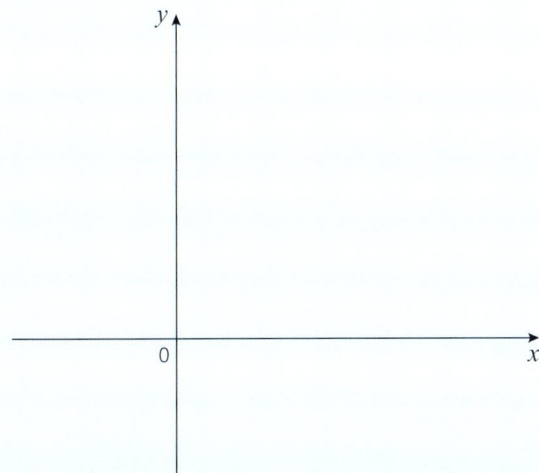

10 Solve the equation $10^{3x-2} = 8$, giving your answer correct to 2 decimal places. [3]

11 Write $\dfrac{\log_5 a + \log_5 b}{(\log_5 c)(\log_c 5)}$ as a single logarithm to base 5. [2]

7 Straight line graphs

We will always have STEM with us. Some things will drop out of the public eye and will go away, but there will always be science, engineering, and technology. And there will always, always be mathematics.

Katherine Johnson (1918–2020)

Discussion point

If you do a bungee jump, you will want to be certain that the rope won't stretch too far.

In an experiment a rope is tested by hanging different loads on it and measuring its length. The measurements are plotted on this graph.

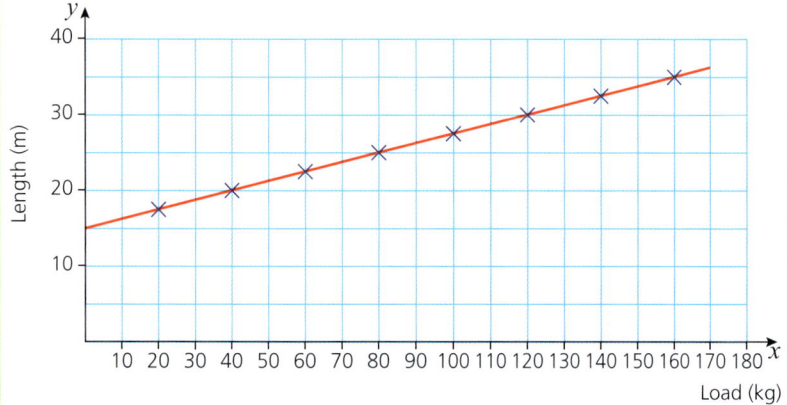

You will have met straight line graphs frequently in abstract algebraic problems, but they can also be used to find information in a practical situation such as this.

What does the graph tell you about the rope?

The straight line y = mx + c

When a load of 200g is attached to a spring, its stretched length is 40cm. With a load of 300g, its length is 50cm. Assuming that the extension is proportional to the load, draw a graph to show the relationship between the load and the length of the spring and use it to find the natural length of the spring.

Since the load is the variable that can be directly controlled, it is plotted on the horizontal axis and the length of the spring on the vertical axis. Plotting the points (200, 40) and (300, 50) and joining them with a straight line gives the graph below.

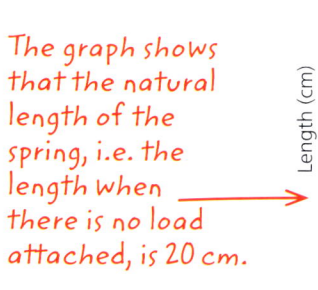

The graph shows that the natural length of the spring, i.e. the length when there is no load attached, is 20 cm.

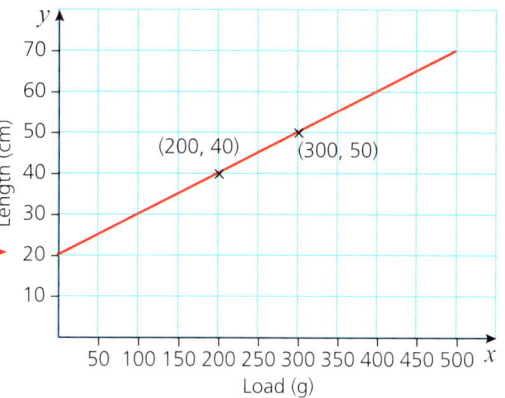

The information that the extension is proportional to the load tells you that the graph of the relationship will be a straight line.

The straight line $y = mx + c$

When the equation of a straight line is written in the form $y = mx + c$, m represents the gradient of the line and the line crosses the y-axis at $(0, c)$.

You can use this to find the equation of a straight line given the graph.

➡ Worked example

Find the equation of this straight line.

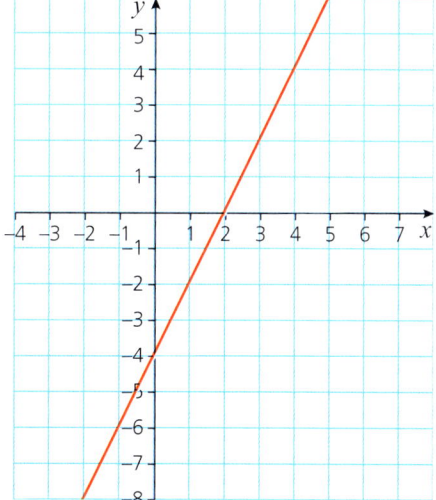

119

7 STRAIGHT LINE GRAPHS

Solution

The line crosses the y-axis at $(0, -4)$ so $c = -4$.

To find the gradient of the line, choose two points on the line and call them (x_1, y_1) and (x_2, y_2). The points of intersection with the axes, $(0, -4)$ and $(2, 0)$, are obvious choices.

The gradient of the line joining the points (x_1, y_1) and (x_2, y_2) is given by
$$\text{Gradient} = \frac{y_2 - y_1}{x_2 - x_1}$$

Using the gradient formula:

Gradient $(m) = \dfrac{0 - (-4)}{2 - 0} = 2$

So the equation of the line is $y = 2x - 4$.

→ Worked example

Find the equation of the line shown.

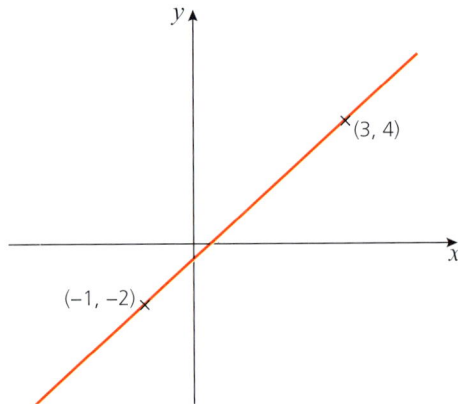

Solution

Substitute each pair of coordinates into the equation $y = mx + c$.

Point $(-1, -2)$: $\quad -2 = m(-1) + c \quad\quad (1)$

Point $(3, 4)$: $\quad\quad\ 4 = m(3) + c \quad\quad\ \ (2)$

Subtract equation (1) from equation (2).

$4 - (-2) = (3m + c) - (-m + c)$

$\Rightarrow\ 6 = 3m + c + m - c$

$\Rightarrow\ 6 = 4m$

$\Rightarrow\ m = 1.5$

Equation 2 is the more straightforward equation because it has no negative signs.

Substitute $m = 1.5$ into equation (2).

$4 = 3(1.5) + c$

$\Rightarrow\ c = -0.5$

So the equation of the line is $y = 1.5x - 0.5$.

The straight line y = mx + c

As well as $y = mx + c$, there are several other formulae for the equation of a straight line. One that you are likely to find useful deals with the situation where you know the gradient of the line, m, and the coordinates of one point on it, (x_1, y_1).

The equation is $y - y_1 = m(x - x_1)$.

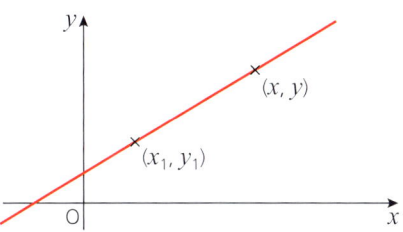

Midpoint of a line

When a line has a fixed length, the midpoint, i.e. the point half way between the two ends of the line, has as its coordinates the average of the individual x- and y-coordinates.

The midpoint of the line joining the points (x_1, y_1) and (x_2, y_2) is given by midpoint $= \left(\frac{x_1 + x_2}{2}, \frac{y_1 + y_2}{2}\right)$.

➜ Worked example

Find the midpoint of the line joining (2, 5) and (4, 13).

Solution

The coordinates of the midpoint are $\left(\frac{2+4}{2}, \frac{5+13}{2}\right) = (3, 9)$.

Length of a line

To find the length of the line joining two points, use Pythagoras' theorem.

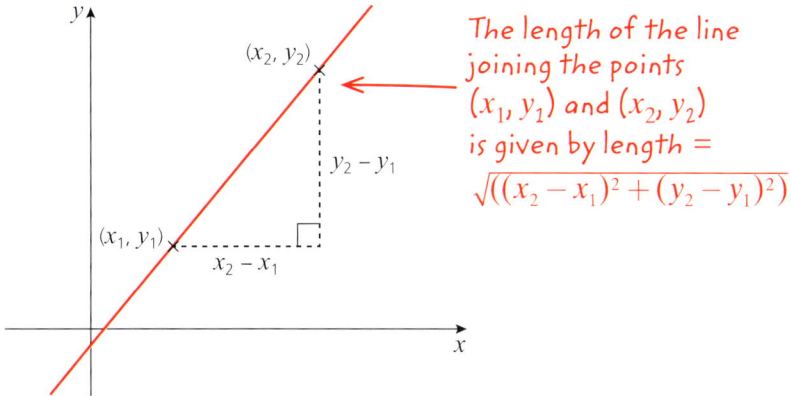

The length of the line joining the points (x_1, y_1) and (x_2, y_2) is given by length $= \sqrt{((x_2 - x_1)^2 + (y_2 - y_1)^2)}$

7 STRAIGHT LINE GRAPHS

 Worked example

Work out the length of the line joining the points $A(-2, 5)$ and $B(2, 2)$.

Solution

You can either:

» sketch the triangle and then use Pythagoras' theorem or
» use the formula given above.

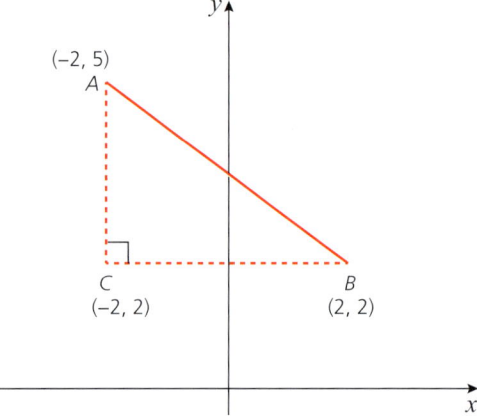

From the triangle: $AC = 3$ units
                     $BC = 4$ units

So $AB^2 = 3^2 + 4^2$
                           $= 25$

$AB = 5$ units

Alternatively, substituting directly into the formula (without drawing a diagram) gives:

length $= \sqrt{((2-(-2))^2 + (2-5)^2)}$

$= 5$ units

Parallel lines

Two lines are **parallel** if they have the same gradient. If you are given the equations of two straight line graphs in the form $y = mx + c$, you can immediately identify whether or not the lines are parallel. For example, $y = 3x - 7$ and $y = 3x + 2$ are parallel since they both have a gradient of 3.

If one or both of the equations are given in a different form, you will need to rearrange them in order to find out whether or not they are parallel.

Worked example

Show that the two lines $y = \frac{1}{2}x - 4$ and $x - 2y - 6 = 0$ are parallel.

Solution

Start by rearranging the second equation into the form $y = mx + c$.
$x - 2y - 6 = 0 \Rightarrow x - 6 = 2y$
$$\Rightarrow 2y = x - 6$$
$$\Rightarrow y = \frac{1}{2}x - 3$$

Both lines have a gradient of $\frac{1}{2}$ so are parallel.

Perpendicular lines

Two lines are **perpendicular** if they intersect at an angle of 90°.

Activity

The diagram shows two congruent right-angled triangles where p and q can take any value.

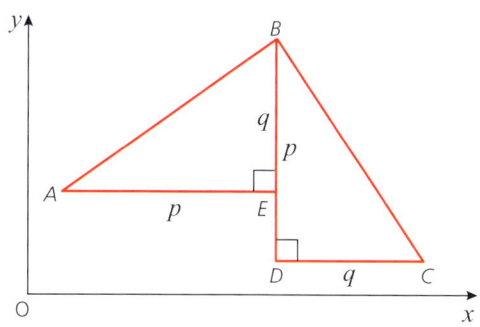

1. Copy the diagram onto squared paper.
2. Explain why $\angle ABC = 90°$.
3. Calculate the gradient of AB (m_1) and the gradient of BC (m_2).
4. Show that $m_1 m_2 = -1$.

7 STRAIGHT LINE GRAPHS

Worked example

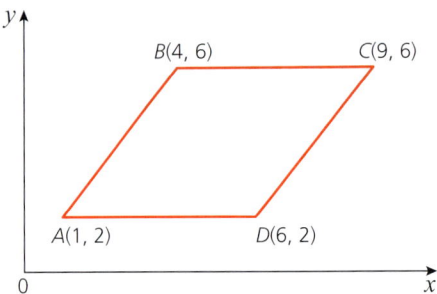

a) Explain why $ABCD$ is a rhombus.

b) Show that the diagonals AC and BD are perpendicular. (This result is always true for a rhombus.)

Solution

a) A rhombus is a parallelogram with all sides equal in length.

AD and BC are both parallel to the x-axis and have length 5 units.

gradient of AB = gradient DC = $\frac{\text{increase in } y}{\text{increase in } x} = \frac{4}{3}$

$AB = DC = \sqrt{3^2 + 4^2} = 5$ units

So $ABCD$ is a rhombus.

b) Using the formula gradient = $\frac{y_2 - y_1}{x_2 - x_1}$

gradient of $AC = \frac{6-2}{9-1} = \frac{1}{2}$

gradient of $BD = \frac{2-6}{6-4} = -2$

$\frac{1}{2} \times (-2) = -1$ so diagonals AC and BD are perpendicular.

Worked example

Find the equation of the perpendicular bisector of the line AB which joins the points $A(5, -2)$ and $B(-7, 4)$.

Solution

The perpendicular bisector of AB passes through the midpoint of AB.

The midpoint of the line joining (x_1, y_1) and (x_2, y_2) is given by $\left(\frac{(x_1 + x_2)}{2}, \frac{(y_1 + y_2)}{2}\right)$

Midpoint of $AB = \left(\frac{5 + (-7)}{2}, \frac{-2 + 4}{2}\right) = (-1, 1)$

Gradient of $AB = \frac{4 - (-2)}{(-7) - 5}$

$= \frac{6}{-12}$

$= -\frac{1}{2}$

The gradient of the line joining (x_1, y_1) and (x_2, y_2) is given by gradient = $\frac{y_2 - y_1}{x_2 - x_1}$

The straight line y = mx + c

For perpendicular lines, $m_1 m_2 = -1$ → Therefore, the gradient of the line perpendicular to AB is 2 and the equation of the perpendicular has the form $y = 2x + c$.

The perpendicular passes through the point $(-1, 1)$. Use this to find c.

Substitute $x = -1$ and $y = 1$ into $y = 2x + c$. →

$1 = 2 \times -1 + c$

$\Rightarrow 1 = -2 + c$

$\Rightarrow c = 3$

So the equation of the perpendicular bisector of AB is $y = 2x + 3$.

Exercise 7.1

1. For each of the following pairs of points A and B, calculate:
   i) the gradient of the line AB
   ii) the gradient of the line perpendicular to AB
   iii) the length of AB
   iv) the coordinates of the midpoint of AB.
   a) $A(4, 3)$ $B(8, 11)$
   b) $A(5, 3)$ $B(10, -8)$
   c) $A(6, 0)$ $B(8, 15)$
   d) $A(-3, -6)$ $B(2, -7)$

2. $A(0, 5)$, $B(4, 1)$ and $C(2, 7)$ are the vertices of a triangle. Show that the triangle is right angled:
   a) by working out the gradients of the sides
   b) by calculating the lengths of the sides.

3. $A(3, 5)$, $B(3, 11)$ and $C(6, 2)$ are the vertices of a triangle.
   a) Work out the perimeter of the triangle.
   b) Sketch the triangle and work out its area using AB as the base.

4. A quadrilateral $PQRS$ has vertices at $P(-2, -5)$, $Q(11, -7)$, $R(9, 6)$ and $S(-4, 8)$.
   a) Work out the lengths of the four sides of $PQRS$.
   b) Find the coordinates of the midpoints of the diagonals PR and QS.
   c) Without drawing a diagram, show that $PQRS$ cannot be a square. What shape is $PQRS$?

5. The points A, B and C have coordinates $(2, 3)$, $(6, 12)$ and $(11, 7)$ respectively.
   a) Draw the triangle ABC.
   b) Show by calculation that the triangle is isosceles and write down the two equal sides.
   c) Work out the midpoint of the third side.
   d) By first calculating appropriate lengths, calculate the area of triangle ABC.

6. A triangle ABC has vertices at $A(3, 2)$, $B(4, 0)$ and $C(8, 2)$.
   a) Show that the triangle is right angled.
   b) Find the coordinates of point D such that $ABCD$ is a rectangle.

7. $P(-2, 3)$, $Q(1, q)$ and $R(7, 0)$ are collinear points (i.e. they lie on the same straight line).
   a) Find the value of Q.
   b) Write down the ratio of the lengths $PQ : QR$.

125

7 STRAIGHT LINE GRAPHS

Exercise 7.1 (cont)

8 A quadrilateral has vertices $A(-2, 8)$, $B(-5, 5)$, $C(5, 3)$ and $D(3, 7)$.
   a) Draw the quadrilateral.
   b) Show by calculation that it is a trapezium.
   c) $ABCE$ is a parallelogram. Find the coordinates of E.

9 In each part, find the equation of the line through the given point that is:
   i) parallel and
   ii) perpendicular to the given line.
   a) $y = 2x + 6$; $(5, -3)$
   b) $x + 3y + 5 = 0$; $(-4, 7)$
   c) $2x = 3y + 1$; $(-1, -6)$

10 Find the equation of the perpendicular bisector of the line joining each pair of points.
   a) $(2, 3)$ and $(8, -1)$
   b) $(-7, 3)$ and $(1, 5)$
   c) $(5, 6)$ and $(4, -3)$

11 P is the point $(2, -1)$ and Q is the point $(8, 2)$.
   a) Write the equation of the straight line joining P and Q.
   b) Find the coordinates of M, the midpoint of PQ.
   c) Write the equation of the perpendicular bisector of PQ.
   d) Write down the coordinates of the points where the perpendicular bisector crosses the two axes.

Relationships of the form $y = ax^n$

When you draw a graph to represent a practical situation, in many cases your points will lie on a curve rather than a straight line. When the relationships are of the form $y = ax^n$ or $y = Ab^x$, you can use logarithms to convert the curved graphs into straight lines. Although you can take the logarithms to any positive base, the forms log and ln are used in most cases.

Worked example

The data in the table were obtained from an experiment. y represents the mass in grams of a substance (correct to 2 d.p.) after a time t minutes.

t	4	9	14	19	24	29
y	3.00	4.50	5.61	6.54	7.35	8.08

Saira wants to find out if these values can be modelled by the function $y = at^n$.

a) By taking logarithms to base 10 of both sides, show that the model can be written as $\log y = n \log t + \log a$.

b) Explain why, if the model is valid, plotting the graph of $\log y$ against $\log t$ will result in a straight line.

c) Plot the graph of $\log y$ against $\log t$ and use it to estimate the values of a and t. Hence express the relationship in the form $y = at^n$.

d) Assuming that this relationship continues for at least the first hour, after how long would there be $10\,\text{g}$ of the substance?

Relationships of the form y = ax^n

Solution

a) $y = at^n \Rightarrow \log y = \log at^n$
$\Rightarrow \log y = \log a + \log t^n$
$\Rightarrow \log y = \log a + n \log t$
$\Rightarrow \log y = n \log t + \log a$

b) Comparing this with the equation $Y = mX + c$ gives $Y = \log y$ and $X = \log t$. This shows that if the model is valid, the graph of $\log y$ (on the vertical axis) against $\log t$ will be a straight line with gradient n and intercept on the vertical axis at $\log a$.

c)

t	4	9	14	19	24	29
y	3.00	4.50	5.61	6.54	7.35	8.08
log t	0.60	0.95	1.15	1.28	1.38	1.46
log y	0.48	0.65	0.75	0.82	0.87	0.91

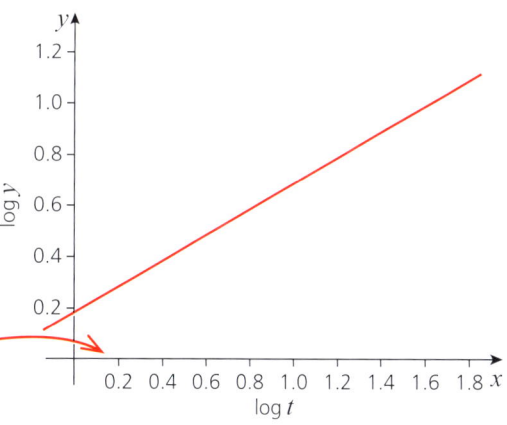

NOTE: You cannot have a break in the horizontal axis because this would lead to an incorrect point of intersection with the y-axis.

Using the points (0, 0.18) and (1.5, 0.92), the gradient of the line is:

gradient = $\frac{y_2 - y_1}{x_2 - x_1}$ ⟶ $\frac{0.92 - 0.18}{1.5 - 0} = 0.493$

This is approximately equal to 0.5, so $n = 0.5$.

Since the values from the graph are only approximate, the results should only be given to 1 or 2 d.p.

Using intercept on the y-axis = $\log a$ ⟶ $0.18 = \log a$

$\Rightarrow a = 1.513 \approx 1.5$

Therefore the relationship is $y = 1.5t^{0.5}$ or $y = 1.5\sqrt{t}$.

Do not go beyond the values in the table unless the question tells you to. If it doesn't, you cannot be sure that the relationship you have found is valid outside of known bounds.

d) There will be 10 g of the substance when $y = 10$
$\Rightarrow 10 = 1.5\sqrt{t}$
$\Rightarrow 100 = 2.25t$
$\Rightarrow t = \frac{100}{2.25}$
$= 44.44$

So, there will be 10 g after about 44 minutes.

7 STRAIGHT LINE GRAPHS

Relationships of the form $y = Ab^x$

These are often referred to as exponential relationships since the variable is the power.

→ Worked example

The table shows the temperature, θ, recorded in degrees Celsius to the nearest degree, of a cup of coffee t minutes after it is poured and milk is added.

t	0	4	8	12	16	20
θ	80	63	50	40	32	25

Seb is investigating whether the relationship between temperature and time can be modelled by an equation of the form $\theta = Ab^t$.

a) By taking logarithms to base e of both sides, show that the model can be written as $\ln \theta = \ln A + t \ln b$.

b) Explain why, if the model is valid, plotting the graph of $\ln \theta$ against t will result in a straight line.

c) Plot the graph of $\ln \theta$ against t and use it to estimate the values of A and b. Hence express the relationship in the form $\theta = Ab^t$.

d) Why will this relationship not continue indefinitely?

Solution

a) $\theta = Ab^t \Rightarrow \ln \theta = \ln Ab^t$

$\Rightarrow \ln \theta = \ln A + \ln b^t$

$\Rightarrow \ln \theta = \ln A + t \ln b$

b) Rewriting $\ln \theta = \ln A + t \ln b$ as $\ln \theta = (\ln b)t + \ln A$ and comparing it with the equation $y = mx + c$ shows that plotting $\ln \theta$ against t will give a straight line with gradient $\ln b$ and intercept on the vertical axis at $\ln A$.

c)
t	0	4	8	12	16	20
θ	80	63	50	40	32	25
$\ln \theta$	4.38	4.14	3.91	3.69	3.47	3.22

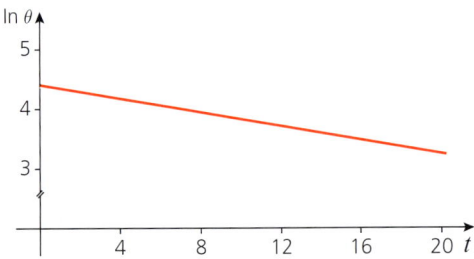

Using the points (0, 4.40) and (20, 3.22), the gradient of the line is:

$$\frac{3.22 - 4.38}{20 - 0} = -0.058$$

$-0.058 = \ln b \Rightarrow b = 0.94$ (2 d.p.)

The intercept on the vertical axis is at $\ln \theta = 4.40$.

From the table, this corresponds $\theta = 80$.

Therefore, the relationship is $\theta = 80 \times 0.94^t$.

d) The relationship will not continue indefinitely since the coffee will not cool below room temperature.

Other relationships of the form $Y = mX + c$

For models of the form $y = ax^n$ and $y = Ab^x$, straight line graphs were used to estimate the values of the constants a, n, A and b. Straight lines can also be used to calculate constants in other mathematical models.

If a mathematical model has the form $Y = mX + c$, where Y is a function of y and X is a function of x, plotting the graph of Y against X will result in a straight line. This straight line can be used to calculate estimates of the constants used in the model.

Worked example

The table shows the amount of gas produced when a particular substance is added during an experiment. y represents the amount of gas in mm³ and x represents the mass of the substance in milligrams.

x	0.8	1	1.2	1.4	1.6	1.8
y	1.42	1.73	2.11	2.55	3.03	3.56

It is known that the variables x and y satisfy the relationship $y^2 = Ax^3 + B$ where A and B are constants.

a) By drawing a suitable straight line graph, find an estimate for the constants A and B.

b) Comment on the suitability of the model for the given data.

Solution

$y^2 = Ax^3 + B$ has the form $Y = mX + c$ where $y^2 = Y$ and $x^3 = X$, therefore you should plot the graph of y^2 against x^3.

7 STRAIGHT LINE GRAPHS

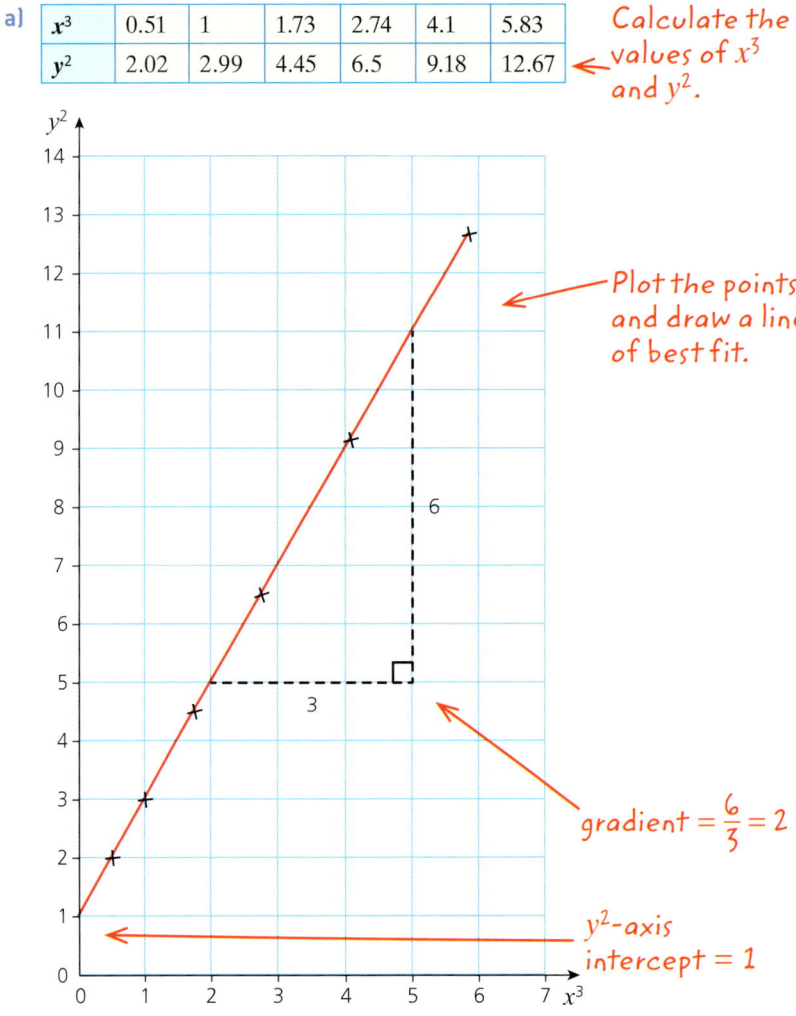

a)

x^3	0.51	1	1.73	2.74	4.1	5.83
y^2	2.02	2.99	4.45	6.5	9.18	12.67

← Calculate the values of x^3 and y^2.

— Plot the points and draw a line of best fit.

gradient = $\frac{6}{3}$ = 2

y^2-axis intercept = 1

Comparing $y^2 = Ax^3 + B$ with $Y = mX + c$, the gradient is A and the y^2-axis intercept is B.
From the graph $A = 2$ and $B = 1$.
This gives the model $y^2 = 2x^3 + 1$.

b) The points form a straight line so the model is a good fit for the given data.

→ Worked example

Ella grows potatoes in pots in her garden. She is experimenting to see whether increasing the amount of fertiliser results in a greater yield. She places one seed potato in each of six similar sized pots and adds varying amounts of fertiliser. Upon harvest, she measures and records the total mass of potatoes per pot. The table shows the data she collected. x represents the amount of fertiliser added in grams and y represents the mass of potatoes in kilograms.

Other relationships of the form Y = mX + c

x	2	3	4	5	6	7
y	0.82	0.91	0.96	1.00	1.03	1.05

Ella believes that the data can be modelled by $y^3 = A\ln x + B$ where A and B are constants.

a) By drawing a suitable straight line graph, find an estimate for the constants A and B.

b) Assuming that this relationship continues, what would be the expected mass of potatoes if an extra 10 grams of fertiliser were added?

Solution

$y^3 = A\ln x + B$ has the form $Y = mX + c$, where $y^3 = Y$ and $\ln x = X$, therefore you should plot the graph of y^3 against $\ln x$.

$\ln x$	0.69	1.10	1.39	1.61	1.79	1.95
y^3	0.55	0.75	0.88	1.00	1.09	1.16

Calculate the values of $\ln x$ and y^3.

a)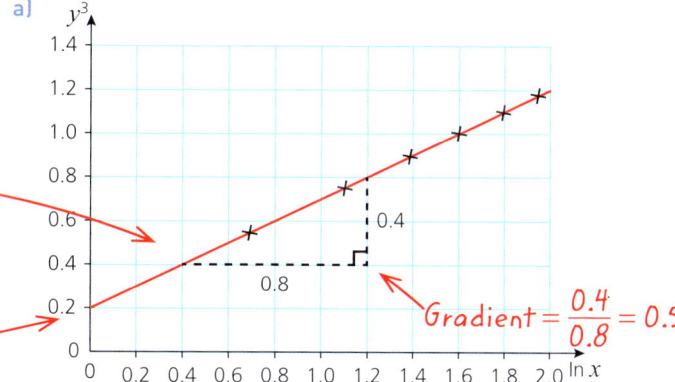

Plot the points and draw a line of best fit

y^3-axis intercept = 0.2

Gradient = $\dfrac{0.4}{0.8} = 0.5$

Comparing $y^3 = A\ln x + B$ with $Y = mX + c$, the gradient is A and the y^3-axis intercept is B.
From the graph $A = 0.5$ and $B = 0.2$
This gives the model $y^3 = 0.5\ln x + 0.2$

b) $x = 10 \Rightarrow y^3 = 0.5\ln 10 + 0.2$

$\Rightarrow y = \sqrt[3]{0.5\ln 10 + 0.2} = 1.11$ kg (2 d.p.)

Exercise 7.2

1 Match the equivalent relationships.
   i) $y = pr^x$
   ii) $y = rp^x$
   iii) $y = px^r$
   iv) $y = xp^r$
   a) $\log y = \log p + r\log x$
   b) $\log y = \log r + x\log p$
   c) $\lg y = \lg p + x\lg r$
   d) $\lg y = \lg x + r\lg p$

2 For each of the following models, k, a and b are constants. Use logarithms to base e to rewrite them in the form $y = mx + c$, stating the expressions equal to x, y, m and c in each case.
   a) $y = ka^x$
   b) $y = kx^a$
   c) $y = ak^x$
   d) $y = ax^k$

7 STRAIGHT LINE GRAPHS

Exercise 7.2 (cont)

3 The table below shows the area, A, in square centimetres of a patch of mould t days after it first appears.

t	1	2	3	4	5	6
A	1.8	2.6	3.6	5.1	7.2	10.3

It is thought that the relationship between A and t is of the form $A = kb^t$.
   a) Show that the model can be written as $\ln A = (\ln b)t + \ln k$.
   b) Plot the graph of $\ln A$ against t and explain why it supports the assumption that $A = kb^t$.
   c) Use your graph to estimate the values of b and k.
   d) Estimate: i) the time when the area of the mould was $6\,\text{cm}^2$
       ii) the area of the mould after 4.5 days.

4 It is thought that the relationship between two variables, a and b, is of the form $b = Pa^n$. An experiment is conducted to test this assumption. The results are shown in the table.

a	2	4	6	8	10	12
b	9.8	12.1	13.7	14.9	16.0	16.9

   a) Show that the model can be written as $\ln b = n \ln a + \ln P$.
   b) Plot the graph of $\ln b$ against $\ln a$ and say why this supports the assumption $b = Pa^n$.
   c) Estimate the values of n and p.

5 With the exception of one, all the results in table satisfy, to one decimal place, the relationship $y = ax^n$.

x	1.3	1.6	1.9	2.2	2.5	2.8
y	5.8	8.2	14.9	21.5	29.6	39.4

   a) Use a suitable logarithmic method to find the values of a and n.
   b) If the values of x are correct, identify the incorrect value of y and estimate the correct value to 1 d.p.

6 The population P (in thousands) of a new town is modelled by the relationship $P = ka^t$ where t is the time in years. Its growth over the first five years is shown in the table below.

Year (t)	1	2	3	4	5
Population (P)	3.6	4.3	5.2	6.2	7.5

   a) Explain why you would expect that the graph of $\ln P$ against t to be a straight line.
   b) Draw up a table of values, plot the graph and use it to estimate values for a and k to 1 d.p.
   c) Using these values, calculate an estimate for the population after 20 years. How reliable is this likely to be?

Other relationships of the form Y = mX + c

7 It is thought that the relationship between the variables p and q is of the form $q^2 = Ap^3 + B$. An experiment is carried out to test the assumption. The results are given below.

p	1.1	1.2	1.3	1.4	1.5	1.6
q	2.51	2.74	2.98	3.25	3.52	3.82

  a) Draw up an appropriate table of values. Then plot a suitable graph and use it to estimate the values of the constants A and B to 1 d.p.
  b) Comment on the suitability of the model for the given data.

8 The table shows the diameter, y cm, of a colony of bacteria at time t hours.

t	0.5	1	1.5	2	2.5	3
y	0.55	0.90	1.20	1.45	1.65	1.82

Suzanne, a microbiologist, models the increase in diameter using the relationship $e^{2y} = At^2 + B$ where A and B are constants.
  a) Draw up an appropriate table of values. Then plot the graph of e^{2y} against t^2.
  b) Use your graph to estimate the values of A and B to 1 d.p.
  c) Use the model to estimate the diameter of the colony after 5 hours.
  d) Comment on the validity of your answer.

9 The rate of a reaction is calculated by recording the amount of oxygen produced at regular time intervals. The table shows the volume of oxygen, y cm^3, that has been produced up to various times, t minutes.

t	1.5	2.5	3.5	4.5	5.5	6.5
y	1.22	1.31	1.36	1.39	1.42	1.44

The data can be modelled using the relationship $y^3 = A \ln t + B$.
  a) Plot a suitable graph and use it to estimate the values of the constants A and B to 1 d.p.
  b) Assuming that this relationship continues for at least the first 15 minutes, after how long would there be 1.5 cm^3 of oxygen?

Past-paper questions

1 **Solutions to this question by accurate drawing will not be accepted.**
The points $A(p, 1)$, $B(1, 6)$, $C(4, q)$ and $D(5, 4)$, where p and q are constants, are the vertices of a kite $ABCD$. The diagonals of the kite, AC and BD, intersect at the point E. The line AC is the perpendicular bisector of BD. Find
  (i) the coordinates of E, [2]
  (ii) the equation of the diagonal AC, [3]
  (iii) the area of the kite $ABCD$. [3]

Cambridge O Level Additional Mathematics (4037)
Paper 21 Q9, June 2014
Cambridge IGCSE Additional Mathematics (0606)
Paper 21 Q9, June 2014

7 STRAIGHT LINE GRAPHS

2 Solutions to this question by accurate drawing will not be accepted.

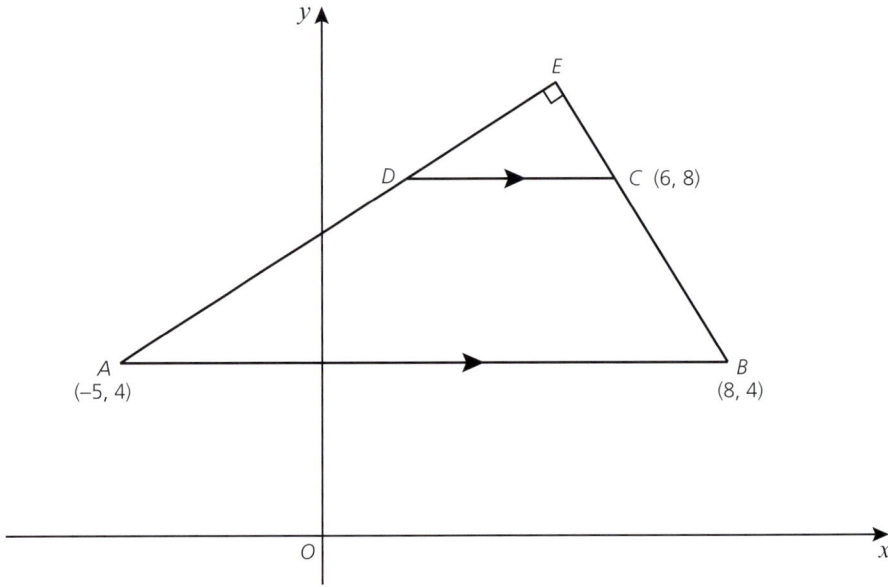

The vertices of the trapezium $ABCD$ are the points $A(-5, 4)$, $B(8, 4)$, $C(6, 8)$ and D. The line AB is parallel to the line DC. The lines AD and BC are extended to meet at E and angle $AEB = 90°$.
(i) Find the coordinates of D and of E. [6]
(ii) Find the area of the trapezium $ABCD$. [2]

Cambridge O Level Additional Mathematics (4037)
Paper 12 Q7, November 2012
Cambridge IGCSE Additional Mathematics (0606)
Paper 12 Q7, November 2012

3 Solutions to this question by accurate drawing will not be accepted.

The points $A(-3, 2)$ and $B(1, 4)$ are vertices of an isosceles triangle ABC, where angle $B = 90°$.
(i) Find the length of the line AB. [1]
(ii) Find the equation of the line BC. [3]
(iii) Find the coordinates of each of the two possible positions of C. [6]

Cambridge O Level Additional Mathematics (4037)
Paper 12 Q10, November 2013
Cambridge IGCSE Additional Mathematics (0606)
Paper 12 Q10, November 2013

Other relationships of the form Y = mX + c

Now you should be able to:
- ★ use the equation of a straight line
- ★ know and use the condition for two lines to be parallel or perpendicular
- ★ solve problems involving the midpoint and length of a line, including finding and using the equation of a perpendicular bisector
- ★ transform given relationships to and from straight line form, including determining unknown constants by calculating the gradient or intercept of the transformed graph.

Key points

- ✔ An equation of the form $y = mx + c$ represents a straight line that has gradient m and intersects the y-axis at $(0, c)$.
- ✔ The midpoint of the line joining the points (x_1, y_1) and (x_2, y_2) is given by:

  midpoint = $\left(\dfrac{x_1 + x_2}{2}, \dfrac{y_1 + y_2}{2}\right)$.
- ✔ The length of the line joining the points (x_1, y_1) and (x_2, y_2) is given by:

  length = $\sqrt{(x_2 - x_1)^2 + (y_2 - y_1)^2}$.
- ✔ Two lines are parallel if they have the same gradient.
- ✔ Two lines are perpendicular if they intersect at an angle of 90°.
- ✔ When the gradients of two parallel lines are given by m_1 and m_2, $m_1 m_2 = -1$.
- ✔ Logarithms can be used to describe the relationship between two variables in the following cases:

  i $y = ax^n$ Taking logs, $y = ax^n$ is equivalent to $\log y = \log a + n \log x$. Plotting $\log y$ against $\log x$ gives a straight line of gradient n that intersects the vertical axis at the point $(0, \log a)$.

  ii $y = Ab^x$ Taking logs, $= Ab^x$ is equivalent to $\log y = \log A + x \log b$. Plotting $\log y$ against x gives a straight line of gradient $\log b$ that intersects the vertical axis at the point $(0, \log A)$.

8 Coordinate geometry of the circle

The description of right lines and circles, upon which geometry is founded, belongs to mechanics. Geometry does not teach us to draw these lines, but requires them to be drawn.

Isaac Newton (1642–1727)

Discussion point

In order to use a mapping app on a mobile phone, it is important to know your precise location. To find your precise location, the mobile phone uses the Global Positioning System (GPS).

The GPS receiver in the mobile phone can communicate with satellites orbiting the Earth; the GPS receiver knows where the satellites are and its distance from them. It uses this information to calculate your location using a method called 'trilateration'. Trilateration can be explained by looking at the intersections of circles.

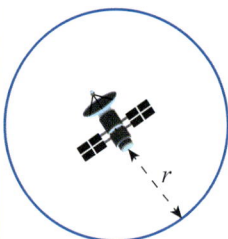

Given that the GPS receiver knows the distance, r, between it and a satellite, when it connects with a single satellite, your position may be any point on the circle with radius r.

So, when it connects with multiple satellites, your precise location can be determined by looking for the intersection of the different circles.

Your location is the point at which these three circles intersect. →

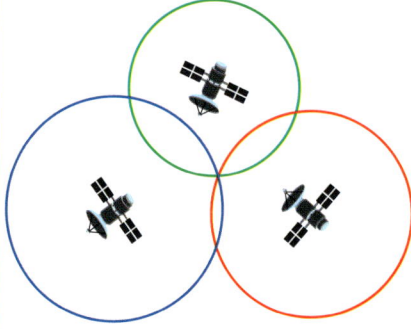

Why would two satellites not be enough to determine your exact location?

Equation of a circle

A circle can be described as the 'locus' of the points in a plane which are a fixed distance from a given point. The fixed distance is the radius of the circle and the given point is its centre. The radius and the centre are used to derive the equation of a circle.

Circles with centre (a,b)

The diagram shows a circle with centre $C(5, 6)$ and radius 4.

$P(x, y)$ is a general point on the circle.

To find the equation of a circle, use Pythagoras' theorem.

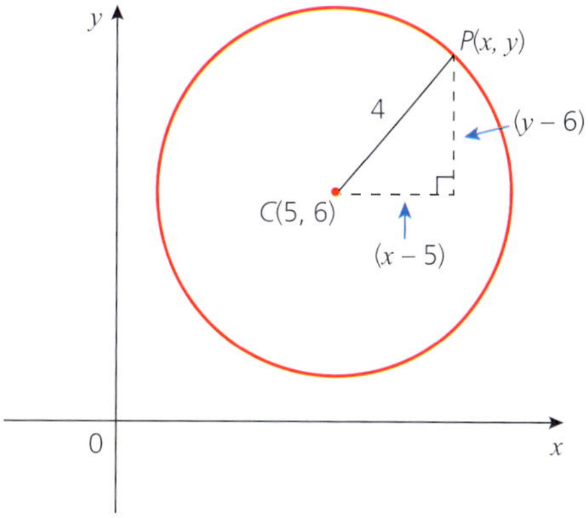

$$(x - 5)^2 + (y - 6)^2 = 4^2$$

$$\Rightarrow (x - 5)^2 + (y - 6)^2 = 16$$

This is the equation of the circle.

This result can be generalised as follows.

The circle with centre (a, b) and radius r has the equation

$$(x - a)^2 + (y - b)^2 = r^2$$

Using this result, you can see that the circle with centre $(0, 0)$ and radius r has the equation

$$x^2 + y^2 = r^2$$

➡ Worked example

Find the centre and the radius of the circle $(x + 4)^2 + y^2 = 49$.

Solution

Comparing this with the general equation for a circle with centre (a, b) and radius r,

$$(x - a)^2 + (y - b)^2 = r^2$$

this gives $a = -4$, $b = 0$ and $r = 7$.

$\Rightarrow$ The centre is $(-4, 0)$ and the radius is 7.

8 COORDINATE GEOMETRY OF THE CIRCLE

➡ Worked example

Find the equation of the circle with centre $(-1, 3)$ that passes through the point $(4, 15)$.

Solution

Use the points given to find the radius:

Use Pythagoras' theorem

$$r^2 = (4-(-1))^2 + (15-3)^2$$
$$\Rightarrow r^2 = 5^2 + 12^2$$
$$\Rightarrow r = \sqrt{5^2 + 12^2}$$
$$\Rightarrow r = 13$$

So the radius is 13.

Then, using $(x-a)^2 + (y-b)^2 = r^2$, you get the equation of the circle:

$$(x+1)^2 + (y-3)^2 = 169$$

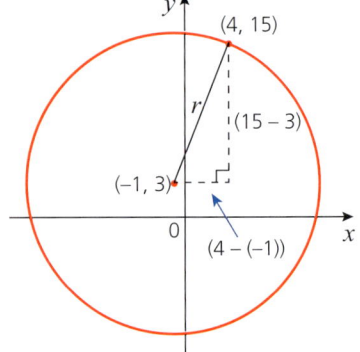

You may find it useful to draw a sketch.

Note

By multiplying out brackets, this equation can also be written in an expanded form as

$$(x+1)^2 + (y-3)^2 = 169$$
$$\Rightarrow x^2 + 2x + 1 + y^2 - 6y + 9 = 169$$
$$\Rightarrow x^2 + y^2 + 2x - 6y - 159 = 0$$

➡ Worked example

Show that the circle with equation $x^2 + y^2 + 8x - 12y + 3 = 0$ can be written in the form $(x-a)^2 + (y-b)^2 = r^2$, where a, b and r are constants to be found.

Solution

Collect the x terms and the y terms together.

$$x^2 + y^2 + 8x - 12y + 3 = 0$$
$$\Rightarrow x^2 + 8x + y^2 - 12y = -3$$

Complete the square on the x terms and on the y terms.

$$\Rightarrow (x+4)^2 - 16 + (y-6)^2 - 36 = -3$$
$$\Rightarrow (x+4)^2 + (y-6)^2 = -3 + 16 + 36$$
$$\Rightarrow (x+4)^2 + (y-6)^2 = 49$$
$$\Rightarrow (x+4)^2 + (y-6)^2 = 7^2$$

Therefore $a = -4$, $b = 6$ and $r = 7$.

Equation of a circle

> **Note**
>
> The equation of a circle can be written in the expanded form $x^2 + y^2 + 2gx + 2fy + c = 0$, where the centre is $(-g, -f)$ and the radius is $\sqrt{g^2 + f^2 - c}$.

➡ Worked example

$2g = 6, 2f = -10,$
$c = -2$

A circle has equation $x^2 + y^2 + 6x - 10y - 2 = 0$.

a) State the centre and the radius of the circle.

b) Give the equation of the circle in the form $(x-a)^2 + (y-b)^2 = r^2$.

Solution

a) Compare with the expanded equation for a circle $x^2 + y^2 + 2gx + 2fy + c = 0$, with centre $(-g, -f)$ and radius $\sqrt{g^2 + f^2 - c}$.

$2g = 6$ and $2f = -10$ and $c = -2$

$\Rightarrow g = 3$ $\Rightarrow f = -5$

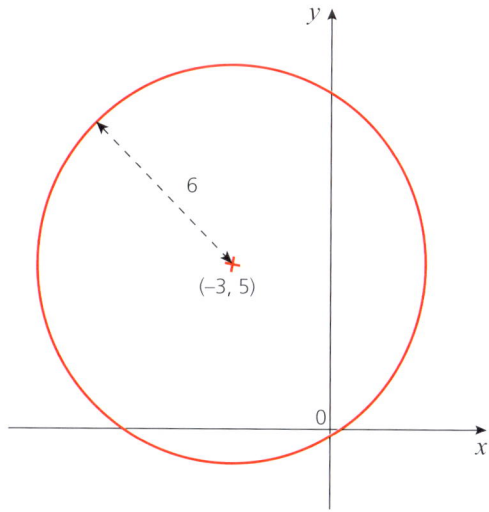

So, the centre is $(-3, 5)$ and the radius is $\sqrt{3^2 + (-5)^2 - (-2)} = 6$.

b) You know that the centre is $(-3, 5)$ and the radius is 6.

Comparing with the general equation for a circle $(x-a)^2 + (y-b)^2 = r^2$, you get the equation

$(x - (-3))^2 + (y - 5)^2 = 6^2$

$\Rightarrow (x + 3)^2 + (y - 5)^2 = 36$

8 COORDINATE GEOMETRY OF THE CIRCLE

> **Worked example**
>
> The points P and Q are $(10, -11)$ and $(-2, 5)$ respectively. The line PQ is the diameter of a circle. Find the equation of the circle in the form $(x - a)^2 + (y - b)^2 = r^2$.

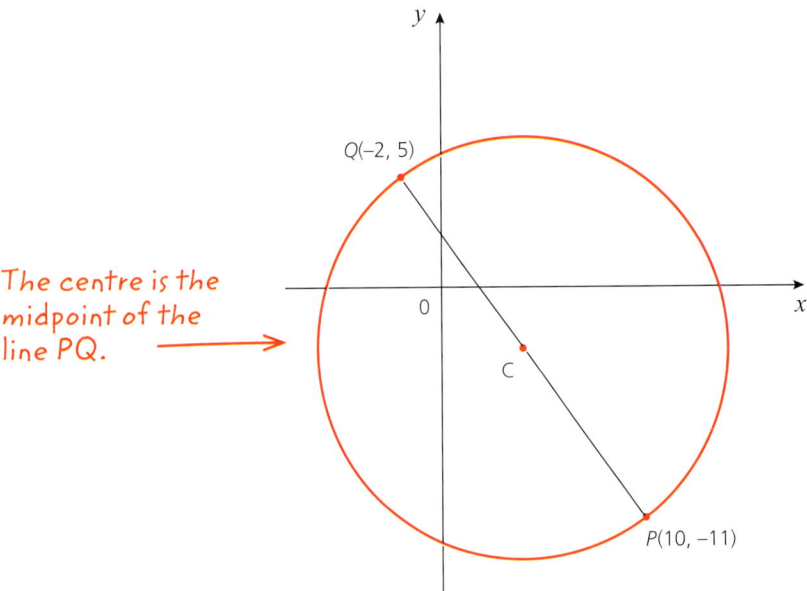

The centre is the midpoint of the line PQ.

The midpoint of the line joining (x_1, y_1) and (x_2, y_2) is given by $\left(\dfrac{(x_1 + x_2)}{2}, \dfrac{(y_1 + y_2)}{2}\right)$

The length of the line joining (x_1, y_1) and (x_2, y_2) is given by $\sqrt{(x_2 - x_1)^2 + (y_2 - y_1)^2}$

Solution

The centre of the circle is the midpoint of the line PQ.

Midpoint of $PQ = \left(\dfrac{10 + -2}{2}, \dfrac{-11 + 5}{2}\right) = (4, -3)$

The radius is half of the length of the diameter.

Radius $= \dfrac{\sqrt{(-2 - 10)^2 + (5 - (-11))^2}}{2} = 10$

Then, using $(x - a)^2 + (y - b)^2 = r^2$, the equation of the circle is

$(x - 4)^2 + (y + 3)^2 = 100$

Equation of a circle

Exercise 8.1

1. Find the equations of the following circles:
   a) centre (2, 7), radius 3
   b) centre (6, 0), radius 8
   c) centre (−2, 5), radius 1
   d) centre (3, −10), radius 5
   e) centre (−5, −9), radius 2

2. For each of the following circles state:
   a) the radius
   b) the coordinates of the centre.
      i) $x^2 + y^2 = 16$
      ii) $(x-1)^2 + y^2 = 64$
      iii) $(x+6)^2 + (y-5)^2 = 1$
      iv) $(x-1)^2 + (y+1)^2 = 25$
      v) $(x+4)^2 + (y+4)^2 = 36$

3. Sketch the circles with the following equations:
   a) $x^2 + y^2 = 36$
   b) $x^2 + (y+1)^2 = 25$
   c) $(x-4)^2 + (y-7)^2 = 9$

4. Find the equation of the circle with centre (1, −2) that passes through (5, 1).

5. Find the equation of the circle with centre (−3, −6) that passes through (−11, 9).

6. The points S and T are (−2, −1) and (4, 7) respectively. The line ST is the diameter of a circle.
   a) Find the coordinates of the centre of the circle.
   b) Calculate the radius of the circle.
   c) State the equation of the circle.
   d) Show that the circle passes through the point (1, −2).

7. a) Show that the circle with equation $x^2 + y^2 - 6x + 14y + 54 = 0$ can be written in the form $(x-a)^2 + (y-b)^2 = r^2$, where a, b and r are constants to be found.
   b) Hence state the radius and the coordinates of the centre of the circle.

8. Find the radius and the coordinates of the centre of the following circles:
   a) $x^2 + y^2 + 10x - 56 = 0$
   b) $x^2 + y^2 + 2x + 2y - 3 = 0$
   c) $x^2 + y^2 = 10x + 16y - 81$

9. The points (−7, 14) and (3, 10) mark the ends of the diameter of a circle. Find the equation of the circle, writing it in the form $(x-a)^2 + (y-b)^2 = r^2$.

10. For the circle with equation $x^2 + y^2 + 2gx + 2fy + c = 0$, prove that
    a) the centre is $(-g, -f)$
    b) the radius is $\sqrt{g^2 + f^2 - c}$.

8 COORDINATE GEOMETRY OF THE CIRCLE

The intersection of a circle and a straight line

A straight line may intersect any given circle at two distinct points or at one point. Alternatively, the straight line may not intersect the circle at all.

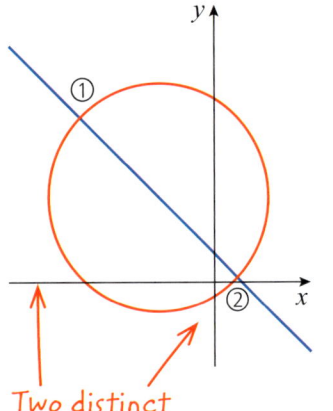

Two distinct points of intersection

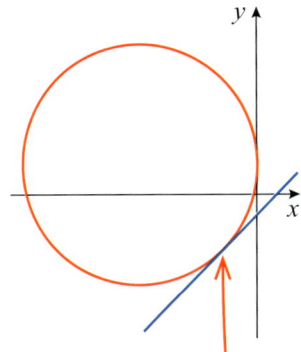

The line is a tangent to the circle

One point of intersection

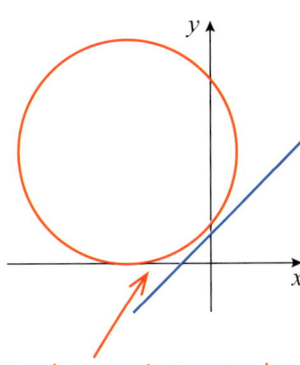

The line and the circle do not intersect.

➡ Worked example

Find the coordinates of the points where the line $y = x + 1$ intersects the circle $(x + 4)^2 + (y - 1)^2 = 1$.

Solution

Solve the equations simultaneously.

$$(x + 4)^2 + (y - 1)^2 = 16$$
$$y = x + 1$$

Substitute $(x + 1)$ for y in the equation of the circle then solve for x.

$$\Rightarrow (x + 4)^2 + ((x + 1) - 1)^2 = 16$$
$$\Rightarrow (x + 4)^2 + x^2 = 16$$
$$\Rightarrow x^2 + 8x + 16 + x^2 = 16$$
$$\Rightarrow 2x^2 + 8x + 16 = 16$$

Solve the quadratic equation.

$$\Rightarrow 2x^2 + 8x = 0$$
$$\Rightarrow x^2 + 4x = 0$$

There are two x-values so two points of intersection.

$$\Rightarrow x(x + 4) = 0$$
$$\Rightarrow x = 0 \text{ or } x = -4$$
$$\Rightarrow \text{when } x = 0, y = 1, \text{ and when } x = -4, y = -3.$$

So the line intersects the circle at $(0, 1)$ and $(-4, -3)$.

The intersection of a circle and a straight line

Worked example

Show that the line $x + y = 1$ is a tangent to the circle $(x + 4)^2 + (y - 3)^2 = 2$.

Solution

Rearrange to get $y = 1 - x$ then substitute into the equation of the circle.

$$(x + 4)^2 + (y - 3)^2 = 2$$
$$x + y = 1$$
$$\Rightarrow (x + 4)^2 + ((1 - x) - 3)^2 = 2$$
$$\Rightarrow (x + 4)^2 + (-x - 2)^2 = 2$$
$$\Rightarrow x^2 + 8x + 16 + x^2 + 4x + 4 = 2$$
$$\Rightarrow 2x^2 + 12x + 20 = 2$$
$$\Rightarrow 2x^2 + 12x + 18 = 0$$

Solve the quadratic equation.

$$\Rightarrow x^2 + 6x + 9 = 0$$
$$\Rightarrow (x + 3)^2 = 0$$

Repeated root

$$\Rightarrow x = -3$$

When $x = -3$, $y = 4$.

As there is a repeated root, there is only one point of intersection. It is at $(-3, 4)$.

Hence the line $x + y = 1$ is a tangent to the circle.

Note

The discriminant can be useful when trying to determine whether a line intersects a circle.

Remember, for the quadratic $ax^2 + bx + c = 0$, if
- $b^2 - 4ac > 0$ there are 2 real roots
- $b^2 - 4ac = 0$ there is 1 repeated root
- $b^2 - 4ac < 0$ there are no real roots.

Worked example

Show that the line $x + 2y + 6 = 0$ does not intersect the circle $x^2 + y^2 + 2x - 3y - 5 = 0$.

Solution

Attempt to solve the equations simultaneously.

$$x^2 + y^2 + 2x - 3y - 5 = 0$$
$$x + 2y + 6 = 0$$
$$\Rightarrow (-2y - 6)^2 + y^2 + 2(-2y - 6) - 3y - 5 = 0$$

Rearrange $x + 2y + 6 = 0$ to get $x = -2y - 6$ then substitute into the equation of the circle.

143

8 COORDINATE GEOMETRY OF THE CIRCLE

Now use the discriminant $b^2 - 4ac$.

$\Rightarrow 4y^2 + 24y + 36 + y^2 - 4y - 12 - 3y - 5 = 0$

$\Rightarrow 5y^2 + 17y + 19 = 0$

$b^2 - 4ac = 17^2 - 4 \times 5 \times 19$

$= 289 - 380$

$= -91$

$b^2 - 4ac < 0$

You can also attempt to solve the quadratic $5y^2 + 17y + 19 = 0$ using the quadratic formula. You will find that the quadratic has no real roots.

The quadratic has no real roots so the line does not intersect the circle.

Discussion point

In the example above, $x + 2y + 6 = 0$ was rearranged to get $x = -2y - 6$ before it was substituted into the equation of the circle. Why do you think the equation was rearranged to make x the subject rather than y?

Tangents to a circle

A radius of a circle is perpendicular to the tangent at the point at which they meet.

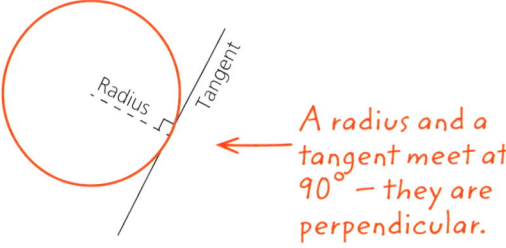

A radius and a tangent meet at $90°$ — they are perpendicular.

Remember, for perpendicular lines with gradients m_1 and m_2

$m_1 m_2 = -1$

You can use these properties, along with the equation of a circle, to find the equation of a tangent at any given point.

➜ Worked example

A circle with equation $(x + 5)^2 + (y - 10)^2 = 125$ has centre C. The circle has a tangent at the point $P(5, 5)$.

a) Verify that the circle passes through the point $(5, 5)$.

b) Find the gradient of the radius CP.

c) State the gradient of the tangent.

d) Write the equation of the tangent.

Tangents to a circle

Substitute x = 5 and y = 5 into $(x+5)^2 + (y-10)^2$ to check that you get 125.

Solution

a) $(5+5)^2 + (5-10)^2$
$= (10)^2 + (-5)^2$
$= 100 + 25$
$= 125$ as required

So the circle does pass through $(5, 5)$.

From the equation of the circle

b) A radius of the circle joins its centre, $C(-5, 10)$, with the point $P(5, 5)$.

gradient of the radius $= \dfrac{5-10}{5-(-5)}$

$= \dfrac{-5}{10}$

$= -\dfrac{1}{2}$

The gradient of the line joining (x_1, y_1) and (x_2, y_2) is given by gradient $= \dfrac{y_2 - y_1}{x_2 - x_1}$

For perpendicular lines, $m_1 m_2 = -1$

c) The radius is perpendicular to the tangent at $(5, 5)$.

$\Rightarrow$ gradient of the tangent $= -\dfrac{1}{-\dfrac{1}{2}}$

$= 2$

The tangent is a straight line. Straight lines have the form $y = mx + c$.

d) The equation of the tangent has the form $y = 2x + c$.
The tangent passes through the point $(5, 5)$; use this to find c.

$5 = 2 \times 5 + c$
$\Rightarrow 5 = 10 + c$
$\Rightarrow c = -5$

So the equation of the tangent at the point $(5, 5)$ is $y = 2x - 5$.

Substitute $x = 5$ and $y = 5$ into $y = 2x + c$

→ Worked example

A circle has equation $x^2 + y^2 - 6x - 10y - 6 = 0$.

a) Show that the circle passes through the point $(9, 7)$.
b) Find the equation of the tangent to the circle at the point $(9, 7)$.
c) Give your answer in the form $ax + by + c = 0$.

Solution

Substitute $x = 9$ and $y = 7$ into $x^2 + y^2 - 6x - 16y - 27$ to check that you get 0.

a) $9^2 + 7^2 - 6 \times 9 - 10 \times 7 - 6$
$= 81 + 49 - 54 - 70 - 6$
$= 0$

So the circle does pass through $(9, 7)$.

145

8 COORDINATE GEOMETRY OF THE CIRCLE

b) First you need to find the centre of the circle.

Compare with the expanded form $x^2 + y^2 + 2gx + 2fy + c = 0$ where the centre is given by $(-g, -f)$.

$$x^2 + y^2 - 6x - 10y - 6 = 0$$

$2g = -6$ and $2f = -10$

$\Rightarrow g = -3$ $\Rightarrow f = -5$

Hence the centre of the circle is $(3, 5)$.

A radius of the circle joins its centre with the point $(9, 7)$.

The gradient of the line joining (x_1, y_1) and (x_2, y_2) is given by gradient $= \dfrac{y_2 - y_1}{x_2 - x_1}$

gradient of the radius $= \dfrac{7-5}{9-3}$

$= \dfrac{2}{6}$

$= \dfrac{1}{3}$

$\Rightarrow$ gradient of the tangent $= -\dfrac{1}{\frac{1}{3}}$ *For perpendicular lines, $m_1 m_2 = -1$*

$= -3$

Therefore, the equation of the tangent has the form $y = -3x + c$.

The tangent passes through the point $(9, 7)$; use this to find c.

Substitute $x = 9$ and $y = 7$ into $y = -3x + c$

$7 = -3 \times 9 + c$

$\Rightarrow 7 = -27 + c$

$\Rightarrow c = 34$

So the equation of the tangent at the point $(9, 7)$ is $y = -3x + 34$.

c) Now rearrange into the form $ax + by + c = 0$.

$y = -3x + 34$

$\Rightarrow 3x + y - 34 = 0$

Exercise 8.2

1 For each pair of equations, determine if the line intersects the circle, is a tangent to the circle or does not meet the circle. Give the coordinates for any point where the line and circle intersect or touch.

a) $x^2 + y^2 = 20$
$y = x + 2$

b) $(x + 4)^2 + y^2 = 35$
$y = 2x - 6$

c) $(x - 3)^2 + (y + 5)^2 = 18$
$y = -x + 4$

d) $(x + 6)^2 + (y + 12)^2 = 45$
$2y = x - 18$

e) $x^2 + y^2 - 3x + 9y - 1 = 0$
$x + 4y - 5 = 0$

f) $x^2 + y^2 + 2x - 4y - 164 = 0$
$5x + 12y = 188$

2 Show that the line that passes through the points (0, −6) and (1.5, 0) does not intersect the circle $(x - 5)^2 + (y + 7)^2 = 20$.

3 Prove that the line that passes through the points (−13, −2) and (−1, 7) is a tangent to the circle in the diagram below.

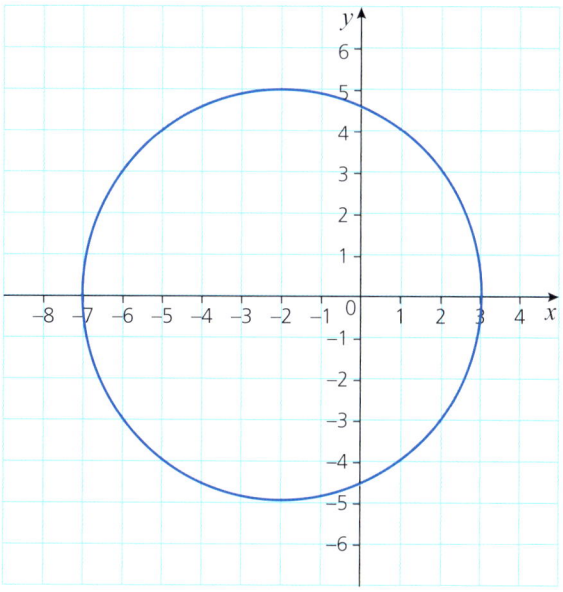

4 The line $y = 3x + 5$ intersects the circle $(x + 1)^2 + (y - 2)^2 = 10$ at the points P and Q.
  a) Find the coordinates of the points P and Q.
  b) Find the midpoint of the line PQ.
  c) Show that PQ is a diameter of the circle.

5 A circle has equation $(x + 2)^2 + (y - 6)^2 = 36$.
  a) Sketch the circle.
  b) State the equations of the tangents to the circle that are parallel to the x-axis.
  c) State the equations of the tangents to the circle that are parallel to the y-axis.

6 A circle with equation $(x + 5)^2 + (y - 3)^2 = 10$ has centre C. The circle has a tangent at the point $P(-2, 4)$.
  a) Find the gradient of the radius CP.
  b) State the gradient of the tangent.
  c) Write the equation of the tangent.

7 Find the equations of the tangents to the following circles at the given points.
  a) $(x - 1)^2 + y^2 = 5$ at $(3, -1)$
  b) $(x + 3)^2 + (y - 2)^2 = 8$ at $(-1, 0)$
  c) $(x - 2)^2 + (y + 5)^2 = 20$ at $(4, -1)$
  d) $x^2 + y^2 + 8x - 4y + 10 = 0$ at $(-1, 1)$
  e) $x^2 + y^2 - 6x - 16y - 27 = 0$ at $(-3, 0)$

8 COORDINATE GEOMETRY OF THE CIRCLE

Exercise 8.2 (cont)

8 A circle has centre $C(5, 6)$. The line l_1 is a tangent to the circle at the point $P(1, 2)$. A second line l_2 has a gradient of 2 and passes through the centre of the circle. l_1 and l_2 intersect at the point Q.
   a) Find the equation of the line l_1.
   b) State the equation of the circle.
   c) Find the coordinates of the point Q.

The intersection of two circles

Two circles may intersect at two distinct points, touch at one point or not meet at all.

Two points of intersection

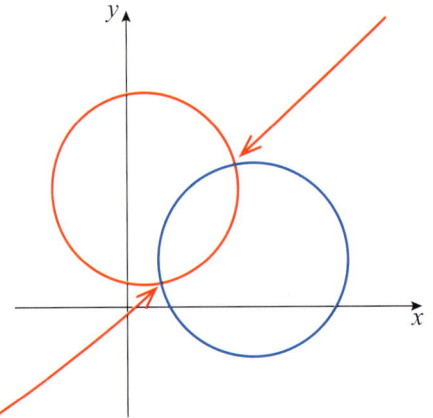
These circles intersect at two distinct points.

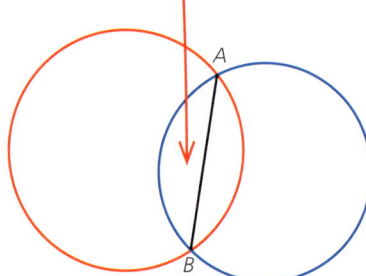
When two circles intersect at two distinct points A and B, the line AB is a common chord.

> **Worked example**
>
> Two circles with equations $(x - 1)^2 + y^2 = 25$ and $(x - 11)^2 + (y - 5)^2 = 100$ intersect at two distinct points A and B.
>
> a) Find the coordinates of the points of intersection.
> b) Find the equation of the common chord AB.
>
> **Solution**
>
> a) $(x - 1)^2 + y^2 = 25 \Rightarrow x^2 + y^2 - 2x - 24 = 0$
>
> $(x - 11)^2 + (y - 5)^2 = 100 \Rightarrow x^2 + y^2 - 22x - 10y + 46 = 0$
>
> Now, subtracting the second equation from the first gives
>
> $(x^2 + y^2 - 2x - 24) - (x^2 + y^2 - 22x - 10y + 46) = 0$
>
> $\Rightarrow 20x + 10y - 70 = 0$
>
> $\Rightarrow 10y = -20x + 70$
>
> $\Rightarrow y = -2x + 7$

Write the equations in expanded form.

It does not matter which way around you do this subtraction; subtracting the first equation from second will give the same result.

Rearrange to make either x or y the subject.

The intersection of two circles

You can substitute into either of the circle equations.

Substituting $y = -2x + 7$ into the first circle equation gives an equation in terms of x.

$x^2 + (-2x + 7)^2 - 2x - 24 = 0$

$\Rightarrow x^2 + 4x^2 - 28x + 49 - 2x - 24 = 0$

$\Rightarrow 5x^2 - 30x + 25 = 0$ ← *Solving this quadratic will give the two x-coordinates for the points of intersection.*

$\Rightarrow x^2 - 6x + 5 = 0$

$\Rightarrow (x-1)(x-5) = 0$

So $x = 1$ or $x = 5$

Substitute the x-values into the linear equation $y = -2x + 7$ from above.

When $x = 1$, $y = -2 \times 1 + 7 = 5$

When $x = 5$, $y = -2 \times 5 + 7 = -3$

Therefore, the two circles intersect at the points $(1, 5)$ and $(5, -3)$.

b) The equation of the common chord AB is the equation of the line which joins the points of intersection $(1, 5)$ and $(5, -3)$.

The gradient of the line joining (x_1, y_1) and (x_2, y_2) is given by gradient $= \dfrac{y_2 - y_1}{x_2 - x_1}$

gradient of the chord $AB = \dfrac{-3 - 5}{5 - 1}$

$= \dfrac{-8}{4} = -2$

So the equation of the chord AB has the form $y = -2x + c$

$\Rightarrow 5 = -2 \times 1 + c$ ← *Substitute the x- and y-values from one of the points of intersection in order to find c. Here, the point $(1, 5)$ has been used.*

$\Rightarrow 5 = -2 + c$

$\Rightarrow c = 7$

Hence the equation of the common chord AB is $y = -2x + 7$.

You may have noticed in the worked example above that the equation of the common chord, $y = -2x + 7$, occurred in the working in part **a**, when the equation of one circle was subtracted from the other. This shortcut is shown in the following worked example.

➡ Worked example

Two circles with equations $x^2 + y^2 + 16x - 20y - 36 = 0$ and $x^2 + y^2 - 24x + 44 = 0$ intersect at two distinct points A and B. Find the equation of the common chord AB.

Solution

$x^2 + y^2 + 16x - 20y - 36 = 0$

$x^2 + y^2 - 24x + 44 = 0$

It does not matter which way around you do this subtraction; subtracting the first equation from second will give the same result. → Subtracting the second equation from the first gives

$(x^2 + y^2 + 16x - 20y - 36) - (x^2 + y^2 - 24x + 44 = 0) = 0$

$\Rightarrow 40x - 20y - 80 = 0$

$\Rightarrow -20y = -40x + 80$

$\Rightarrow y = 2x - 4$

You can give the equation in the form $y = mx + c$ unless the question says otherwise. → So, the equation of the common chord AB is $y = 2x - 4$.

8 COORDINATE GEOMETRY OF THE CIRCLE

> **Discussion point**
>
> For the worked example above, what are the coordinates of A and B?

One point of intersection

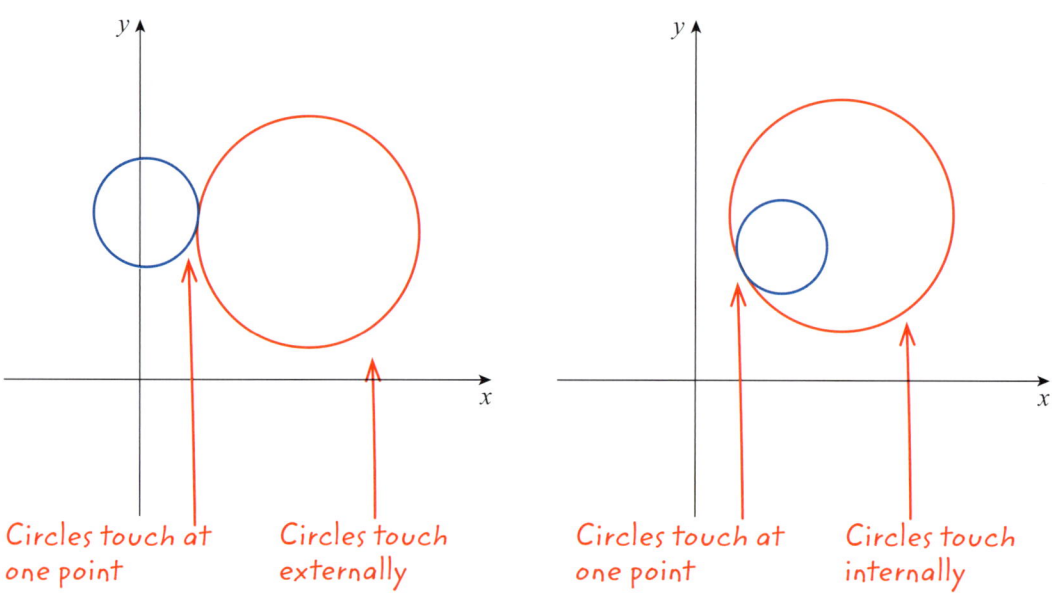

Circles touch at one point

Circles touch externally

Circles touch at one point

Circles touch internally

→ Worked example

Show that the circles $x^2 + y^2 + 6x - 16y + 57 = 0$ and $x^2 + y^2 - 16x - 16y + 79 = 0$ touch at just one point. Find the coordinates of the point at which they touch.

Solution

$$x^2 + y^2 + 6x - 16y + 57 = 0$$
$$x^2 + y^2 - 16x - 16y + 79 = 0$$

Subtracting the second equation from the first gives *(It does not matter which way around you do this subtraction; subtracting the first equation from second will give the same result.)*

$$(x^2 + y^2 + 6x - 16y + 57) - (x^2 + y^2 - 16x - 16y + 79) = 0$$
$$\Rightarrow 22x - 22 = 0$$
$$\Rightarrow 22x = 22$$ *(Rearrange to make x the subject)*
$$\Rightarrow x = 1$$

Substituting $x = 1$ into the first circle equation gives an equation in terms of y *(You can substitute into either of the circle equations)*

$$1^2 + y^2 + 6 \times 1 - 16y + 57 = 0$$
$$\Rightarrow 1 + y^2 + 6 - 16y + 57 = 0$$
$$\Rightarrow y^2 - 16y + 64 = 0$$
$$\Rightarrow (y - 8)^2 = 0$$

So $y = 8$

(Solving the quadratic for y shows that you get a repeated root)

The intersection of two circles

As you get a repeated root, there can only be one point of intersection between the circles. Hence, the circles touch at just one point.
Using the linear equation $x = 1$ from above gives you the coordinates $(1, 8)$.
The circles touch at the point $(1, 8)$.

> **Note**
>
> You can determine that two circles touch at one point by considering their radii and the distance between their centres.
>
>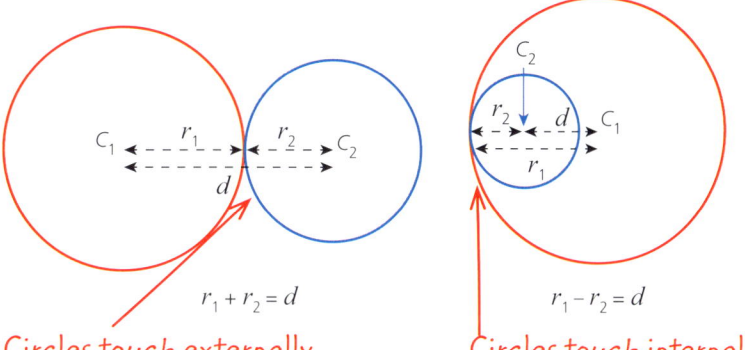
>
> Circles touch externally   Circles touch internally
>
> $r_1 + r_2 = d$ $r_1 - r_2 = d$
>
> If the distance between the centres (d) is equal to the sum of the radii ($r_1 + r_2$) or is equal to the difference of the radii ($r_1 - r_2$) then the circles will touch.

➡ Worked example

Without calculating their point of intersection, show that the circles $x^2 + (y - 3)^2 = 25$ and $(x - 9)^2 + (y - 15)^2 = 100$ touch at one point.

Solution

The circle with equation $x^2 + (y - 3)^2 = 25$ has centre $(0, 3)$ and radius 5.

The circle with equation $(x - 9)^2 + (y - 15)^2 = 100$ has centre $(9, 15)$ and radius 10.

The sum of the radii is $5 + 10 = 15$

The distance between the centres is $\sqrt{(9 - 0)^2 + (15 - 3)^2}$
$$= \sqrt{9^2 + 12^2}$$
$$= \sqrt{225}$$
$$= 15$$

The length of the line joining (x_1, y_1) and (x_2, y_2) is given by $\sqrt{(x_2 - x_1)^2 + (y_2 - y_1)^2}$

Sum of the radii = distance between the centres.
So, the circles must touch at exactly one point.
Since $5 + 10 = 15$, the circles touch externally.

8 COORDINATE GEOMETRY OF THE CIRCLE

Circles do not intersect

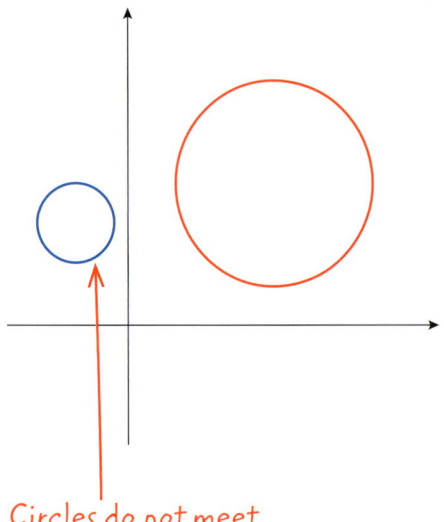

Circles do not meet

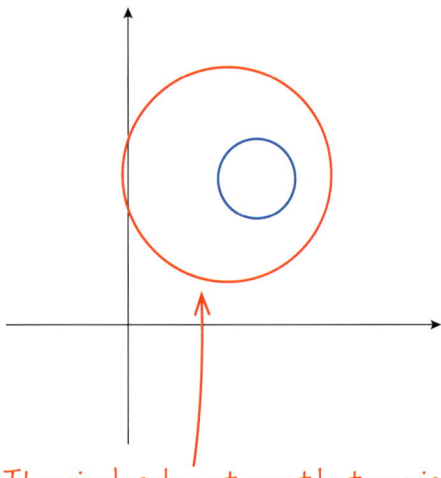

The circles do not meet but one is inside the other.

> **Note**
>
> The discriminant can be useful when trying to determine whether a line intersects a circle.
>
> Remember, for the quadratic $ax^2 + bx + c = 0$, if
> - $b^2 - 4ac > 0$ there are 2 real roots
> - $b^2 - 4ac = 0$ there is 1 repeated root
> - $b^2 - 4ac < 0$ there are no real roots.

→ Worked example

Show that the circles $x^2 + y^2 + 14x + 8y + 4 = 0$ and $x^2 + y^2 + 6x + 4y + 8 = 0$ do not intersect.

Solution

$$x^2 + y^2 + 14x + 8y + 4 = 0$$
$$x^2 + y^2 + 6x + 4y + 8 = 0$$

Attempt to find the points of intersection as in the previous examples. → Subtracting the second equation from the first gives

$$(x^2 + y^2 + 14x + 8y + 4) - (x^2 + y^2 + 6x + 4y + 8) = 0$$
$$\Rightarrow 8x + 4y - 4 = 0$$
$$\Rightarrow 4y = -8x + 4$$
$$\Rightarrow y = -2x + 1$$

The intersection of two circles

You can substitute into either of the circle equations. → Substituting $y = -2x + 1$ into the first circle equation gives an equation in terms of x

$x^2 + (-2x + 1)^2 + 14x + 8(-2x + 1) + 4 = 0$
$\Rightarrow x^2 + 4x^2 - 4x + 1 + 14x - 16x + 8 + 4 = 0$
$\Rightarrow 5x^2 - 6x + 13 = 0$

Now use the discriminant. →
$b^2 - 4ac = (-6)^2 - 4 \times 5 \times 13$
$= 36 - 260$
$= -224$
$b^2 - 4ac < 0$

You can also attempt to solve the quadratic $5x^2 - 6x + 13 = 0$ using the quadratic formula. You will find that the quadratic has no real roots.

The quadratic has no real roots so the circles do not intersect at any point.

> **Note**
>
> If two circles have no points of intersection, you can use the radii and the distance between the centres to determine if the circles are completely separate or if one circle is inside the other.
>
>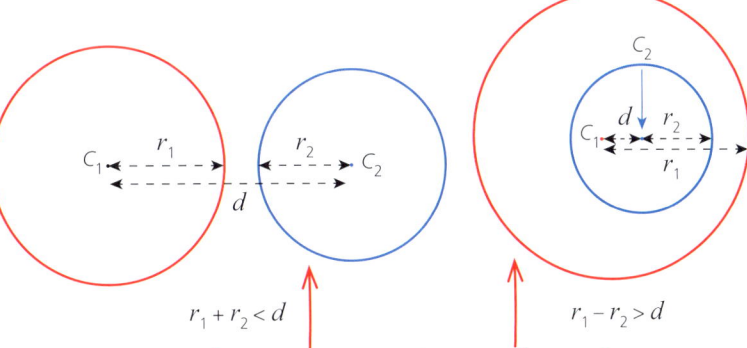
>
> $r_1 + r_2 < d$ $r_1 - r_2 > d$
>
> **Circles completely separate**   **One circle inside the other**
>
> So, when there are two circles that do not meet:
> - The circles are completely separate if the sum of the radii $(r_1 + r_2)$ is less than the distance between the centres (d); $r_1 + r_2 < d$.
> - One circle will be inside the other circle if the difference between the radii $(r_1 - r_2)$ is greater than the distance between the centres (d); $r_1 - r_2 > d$.

Exercise 8.3

1. For each pair of equations, determine if the circles intersect at two distinct points, touch at one point or do not meet at all. Give the coordinates of any points where the circles intersect or touch.

   a) $x^2 + y^2 - 6x + 10y + 31 = 0$
   $x^2 + y^2 - 4x - 2y - 11 = 0$

   b) $x^2 + y^2 + 2x - 6y - 90 = 0$
   $x^2 + y^2 - 40x - 6y + 120 = 0$

   c) $(x - 2)^2 + y^2 = 16$
   $(x - 2)^2 + (y - 9)^2 = 25$

   d) $x^2 + y^2 - 8x + 4y - 5 = 0$
   $x^2 + y^2 - 22x + 6y + 105 = 0$

8 COORDINATE GEOMETRY OF THE CIRCLE

Exercise 8.3 (cont)

2 The circles with equations $(x-16)^2 + (y+1)^2 = 100$ and $x^2 + y^2 - 8x - 10y + 1 = 0$ intersect at two distinct points A and B.
   a) Find the two points of intersection.
   b) State the equation of the common chord AB.

3 The circles with equations $(x-3)^2 + (y-6)^2 = 50$ and $(x-8)^2 + (y-1)^2 = 100$ intersect at the point $A(8, 11)$ and at the point B.
   a) Find the coordinates of the point B.
   b) State the equation of the common chord AB.
   c) Find the midpoint of the chord AB.
   d) Show that the line joining the centres of the two circles is the perpendicular bisector of the chord AB.

4 Two circles have equations $(x-3)^2 + (y-8)^2 = 25$ and $(x+3)^2 + y^2 = 225$.
   a) Show that the circles touch at only one point.
   b) Find the distance between the centres of the two circles.
   c) Using your answer to part **b**, deduce whether one circle is inside the other.

5 Two circles have equations $x^2 + (y-2)^2 = 64$ and $x^2 + (y-4)^2 = 16$.
   a) Verify that the circles have no points of intersection.
   b) Write down their centres and radii.
   c) Draw the circles on a diagram. State whether one circle is or is not inside the other.

Practice questions

1 The points $(-1, -1)$ and $(5, 7)$ mark the ends of a diameter of a circle.
   a) Find the equation of the circle, writing it in the form $(x-a)^2 + (y-b)^2 = r^2$. [3]
   b) A diameter of the circle goes through the point $(6, 0)$. Find the coordinates of the other end of this diameter. [2]

2 The diagram shows the circle $x^2 + y^2 - 6x + 4y - 12 = 0$ and the lines l_1, $y = 2x - 3$, and l_2, $y = 9 - 2x$. The lines intersect at point C and meet the circle at points A and B.

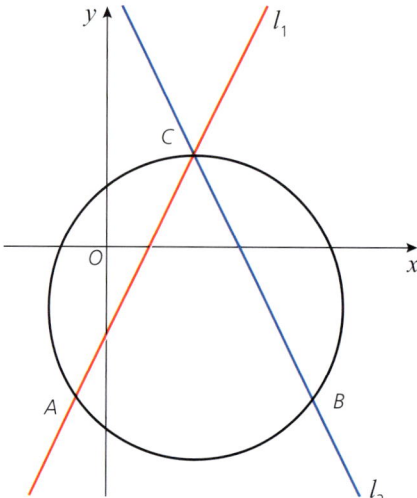

The intersection of two circles

    a) Find the coordinates of the point of intersection of l_1 and l_2 and verify that this point lies on the circumference of the circle. [4]
    b) Give the coordinates of the points A and B. [5]
    c) Find the area of the triangle ABC. [2]

3 Two circles have equations $(x+2)^2 + (y-3)^2 = 36$ and $(x-10)^2 + (y+2)^2 = 25$.
    a) Verify that the circles have no points of intersection. [5]
    b) i Find the distance between the centres of the two circles. [1]
       ii Deduce whether one circle is inside the other. Explain your answer fully. [2]

Now you should be able to:
★ know and use the equation of a circle with radius r and centre (a, b)
★ solve problems involving the intersection of a circle and a straight line
★ solve problems involving tangents to a circle
★ solve problems involving the intersection of two circles

Key points

✔ The circle with centre $(0, 0)$ and radius r has the equation $x^2 + y^2 = r^2$.
✔ The circle with centre (a, b) and radius r has the equation $(x-a)^2 + (y-b)^2 = r^2$.
✔ The equation of a circle can be written in the form $x^2 + y^2 + 2gx + 2fy + c = 0$, where the centre is $(-g, -f)$ and the radius is $\sqrt{g^2 + f^2 - c}$.
✔ A straight line may intersect any given circle at two distinct points or at one point. Alternatively, the straight line may not intersect the circle at all.

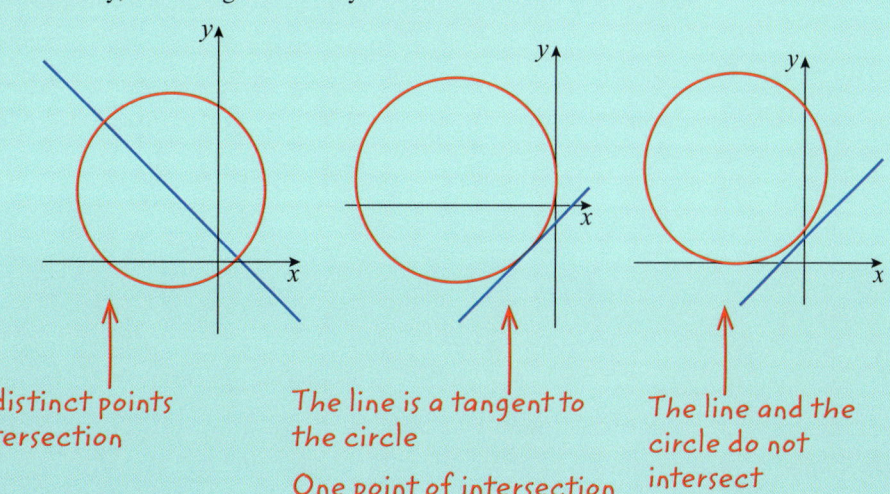

Two distinct points of intersection

The line is a tangent to the circle
One point of intersection

The line and the circle do not intersect

8 COORDINATE GEOMETRY OF THE CIRCLE

✔ Two circles may:
- have two distinct points of intersection

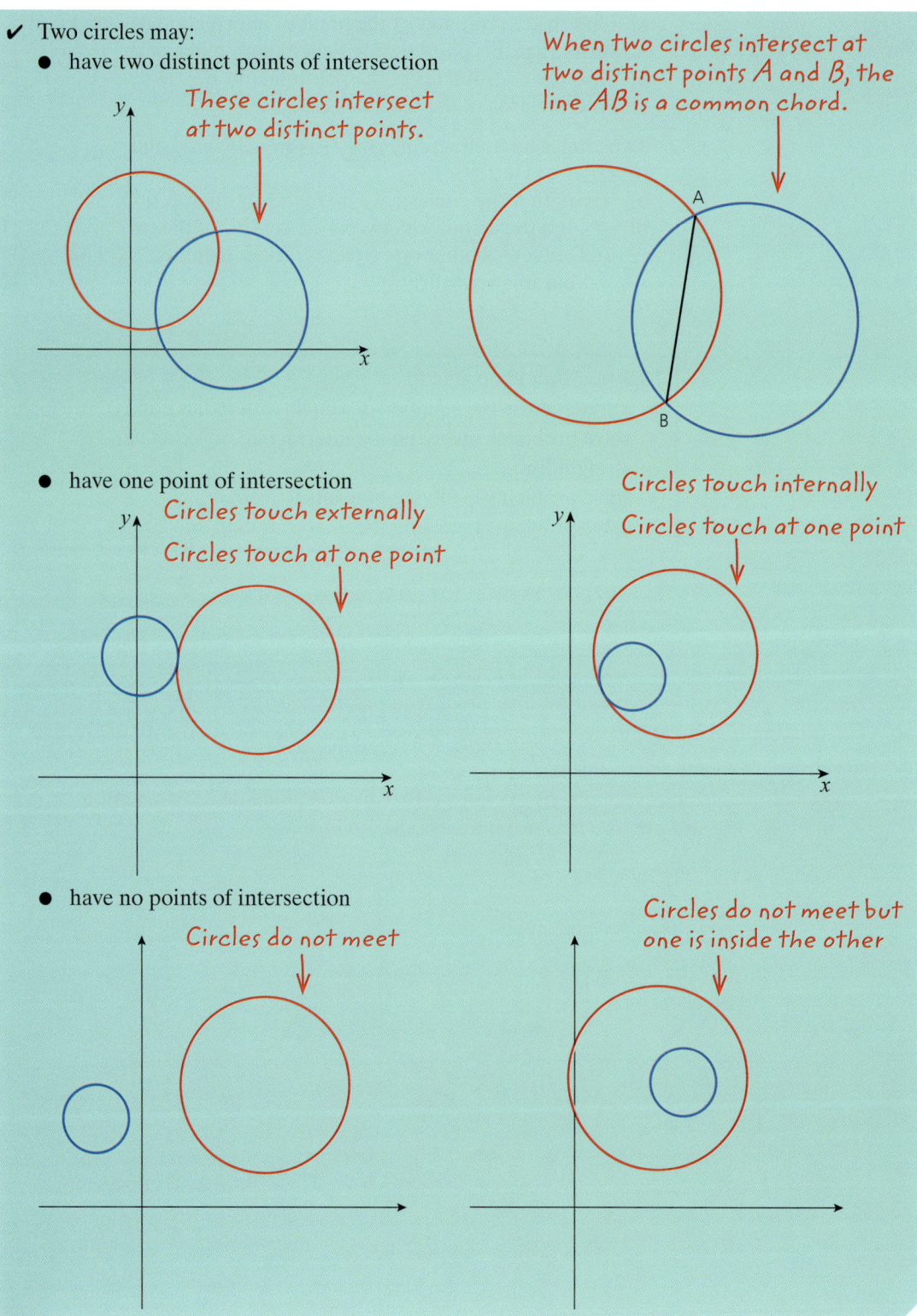

These circles intersect at two distinct points.

When two circles intersect at two distinct points A and B, the line AB is a common chord.

- have one point of intersection

Circles touch externally
Circles touch at one point

Circles touch internally
Circles touch at one point

- have no points of intersection

Circles do not meet

Circles do not meet but one is inside the other

9 Circular measure

A circle is the reflection of eternity. It has no beginning and no end.

Maynard James Keenan (born 1964)

Discussion point

This is the Singapore Flyer. It has a radius of 75 metres. It takes about 30 minutes to complete one rotation, travelling at a constant speed. How fast do the capsules travel?

The tradition of measuring angles in degrees, and there being 360 degrees in one revolution, is thought to have come about because much of early mathematics was connected to astronomy, and the shepherd-astronomers of Sumeria believed that there were 360 days in a year.

The following notation is used in this chapter:

C represents the **circumference** of the circle – the distance round the circle.

r represents the **radius** of the circle – the distance from the centre to any point on the circumference.

θ (the Greek letter theta) is used to represent the **angle** that an arc subtends at the centre of the circle.

A represents **area** – this may be the area of a whole circle or a sector.

9 CIRCULAR MEASURE

Arc length and area of a sector

A **sector** of a circle looks similar to a piece of cake – it is the shape enclosed by an arc of the circle and two radii. If the angle at the centre is less than 180° it is called a **minor sector**, and if it is between 180° and 360° it is called a **major sector**.

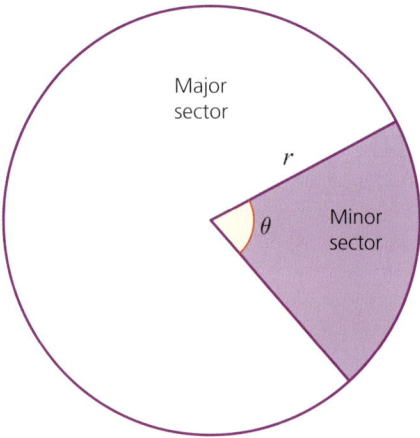

Using ratios:
$$\frac{\text{arc length}}{\text{circumference of the circle}} = \frac{\text{area of the sector}}{\text{area of the circle}} = \frac{\theta}{360}$$

➡ Worked example

For each sector, calculate:

i   the arc length   ii   the area   iii   the perimeter.

a)

b)

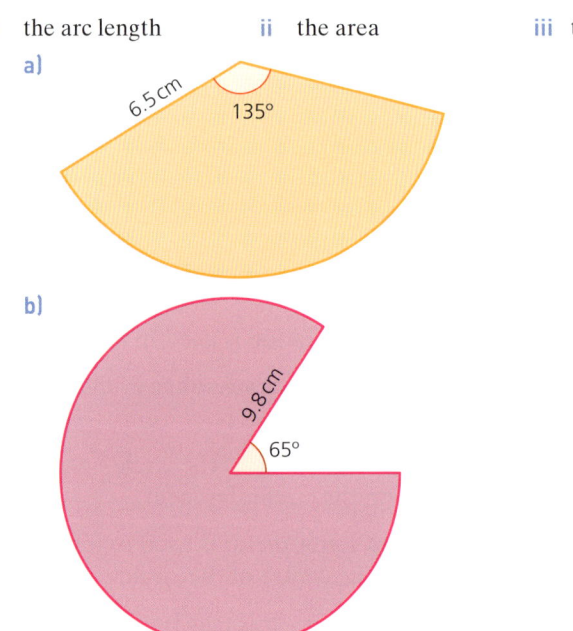

Radian measure

Solution

a) i $\quad \dfrac{\text{arc length}}{2\pi r} = \dfrac{\theta}{360} \quad \Rightarrow \quad \text{arc length} = \dfrac{135}{360} \times 2 \times \pi \times 6.5$

$\quad = 15.3 \text{ cm (3 s.f.)}$

ii $\quad \dfrac{\text{area}}{\pi r^2} = \dfrac{\theta}{360} \quad \Rightarrow \quad \text{area} = \dfrac{135}{360} \times \pi \times 6.5^2$

$\quad = 49.8 \text{ cm}^2 \text{ (3 s.f.)}$

iii perimeter = arc length + 2 × radius

$\quad = 15.3 + 2(6.5) = 28.3 \text{ cm (3 s.f.)}$

b) The angle of this sector is 360 − 65 = 295°

i $\quad \dfrac{\text{arc length}}{2\pi r} = \dfrac{\theta}{360} \quad \Rightarrow \quad \text{arc length} = \dfrac{295}{360} \times 2 \times \pi \times 9.8$

$\quad = 50.5 \text{ cm (3 s.f.)}$

ii $\quad \dfrac{\text{area}}{\pi r^2} = \dfrac{\theta}{360} \quad \Rightarrow \quad \text{area} = \dfrac{295}{360} \times \pi \times 9.8^2$

$\quad = 247 \text{ cm}^2 \text{ (3 s.f.)}$

iii perimeter = arc length + 2 × radius

$\quad = 50.5 + 2(9.8) = 70.1 \text{ cm (3 s.f.)}$

➔ Worked example

A sector of a circle of radius 8 cm has an area of 25 cm². Work out the angle at the centre.

Solution

Using sector area $= \dfrac{\theta}{360} \times \pi r^2$

$25 = \dfrac{\theta}{360} \times \pi \times 8^2$

$\Rightarrow \theta = \dfrac{25 \times 360}{\pi \times 64}$

$= 44.8° \text{ (3 s.f.)}$

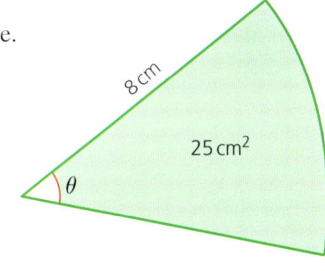

Radian measure

Radian measure is used extensively in mathematics because it simplifies many angle calculations. One radian (rad) is the angle in a sector when the arc length is equal to the radius. 1 rad is approximately 57.3°.

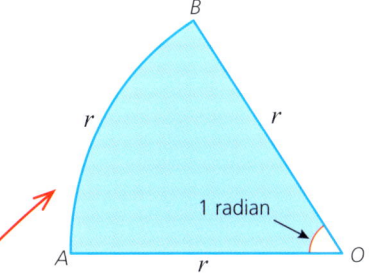

1 radian can also be written as 1^c.

9 CIRCULAR MEASURE

Note
- An angle given as a fraction of π is assumed to be in radians.
- If an angle is a simple fraction of 180°, its equivalent value in radians is usually expressed as a fraction of π.

Since the circumference of a circle is of length $2\pi r$, there are 2π arcs of length r round the circumference. This means that there are 2π radians in 360°.

Degrees	Radians
360	2π
180	π
90	$\frac{\pi}{2}$
60	$\frac{\pi}{3}$
45	$\frac{\pi}{4}$
30	$\frac{\pi}{6}$

1 degree is the same as $\frac{\pi}{180}$ radians, therefore:

» multiply by $\frac{\pi}{180}$ to convert degrees to radians

» multiply by $\frac{180}{\pi}$ to convert radians to degrees.

Worked example

a) Express the following in radians: i 75°   ii 49°

b) Express the following in degrees: i $\frac{\pi}{10}$ radians   ii 1.25 radians

Solution

a) i $75° = 75 \times \frac{\pi}{180} = \frac{5\pi}{12}$ radians

   ii $49° = 49 \times \frac{\pi}{180} = 0.855$ radians (3 s.f.)

b) i $\frac{\pi}{10}$ radians $= \frac{\pi}{10} \times \frac{180}{\pi} = 18°$

   ii 1.25 radians $= 1.25 \times \frac{180}{\pi} = 71.6°$ (3 s.f.)

Using your calculator

Your calculator has modes for degrees and for radians, so always make sure that it is on the correct setting for any calculations that you do. There is usually a button marked DRG for **d**egrees, **r**adians and **g**rad (you will not use grad at this stage).

To find the value of sin 2.3^c, set your calculator to the radian mode and enter sin 2.3 followed by = or EXE, depending on your calculator. You should see the value 0.74570… on your screen.

Arc length and area of a sector in radians

Using the definition of a radian, an angle of 1 radian at the centre of a circle corresponds to an arc length equal to the radius r of the circle. Therefore an angle of θ radians corresponds to an arc length of $r\theta$.

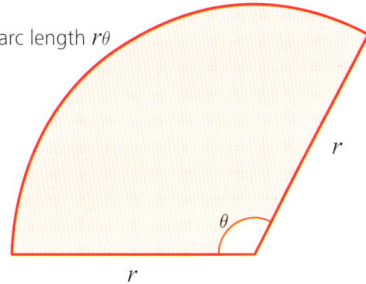

arc length $r\theta$

It is accepted practice to write $r\theta$, with the Greek letter at the end, rather than θr.

The area of this sector is the fraction $\dfrac{\theta}{2\pi}$ of the area of the circle (since 2π is the radian equivalent of 360°).

This gives the formula:

$$\text{area of a sector} = \frac{\theta}{2\pi} \times \pi r^2 = \tfrac{1}{2}r^2\theta.$$

➜ Worked example

Calculate the arc length, area and perimeter of this sector.

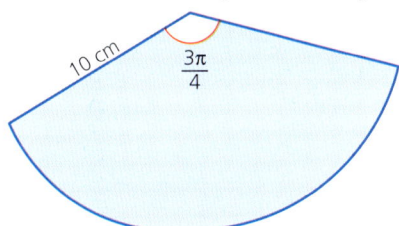

10 cm, $\dfrac{3\pi}{4}$

Solution

arc length $= 10 \times \dfrac{3\pi}{4}$ — Using arc length $= r\theta$

$= \dfrac{15\pi}{2}$ cm

sector area $= \dfrac{1}{2} \times 10^2 \times \dfrac{3\pi}{4}$ — Using area of sector $= \tfrac{1}{2}r^2\theta$

$= \dfrac{75\pi}{2}$ cm² — Using perimeter = arc length + 2 × radius

perimeter $= \dfrac{15\pi}{2} + 2 \times 10 = \dfrac{15\pi}{2} + 20$ cm

9 CIRCULAR MEASURE

Exercise 9.1

1. Express each angle in radians, leaving your answer in terms of π if appropriate:
   a) $120°$ b) $540°$ c) $22°$ d) $150°$ e) $37.5°$

2. Express each angle in degrees, rounding your answer to 3 s.f. where necessary:
   a) $\dfrac{2\pi}{3}$ b) $\dfrac{5\pi}{9}$ c) 3^c d) $\dfrac{\pi}{7}$ e) $\dfrac{3\pi}{8}$

3. The table gives information about some sectors of circles.
   Copy and complete the table. Leave your answers as a multiple of π where appropriate.

Radius, r (cm)	Angle at centre in degrees	Angle at centre in radians	Arc length, s (cm)	Area, A (cm²)
8	120			
10			5	
	60		6	
6				12
	75			20

4. The table gives information about some sectors of circles. Copy and complete the table. Leave your answers as a multiple of π where appropriate.

Radius, r (cm)	Angle at centre in radians	Arc length, s (cm)	Area, A (cm²)
10	$\dfrac{\pi}{3}$		
12		24	
	$\dfrac{\pi}{4}$	16	
5			25
	$\dfrac{3\pi}{5}$		40

5. OAB is a sector of a circle of radius 6 cm. ODC is a sector of a circle radius 10 cm. Angle AOB is $\dfrac{\pi}{3}$.
   Express in terms of π:
   a) the area of $ABCD$
   b) the perimeter of $ABCD$.

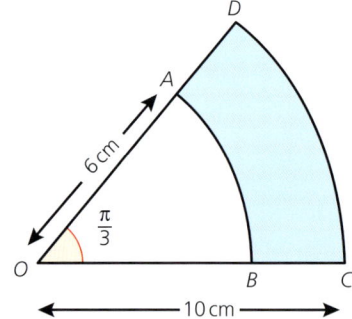

Radian measure

The shaded area is called a **segment** of a circle.

6   a) Work out the area of the sector AOB.
    b) Calculate the area of the triangle AOB.
    c) Work out the shaded area.

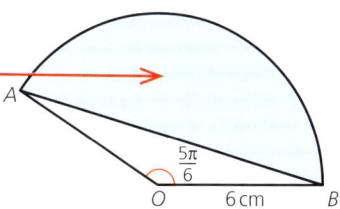

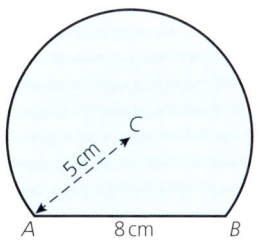

7   The diagram shows the cross-section of a paperweight. The paperweight is a sphere of radius 5 cm with the bottom cut off to create a circular flat base with diameter 8 cm.
    a) Calculate the obtuse angle ACB in radians.
    b) Work out the area of cross-section of the paperweight.

8   The perimeter of the sector in the diagram is $6\pi + 16$ cm.
    Calculate:
    a) angle AOB
    b) the exact area of sector AOB
    c) the exact area of the triangle AOB
    d) the exact area of the shaded segment.

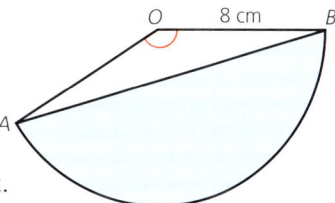

Past-paper questions

1

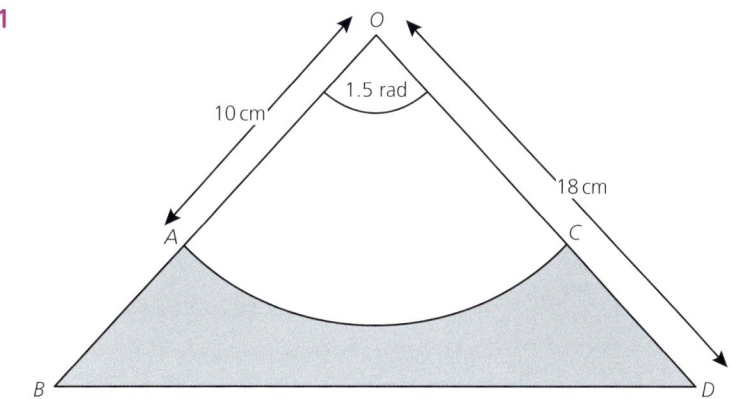

The diagram shows an isosceles triangle OBD in which $OB = OD = 18$ cm and angle $BOD = 1.5$ radians. An arc of the circle, centre O and radius 10 cm, meets OB at A and OD at C.
(i) Find the area of the shaded region. [3]
(ii) Find the perimeter of the shaded region. [4]

Cambridge O Level Additional Mathematics (4037)
Paper 12 Q8, November 2012
Cambridge IGCSE Additional Mathematics (0606)
Paper 12 Q8, November 2012

9 CIRCULAR MEASURE

2 The diagram shows a circle, centre O, radius 8 cm. Points P and Q lie on the circle such that the chord $PQ = 12$ cm and angle $POQ = \theta$ radians.

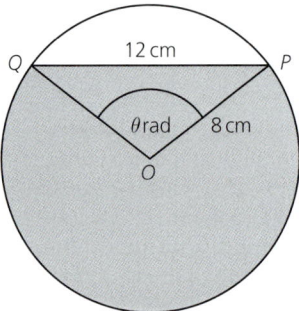

(i) Show that $\theta = 1.696$, correct to 3 decimal places. [2]
(ii) Find the perimeter of the shaded region. [3]
(iii) Find the area of the shaded region. [3]

Cambridge O Level Additional Mathematics (4037)
Paper 12 Q7, June 2014
Cambridge IGCSE Additional Mathematics (0606)
Paper 12 Q7, June 2014

3

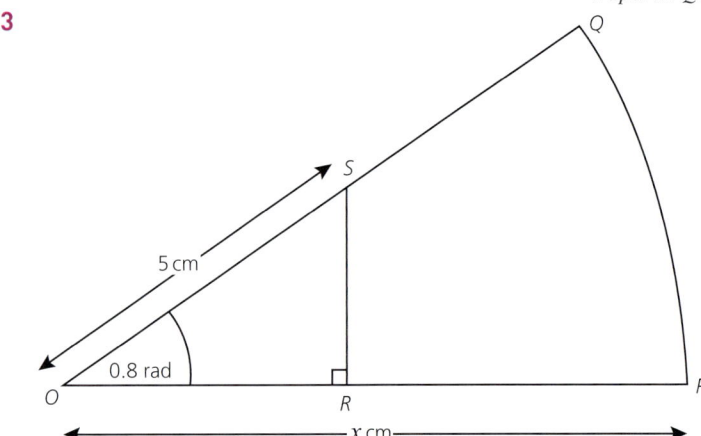

The diagram shows a sector OPQ of a circle with centre O and radius x cm. Angle POQ is 0.8 radians. The point S lies on OQ such that $OS = 5$ cm. The point R lies on OP such that angle ORS is a right angle. Given that the area of triangle ORS is one-fifth of the area of sector OPQ, find

(i) the area of sector OPQ in terms of x and hence show that the value of x is 8.837 correct to 4 significant figures, [5]
(ii) the perimeter of $PQSR$, [3]
(iii) the area of $PQSR$. [2]

Cambridge O Level Additional Mathematics (4037)
Paper 22 Q11, November 2014
Cambridge IGCSE Additional Mathematics (0606)
Paper 22 Q11, November 2014

Now you should be able to:
★ solve problems involving the arc length and sector area of a circle, including knowledge and use of radian measure.

Key points

✔ Angles are measured either in degrees or radians.
  $180° = \pi$ radians
✔ The angle at the centre of the circle subtended by an arc that is the same length as the radius is 1 radian.
✔ The formulae for area of a circle ($A = \pi r^2$) and circumference of a circle ($C = 2\pi r$) are the same whether the angle is measured in degrees or radians.
✔ You will need to learn these formulae.

	Radians
Area	πr^2
Circumference	$2\pi r$
Arc length (θ at centre)	$r\theta$
Sector area (θ at centre)	$\frac{1}{2}r^2\theta$

10 Trigonometry

The laws of nature are written in the language of mathematics ... the symbols are triangles, circles and other geometrical figures without whose help it is impossible to comprehend a single word.

Galileo Galilei (1564–1642)

Discussion point

How can you estimate the angle the sloping sides of this pyramid make with the horizontal?

Using trigonometry in right-angled triangles

The simplest definitions of the trigonometrical functions are given in terms of the ratios of the sides of a right-angled triangle, for values of the angle θ between 0° and 90°.

The Greek letter θ (theta) is often used to denote an angle. The Greek letters α (alpha) and β (beta) are also commonly used for this purpose.

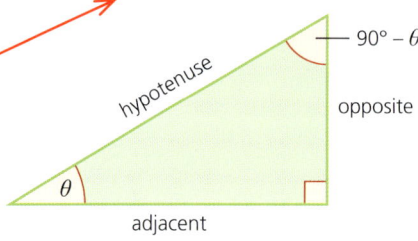

Using trigonometry in right-angled triangles

> Taking the first letters of each part gives the word 'sohcahtoa', which may help you to remember the formulae.

In a triangle:

$$\sin \theta = \frac{\text{opposite}}{\text{hypotenuse}} \qquad \cos \theta = \frac{\text{adjacent}}{\text{hypotenuse}} \qquad \tan \theta = \frac{\text{opposite}}{\text{adjacent}}.$$

sin is an abbreviation of sine, cos of cosine and tan of tangent. The previous diagram shows that:

$$\sin \theta = \cos(90° - \theta) \text{ and } \cos \theta = \sin(90° - \theta).$$

➜ Worked example

Work out the length of x in each triangle. Give your answers correct to three significant figures.

a)
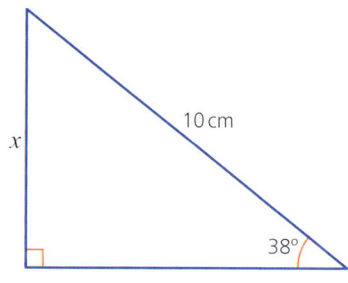

Solution

a) $\frac{x}{10} = \sin 38°$

$\Rightarrow \quad x = 10 \sin 38°$

$\quad x = 6.16 \text{ cm}$

b)
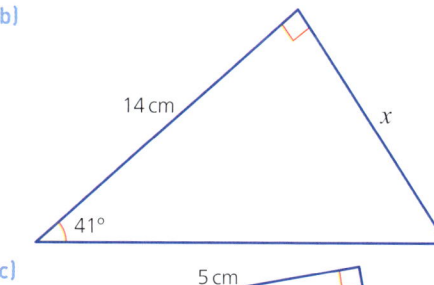

b) $\frac{x}{14} = \tan 41°$

$\Rightarrow \quad x = 14 \tan 41°$

$\Rightarrow \quad x = 12.2 \text{ cm}$

c)
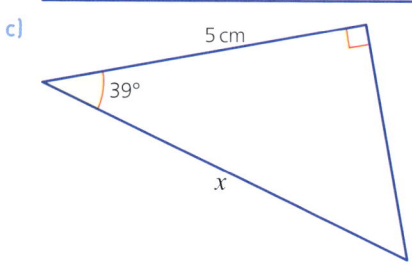

c) $\frac{5}{x} = \cos 39°$

$\Rightarrow \quad 5 = x \cos 39°$

$\Rightarrow \quad x = \frac{5}{\cos 39°}$

$\Rightarrow \quad x = 6.43 \text{ cm}$

10 TRIGONOMETRY

➜ Worked example

Work out the angle marked θ in each triangle. Give your answers correct to one decimal place.

a)
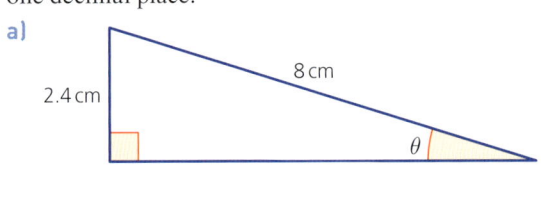

Solution

a) $\sin \theta = \dfrac{2.4}{8}$

$\Rightarrow \theta = \sin^{-1} 0.3$

$\Rightarrow \theta = 17.5°$

$\sin^{-1} 0.3$ is shorthand notation for 'the angle θ where $\sin \theta = 0.3$'. $\cos^{-1} 0.3$ and $\tan^{-1} 0.3$ are similarly defined.

b)
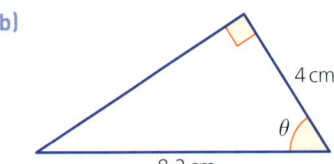

b) $\cos \theta = \dfrac{4}{8.2}$

$\Rightarrow \theta = \cos^{-1} \dfrac{4}{8.2}$

$\Rightarrow \theta = 60.8°$

c)
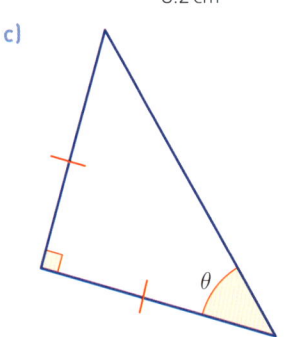

c) The opposite and adjacent sides are equal, so $\tan \theta = 1$
$\Rightarrow \theta = 45°$.

Special cases

Certain angles occur frequently in mathematics and you will find it helpful to know the value of their trigonometrical functions.

The angles 30° and 60°

Triangle ABC is an equilateral triangle with side 2 units, and AD is a line of symmetry.

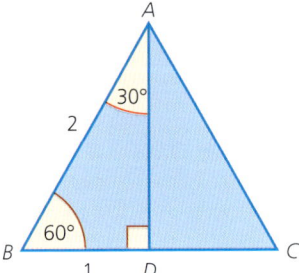

Using trigonometry in right-angled triangles

Using Pythagoras' theorem
$$AD^2 + 1^2 = 2^2 \Rightarrow AD = \sqrt{3}.$$

From triangle ABD,

$\sin 60° = \frac{\sqrt{3}}{2};$ $\cos 60° = \frac{1}{2};$ $\tan 60° = \sqrt{3};$

$\sin 30° = \frac{1}{2};$ $\cos 30° = \frac{\sqrt{3}}{2};$ $\tan 30° = \frac{1}{\sqrt{3}}.$

➔ Worked example

Without using a calculator, find the value of $\sin^2 30° + \sin 60° \cos 30°$.
(Note that $\sin^2 30°$ means $(\sin 30°)^2$.)

Solution

$$\sin^2 30° + \sin 60° \cos 30° = \left(\frac{1}{2}\right)^2 + \frac{\sqrt{3}}{2} \times \frac{\sqrt{3}}{2}$$
$$= \frac{1}{4} + \frac{3}{4}$$
$$= 1$$

> **Note**
>
> The equivalent results using radians are
>
> $\sin \frac{\pi}{3} = \frac{\sqrt{3}}{2};$ $\cos \frac{\pi}{3} = \frac{1}{2};$ $\tan \frac{\pi}{3} = \sqrt{3}$
>
> $\sin \frac{\pi}{6} = \frac{1}{2};$ $\cos \frac{\pi}{6} = \frac{\sqrt{3}}{2};$ $\tan \frac{\pi}{6} = \frac{1}{\sqrt{3}}$

The angle 45°
PQR is a right-angled isosceles triangle with equal sides of length 1 unit.

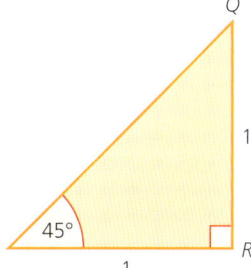

> **Note**
>
> In radians
> $\sin \frac{\pi}{4} = \frac{1}{\sqrt{2}};$
> $\cos \frac{\pi}{4} = \frac{1}{\sqrt{2}};$
> $\tan \frac{\pi}{4} = 1$

Using Pythagoras' theorem, $PQ = \sqrt{2}$.
This gives

$\sin 45° = \frac{1}{\sqrt{2}};$ $\cos 45° = \frac{1}{\sqrt{2}};$ $\tan 45° = 1.$

10 TRIGONOMETRY

> ### ➡ Worked example
>
> Without using a calculator find the value of $\sin^2 \frac{\pi}{4} + \cos^2 \frac{\pi}{4}$.
>
> **Solution**
>
> $$\sin \frac{\pi}{4} = \frac{1}{\sqrt{2}}, \qquad \cos \frac{\pi}{4} = \frac{1}{\sqrt{2}}$$
>
> So $\sin^2 \frac{\pi}{4} + \cos^2 \frac{\pi}{4} = \frac{1}{2} + \frac{1}{2}$
>
> $= 1$
>
> *When an angle is given in terms of π like this, it is in radians. $\frac{\pi}{4}$ radians = 45°*

The angles 0° and 90°

Although you cannot have an angle of 0° in a triangle (because one side would be lying on top of another), you can still imagine what it might look like. In the diagram, the hypotenuse has length 1 unit and the angle at X is very small.

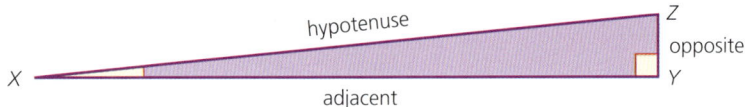

If you imagine the angle at X becoming smaller and smaller until it is zero, you can deduce that

$$\sin 0° = \frac{0}{1} = 0; \qquad \cos 0° = \frac{1}{1} = 1; \qquad \tan 0° = \frac{0}{1} = 0.$$

If the angle at X is 0°, then the angle at Z is 90°, and so you can also deduce that

$$\sin 90° = \frac{1}{1} = 1; \qquad \cos 90° = \frac{0}{1} = 0.$$

Remember that in radians 90° is $\frac{\pi}{2}$.

However, when you come to find $\tan 90°$, there is a problem. The triangle suggests this has value $\frac{1}{0}$, but you cannot divide by zero.

If you look at the triangle XYZ, you will see that what we actually did was to draw it with angle X not zero but just very small, and to argue:

'We can see from this what will happen if the angle becomes smaller and smaller so that it is effectively zero.'

In this case we are looking at the limits of the values of $\sin \theta$, $\cos \theta$ and $\tan \theta$ as the angle θ approaches zero. The same approach can be used to look again at the problem of $\tan 90°$.

If the angle X is not quite zero, then the side ZY is also not quite zero, and $\tan Z$ is 1 (XY is almost 1) divided by a very small number and so is large. The smaller the angle X, the smaller the side ZY and so the larger the value of tan Z. We conclude that in the limit when angle X becomes zero and angle Z becomes 90°, tan Z is infinitely large, and so we say

as $Z \to 90°$, $\tan Z \to \infty$ (infinity).

Read these arrows as 'tends to'.

You can see this happening in the table of values below.

Z	tan Z
80°	5.67
89°	57.29
89.9°	572.96
89.99°	5729.6
89.999°	57 296

When Z actually equals 90°, we say that tan Z is **undefined**.

Positive and negative angles

Unless given in the form of bearings, angles are measured from the x-axis (as shown below). Anticlockwise is taken to be positive and clockwise to be negative.

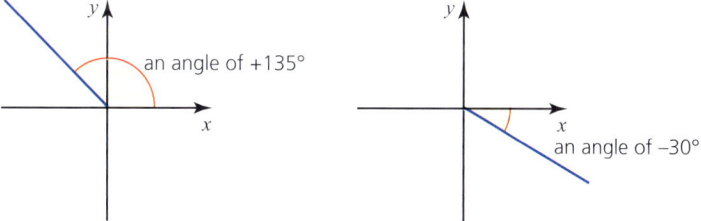

→ Worked example

In the diagram, angles ADB and CBD are right angles, angle $BAD = \frac{\pi}{3}$, $AB = 2l$ and $BC = 3l$.

Calculate the value of θ in radians.

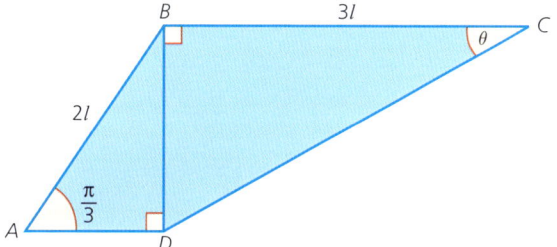

Solution

First, find an expression for BD.

In triangle ABD, $\dfrac{BD}{AB} = \sin\dfrac{\pi}{3}$ ← $AB = 2l$

$\Rightarrow \quad BD = 2l \sin\dfrac{\pi}{3}$

$= 2l \times \dfrac{\sqrt{3}}{2}$

$= \sqrt{3}\,l$

In triangle BCD, $\tan\theta = \dfrac{BD}{BC}$

$= \dfrac{\sqrt{3}\,l}{3l}$

$= \dfrac{1}{\sqrt{3}}$

$\Rightarrow \quad \theta = \tan^{-1}\left(\dfrac{1}{\sqrt{3}}\right)$

$= \dfrac{\pi}{6}$

Exercise 10.1

1. In the triangle PQR, $PQ = 29$ cm, $QR = 21$ cm and $PR = 20$ cm.
   a) Show that the triangle is right-angled.
   b) Write down the values of $\sin Q$, $\cos Q$ and $\tan Q$, leaving your answers as fractions.
   c) Use your answers to part **b** to show that:
      i) $\sin^2 Q + \cos^2 Q = 1$
      ii) $\tan Q = \dfrac{\sin Q}{\cos Q}$

2. Without using a calculator, show that $\cos 60° \sin 30° + \sin 60° \cos 30° = 1$

3. Without using a calculator, show that $\cos^2 60° + \cos^2 45° = \cos^2 30°$

4. Without using a calculator, show that $3\cos^2 \dfrac{\pi}{3} = \sin^2 \dfrac{\pi}{3}$

5. In the diagram, $AB = 12$ cm, angle $BAC = 30°$, angle $BCD = 60°$ and angle $BDC = 90°$.

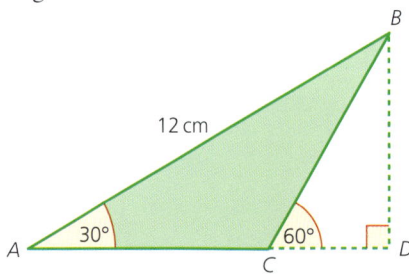

a) Calculate the length of BD.
b) Show that $AC = 4\sqrt{3}$ cm.

6 In the diagram, $OA = 1$ cm, angle $AOB =$ angle $BOC =$ angle $COD = \frac{\pi}{4}$ and angle $OAB =$ angle $OBC =$ angle $OCD = \frac{\pi}{2}$.

  a) Find the length of OD, giving your answer in the form $a\sqrt{2}$.

  b) Show that the perimeter of the pentagon $OABCD$ is $4 + 3\sqrt{2}$ cm.

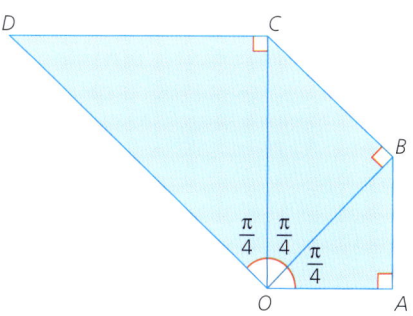

7 In the diagram, $ABED$ is a trapezium with right angles at E and D, and CED is a straight line. The lengths of AB and BC are $(2\sqrt{3})d$ and $2d$ respectively, and angles BAD and CBE are $60°$ and $30°$ respectively.

  a) Find the length of CD in terms of d.

  b) Show that angle $CAD = \tan^{-1}\left(\frac{2}{\sqrt{3}}\right)$.

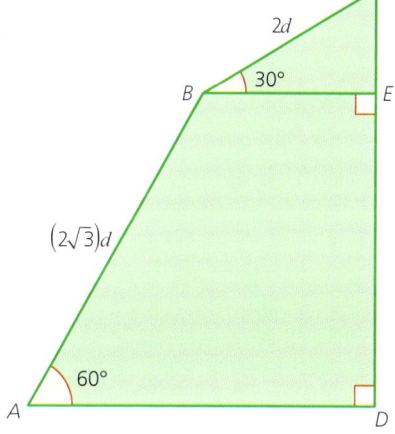

8 In the diagram, ABC is a triangle in which $AB = 6$ cm, $BC = 4$ cm and angle $ABC = \frac{2\pi}{3}$. The line CX is perpendicular to the line ABX.

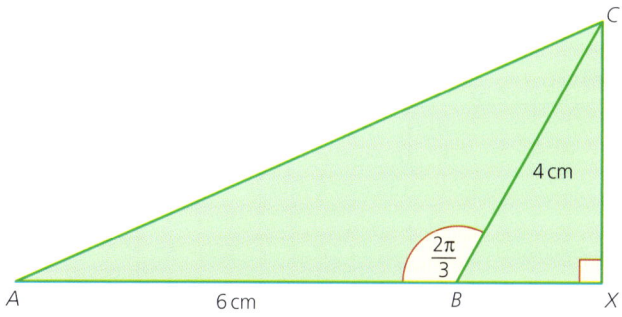

  a) Work out the exact length of BX.

  b) Show that angle $CAB = \tan^{-1}\left(\frac{\sqrt{3}}{4}\right)$.

  c) Show that the exact length of AC is $\sqrt{76}$ cm.

10 TRIGONOMETRY

Reciprocal trigonometrical functions

As well as sin, cos and tan there are three more trigonometrical ratios that you need to be able to use. These are the reciprocals of the three functions you have already met: cosecant (cosec), secant (sec) and cotangent (cot).

$$\text{cosec}\,\theta = \frac{1}{\sin\theta} \qquad \sec\theta = \frac{1}{\cos\theta} \qquad \cot\theta = \frac{1}{\tan\theta} \left(= \frac{\cos\theta}{\sin\theta}\right)$$

Each of these functions is undefined for certain values of θ. For example, $\text{cosec}\,\theta$ is undefined for $\theta = 0°, 180°, 360°…$ since $\sin\theta$ is zero for these values.

➡ Worked example

Solve the following equations for $0° < x < 90°$ rounding your answers to one decimal place where necessary.

a) $\sec x = 2$ b) $\text{cosec}\,x = 2$ c) $\cot x = 2$

Solution

a) $\sec x = 2 \Rightarrow \frac{1}{\cos x} = 2$
$\Rightarrow \cos x = \frac{1}{2}$
$\Rightarrow x = 60°$

b) $\text{cosec}\,x = 2 \Rightarrow \frac{1}{\sin x} = 2$
$\Rightarrow \sin x = \frac{1}{2}$
$\Rightarrow x = 30°$

c) $\cot x = 2 \Rightarrow \frac{1}{\tan x} = 2$
$\Rightarrow \tan x = \frac{1}{2}$
$\Rightarrow x = 26.6°$

Exercise 10.2

'Exact form' means give the answer using fractions and surds.

1. Write each value in exact form. Do not use a calculator.
   a) i) $\sin 30°$ ii) $\cos 30°$ iii) $\tan 30°$
   b) i) $\text{cosec}\,30°$ ii) $\sec 30°$ iii) $\cot 30°$
2. Write each value in exact form. Do not use a calculator.
   a) i) $\sin 45°$ ii) $\cos 45°$ iii) $\tan 45°$
   b) i) $\text{cosec}\,45°$ ii) $\sec 45°$ iii) $\cot 45°$
3. Write each value in exact form. Do not use a calculator.
   a) i) $\sin \frac{\pi}{3}$ ii) $\cos \frac{\pi}{3}$ iii) $\tan \frac{\pi}{3}$
   b) i) $\text{cosec}\,\frac{\pi}{3}$ ii) $\sec \frac{\pi}{3}$ iii) $\cot \frac{\pi}{3}$
4. In the triangle ABC, angle $A = 90°$ and $\sec B = 2$.
   a) Work out the size of angles B and C.
   b) Find $\tan B$.
   c) Show that $1 + \tan^2 B = \sec^2 B$.
5. In the triangle ABC, angle $A = 90°$ and $\text{cosec}\,B = 2$.
   a) Work out the size of angles B and C.
   $AC = 2$ units
   b) Work out the lengths of AB and BC.

6 Given that $\sin\theta = \frac{3}{4}$ and θ is acute, find the values of $\sec\theta$ and $\cot\theta$.
7 In the triangle LMN, angle $M = \frac{\pi}{2}$ and $\cot N = 1$.
   a) Find the angles L and N.
   b) Find $\sec L$, $\csc L$ and $\tan L$.
   c) Show that $1 + \tan^2 L = \sec^2 L$.
8 Malini is 1.5 m tall. At 8 o'clock one evening, her shadow is 6 m long. Given that the angle of elevation of the sun at that moment is α:
   a) show that $\cot\alpha = 4$,
   b) find the value of α.
9 α is an angle in a triangle.
   a) For what values of α are $\sin\alpha$, $\cos\alpha$ and $\tan\alpha$ all positive? Give your answers in both degrees and radians.
   b) Are there any values of α for which $\sin\alpha$, $\cos\alpha$ and $\tan\alpha$ are all negative? Explain your answer.
   c) Are there any values of α for which $\sin\alpha$, $\cos\alpha$ and $\tan\alpha$ are all equal? Explain your answer.

Trigonometrical functions for angles of any size

Is it possible to extend the use of the trigonometrical functions to angles greater than 90°, like $\sin 120°$, $\cos 275°$ or $\tan 692°$? The answer is yes – provided you change the definition of sine, cosine and tangent to one that does not require the angle to be in a right-angled triangle. It is not difficult to extend the definitions, as follows.

First look at the right-angled triangle below, which has hypotenuse of unit length.

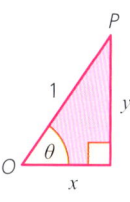
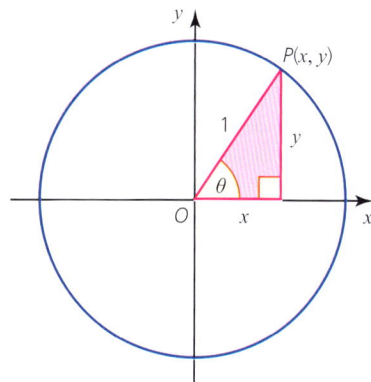

This provides the definitions:

$$\sin\theta = \frac{y}{1} = y; \qquad \cos\theta = \frac{x}{1} = x; \qquad \tan\theta = \frac{y}{x}.$$

Now think of the angle θ being situated at the origin, as in the diagrams above, and allow θ to take any value. The vertex marked P has coordinates (x, y) and can now be anywhere on the unit circle.

10 TRIGONOMETRY

This shows that the definitions above can be applied to *any* angle θ, whether it is positive or negative, and whether it is less than or greater than 90°:

$$\sin\theta = y, \qquad \cos\theta = x, \qquad \tan\theta = \frac{y}{x}.$$

For some angles, x or y (or both) will take a negative value, so the signs of $\sin\theta$, $\cos\theta$ and $\tan\theta$ will vary accordingly.

→ Worked example

The x- and y-axes divide the plane into four regions called quadrants. Draw a diagram showing the quadrants for values of x and y from −1 to 1. Label each quadrant to show which of the trigonometrical functions are positive and which are negative in each quadrant.

Solution

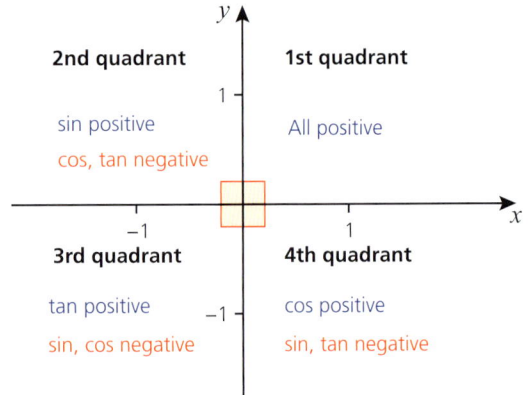

2nd quadrant — sin positive, cos, tan negative
1st quadrant — All positive
3rd quadrant — tan positive, sin, cos negative
4th quadrant — cos positive, sin, tan negative

> **Note**
> Look at this diagram. It gives you a useful aid for remembering the values for which sin, cos and tan are positive and negative.
> A means all are positive.
>
>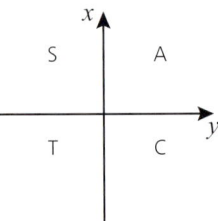
>
> S means sin is positive but the other two, cos and tan, are negative, and so on. Starting from C and working anticlockwise this spells 'CAST'. Consequently, it is often referred to as 'the CAST rule'.

→ Worked example

Find the value of: **a** $\sin 120°$ **b** $\cos 210°$ **c** $\tan 405°$.

Solution

a) 120° is in the second quadrant, so $\sin 120°$ is positive. The line at 120° makes an angle of 60° with the x-axis, so $\sin 120° = +\sin 60° = \dfrac{\sqrt{3}}{2}$

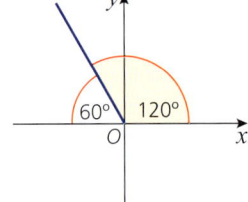

b) 210° is in the third quadrant, so $\cos 210°$ is negative. The line at 210° makes an angle of 30° with the x-axis, so $\cos 210° = -\cos 30° = -\dfrac{\sqrt{3}}{2}$

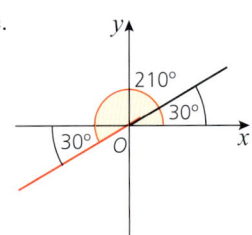

176

Graphs of trigonometrical functions

c) 405° is in the first quadrant, so tan 405° is positive. The line at 405° makes an angle of 45° with the x-axis, so tan 405° = tan 45° = 1.

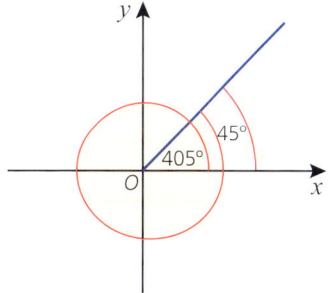

Graphs of trigonometrical functions
The sine and cosine graphs

The diagram on the left below shows angles at intervals of 30° in the unit circle. The resulting coordinates, θ and y, are plotted relative to the axes in the diagram on the right. They have been joined with a continuous curve to give the graph of $\sin \theta$ for $0° \leq \theta \leq 360°$. The resulting wave is called the sine curve.

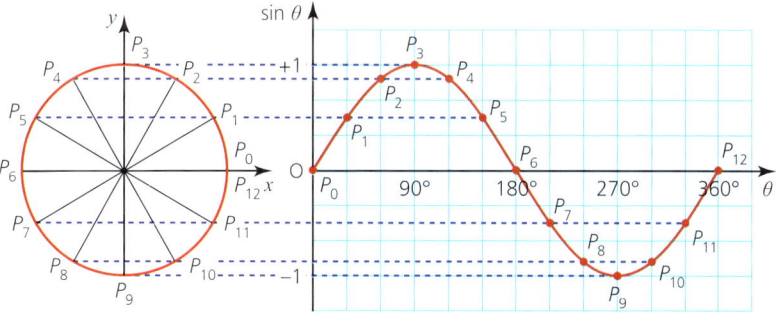

The angle 390° gives the same point P_1 on the circle as the angle 30°, the angle 420° gives point P_2 and so on. You can see that for angles from 360° to 720° the sine wave will simply repeat itself, as shown below. This is true also for angles from 720° to 1080° and so on.

Since the curve repeats itself every 360° the sine function is described as **periodic**, with **period** 360° or 2π radians.

The **amplitude** of such a curve is the largest displacement from the central position, i.e. the horizontal axis.

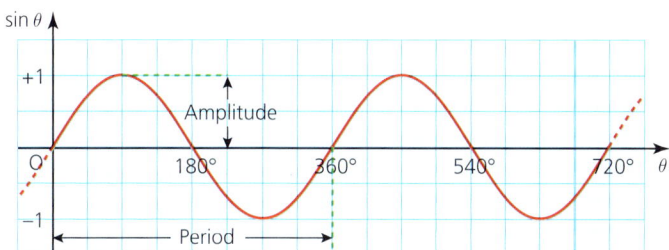

177

You can transfer the *x*-coordinates on to a set of axes in a similar way to obtain the graph of cos θ. This is most easily illustrated if you first rotate the circle through 90° anticlockwise. The diagram shows the circle in this new orientation, together with the resulting graph.

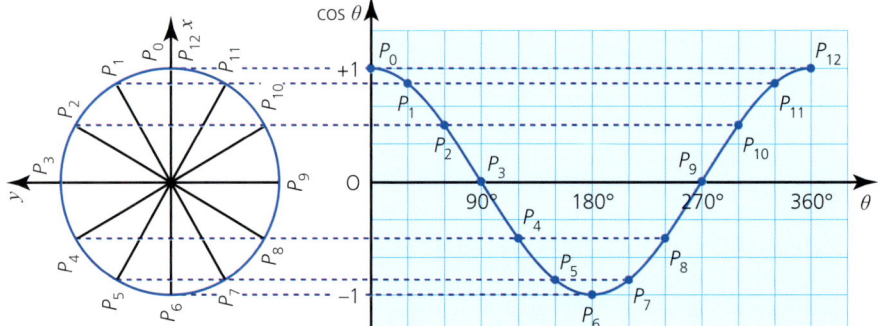

The cosine curve repeats itself for angles in the interval $360° \leq \theta \leq 720°$. This shows that the cosine function is also periodic with a period of 360°.

Notice that the graphs of sin θ and cos θ have exactly the same shape. The cosine graph can be obtained by translating the sine graph 90° to the left, as shown below.

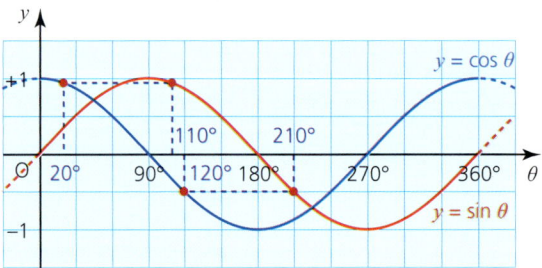

The diagram shows that, for example,

$$\cos 20° = \sin 110°, \cos 90° = \sin 180°, \cos 120° = \sin 210°, \text{ etc.}$$

In general:

$$\cos \theta = \sin(\theta + 90°), \text{ and in radians } \cos \theta = \sin\left(\theta + \frac{\pi}{2}\right).$$

Discussion point

1. What do the graphs of sin θ and cos θ look like for negative angles?
2. Draw the curve of sin θ for $0° \leq \theta \leq 90°$.
   Using only reflections, rotations and translations of this curve, how can you generate the curves of sin θ and cos θ for $0° \leq \theta \leq 360°$?

Solving trigonometrical equations using graphs

> **Note**
>
> The graph of $\tan\theta$ is periodic, like those for $\sin\theta$ and $\cos\theta$, but in this case the period is 180°. Again, the curve for $0 \leq \theta < 90°$ can be used to generate the rest of the curve using rotations and translations.

The tangent graph

The value of $\tan\theta$ can be worked out from the definition $\tan\theta = \frac{y}{x}$ or by using $\tan\theta = \frac{\sin\theta}{\cos\theta}$.

You have already seen that $\tan\theta$ is undefined for $\theta = 90°$. This is also the case for all other values of θ for which $\cos\theta = 0$, namely 270°, 450°, ..., and −90°, −270°, ...

The graph of $\tan\theta$ is shown below.
The dotted lines $\theta = \pm 90°$ and $\theta = 270°$ are **asymptotes**.

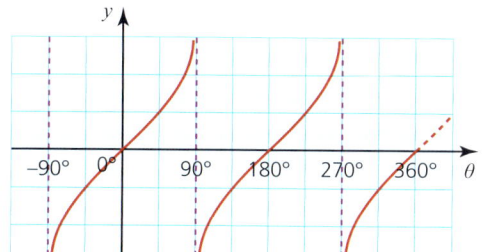

Asymptotes are not part of the curve. The branches get increasingly close to them but never actually touch them.

Solving trigonometrical equations using graphs

You can use these graphs when you solve trigonometric equations.

If you use the inverse function on your calculator to solve the equation $\cos\theta = 0.5$, the answer is given as 60°. However, the graph of $y = \cos\theta$ shows that this equation has infinitely many roots.

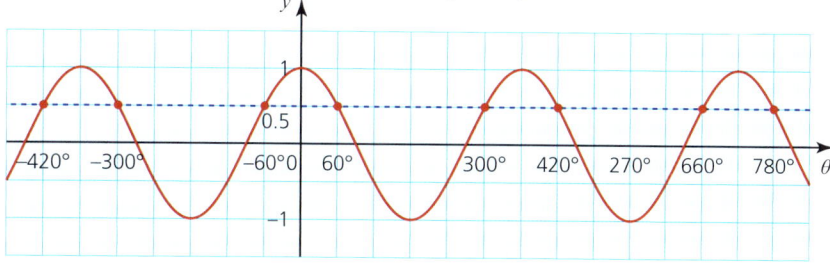

This graph of $y = \cos\theta$ shows that the roots for $\cos\theta = 0.5$ are:
$$\theta = \ldots, -420°, -300°, -60°, 60°, 300°, 420°, 660°, 780°, \ldots$$

The functions cosine, sine and tangent are all many-one mappings, so their inverse mappings are one-many. In other words, the problem 'Find $\cos 60°$' has only one solution (0.5), whilst 'Find θ such that $\cos\theta = 0.5$' has infinitely many solutions.

Remember that a function has to be either one-one or many-one. This means that in order to define inverse functions for cosine, sine and tangent, a restriction must be placed on the domain of each so that it

10 TRIGONOMETRY

becomes a one-one mapping. This is why your calculator will always give the value of the solution between:

$$0° \leq \theta \leq 180° \quad \text{(cos)}$$
$$-90° \leq \theta \leq 90° \quad \text{(sin)}$$
$$-90° < \theta < 90° \quad \text{(tan)}.$$

*Your calculator only gives one of the infinitely many roots. This is called the **principal value**.*

The following diagrams are the graphs of cosine, sine and tangent together with their principal values. The graphs show you that the principal values cover the whole of the range (y values) for each function.

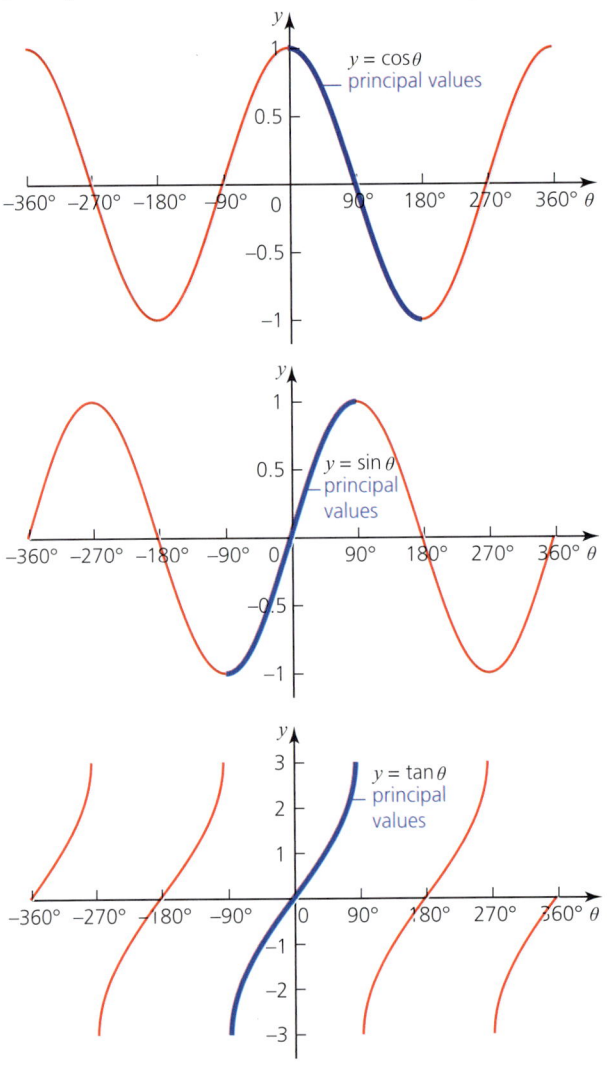

Discussion point

How are the graphs of $\sin\theta$, $\cos\theta$ and $\tan\theta$ changed if θ is measured in radians rather than degrees?

Worked example

Find values of θ in the interval $-360° \leq \theta \leq 360°$ for which $\sin\theta = \frac{\sqrt{3}}{2}$.

Solution

$\sin\theta = \frac{\sqrt{3}}{2} \Rightarrow \theta = \sin^{-1}\left(\frac{\sqrt{3}}{2}\right) = 60°$. The graph of $\sin\theta$ is shown below.

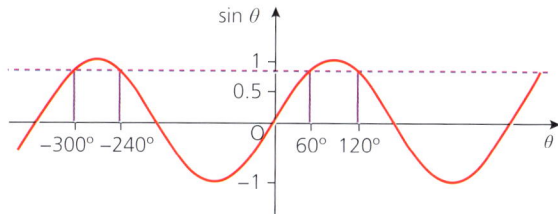

So, the values of θ are $-300°, -240°, 60°, 120°$.

Worked example

Solve the equation $2\tan\theta + 1 = 0$ for $-180° \leq \theta \leq 180°$.

Solution

$2\tan\theta + 1 = 0 \Rightarrow \tan\theta = -\frac{1}{2}$

Using a calculator

$\Rightarrow \theta = \tan^{-1}\left(-\frac{1}{2}\right)$

$= -26.6°$ (1 d.p.)

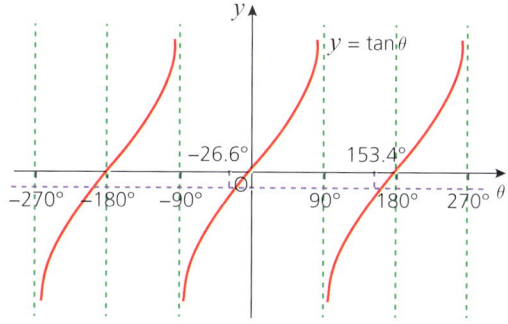

From the graph, the other answer in the range is:

$\theta = -26.6° + 180° = 153.4°$

So, the values of θ are $-26.6°, 153.4°$.

10 TRIGONOMETRY

Exercise 10.3

1. a) Sketch the curve $y = \cos x$ for $0° \leq x \leq 360°$.
   b) Solve the equation $\cos x = 0.5$ for $0° \leq x \leq 360°$, and illustrate the two roots on your sketch.
   c) State two other roots of $\cos x = 0.5$, given that x is no longer restricted to values between $0°$ and $360°$.
   d) Write down, without using your calculator, the value of $\cos 240°$.

2. a) Sketch the curve of $y = \sin x$ for $-2\pi \leq x \leq 2\pi$.
   b) Solve the equation $\sin x = 0.6$ for $-2\pi \leq x \leq 2\pi$, and illustrate all the roots on your sketch.
   c) Sketch the curve $y = \cos x$ for $-2\pi \leq x \leq 2\pi$.
   d) Solve the equation $\cos x = 0.8$ for $-2\pi \leq x \leq 2\pi$, and illustrate all the roots on your sketch.
   e) Explain why some of the roots of $\sin x = 0.6$ are the same as those for $\cos x = 0.8$, and why some are different.

3. Solve the following equations for $0° \leq x \leq 2\pi$.
   a) $\tan x = \sqrt{3}$
   b) $\sin x = 0.5$
   c) $\cos x = -\dfrac{\sqrt{3}}{2}$
   d) $\tan x = \dfrac{1}{\sqrt{3}}$
   e) $\cos x = -0.7$
   f) $\cos x = 0.3$
   g) $\sin x = -\dfrac{1}{3}$
   h) $\sin x = -1$

4. Write the following as integers, fractions, or using square roots. You should not need your calculator.
   a) $\sin 45°$
   b) $\cos 60°$
   c) $\tan 45°$
   d) $\sin 120°$
   e) $\cos 150°$
   f) $\tan 180°$
   g) $\sin 405°$
   h) $\cos(-45°)$
   i) $\tan 225°$

5. In this question all the angles are in the interval $-180°$ to $180°$. Give all answers correct to one decimal place.
   a) Given that $\cos \alpha < 0$ and $\sin \alpha = 0.5$, find α.
   b) Given that $\tan \beta = 0.3587$ and $\sin \beta < 0$, find β.
   c) Given that $\cos \gamma = 0.0457$ and $\tan \gamma > 0$, find γ.

6. a) Draw a sketch of the graph $y = \sin x$ and use it to demonstrate why $\sin x = \sin(180° - x)$.
   b) By referring to the graphs of $y = \cos x$ and $y = \tan x$, state whether the following are true or false.
      i) $\cos x = \cos(180° - x)$
      ii) $\cos x = -\cos(180° - x)$
      iii) $\tan x = \tan(180° - x)$
      iv) $\tan x = -\tan(180° - x)$

7. a) For what values of α are $\sin \alpha$, $\cos \alpha$ and $\tan \alpha$ all positive given that $0° \leq \alpha \leq 360°$?
   b) Are there any values of α for which $\sin \alpha$, $\cos \alpha$ and $\tan \alpha$ are all negative? Explain your answer.
   c) Are there any values of α for which $\sin \alpha$, $\cos \alpha$ and $\tan \alpha$ are all equal? Explain your answer.

8. Solve the following equations for $0° \leq \theta \leq 360°$.
   a) $\cos(\theta - 20°) = \dfrac{1}{2}$
   b) $\tan(\theta + 10°) = \dfrac{\sqrt{3}}{3}$
   c) $\sin(\theta + 80°) = \dfrac{\sqrt{2}}{2}$
   d) $\tan 2\theta = \sqrt{3}$
   e) $\sin\left(\tfrac{1}{2}\theta\right) = \dfrac{1}{2}$
   f) $\cos 2\theta = -\dfrac{\sqrt{3}}{2}$
   g) $\sin 3\theta = \dfrac{1}{2}$
   h) $\sin 2\theta = 0$
   i) $\tan 3\theta = 1$

9. Solve the following equations for $-2\pi \leq x \leq 2\pi$.
   a) $10 \sin x = 1$
   b) $2\cos x - 1 = 0$
   c) $\tan x + 2 = 0$
   d) $5 \sin x + 2 = 0$
   e) $\cos^2 x = 1 - \sin x$

Transformations of trigonometrical graphs

Now that you are familiar with the graphs of the sine, cosine and tangent functions, you can see how to transform these graphs.

$y = a \sin x$ where a is a positive integer

How are the graphs of $y = \sin x$ and $y = 2 \sin x$ related?
To investigate this question, start by drawing graphs of the two functions using a graphical calculator or graph-drawing package.

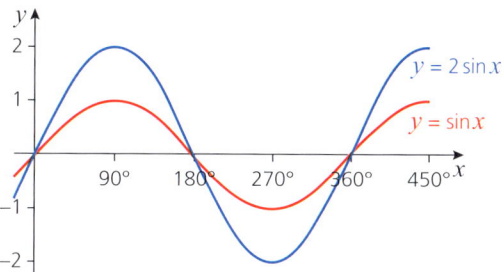

Looking at the graphs:
» $y = \sin x$ has an amplitude of 1 unit and a period of 360°.
» $y = 2 \sin x$ has an amplitude of 2 units and a period of 360°.

The graphs of $y = \sin x$ and $y = 2 \sin x$ illustrate the following general result.

> The graph of $y = a \sin x$ is a sine curve that has an amplitude of a units and a period of 360°.
> The transformation is a **stretch of scale factor a parallel to the y-axis**.

$y = \sin bx$ where b is a simple fraction or integer

What is the relationship between the graph of $y = \sin x$ and $y = \sin 2x$?
Again, start by drawing the graphs. The table of values, calculated to one decimal place, is given below, but you can also draw the graphs using suitable software.

x	0°	30°	60°	90°	120°	150°	180°	210°	240°	270°	300°	330°	360°
$\sin x$	0	0.5	0.9	1.0	0.9	0.5	0	−0.5	−0.9	−1.0	−0.9	−0.5	0
$\sin 2x$	0	0.9	0.9	0	−0.9	−0.9	0	0.9	0.9	0	−0.9	−0.9	0

10 TRIGONOMETRY

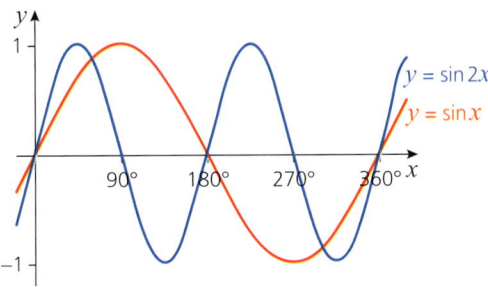

Looking at the graphs:
» $y = \sin x$ has an amplitude of 1 unit and a period of 360°.
» $y = \sin 2x$ has an amplitude of 1 unit and a period of 180°.

Similarly, the graph of $y = \sin 3x$ has a period of $360 \div 3 = 120°$, the graph of $y = \sin\left(\frac{x}{2}\right)$ has a period of $360 \div \frac{1}{2} = 720°$, and so on.

The graphs of $y = \sin x$ and $y = \sin 2x$ illustrate the following general results.

The graph of $y = \sin bx$ is a sine curve that has amplitude 1 unit and period $\left(\frac{360}{b}\right)°$.

The transformation is a **stretch of scale factor $\frac{1}{b}$ parallel to the x-axis**.

$y = \sin x + c$ where c is an integer

How are the graphs of $y = \sin x$ and $y = \sin x + 3$ related?
Again, start by drawing the graphs.

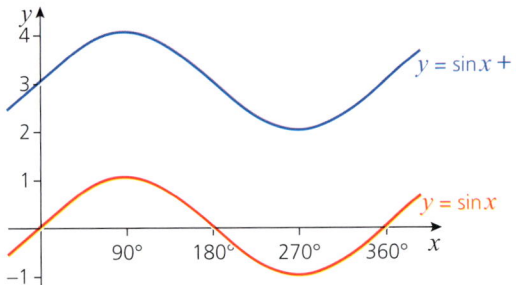

Looking at the graphs, $y = \sin x + 3$ has the same amplitude and period as $y = \sin x$ but is 3 units above it. Similarly, the graph of $y = \sin x - 2$ is 2 units below the graph of $y = \sin x$.

The graphs of $y = \sin x$ and $y = \sin x + 3$ illustrate the following general result.

The graph of $y = \sin x + c$ is the same shape as the graph of $y = \sin x$ but is translated vertically upwards through c units.

The transformation is **a translation of $\begin{pmatrix} 0 \\ c \end{pmatrix}$**.

Transformations of trigonometrical graphs

Combining transformations

The graph of $y = a \sin bx + c$ is a transformation of the graph of $y = \sin x$ effected by:
- a **stretch** parallel to the y-axis, **scale factor** a
- a **stretch** parallel to the x-axis, **scale factor** $\frac{1}{b}$
- a **translation** parallel to the y-axis of c units.

> ### Discussion point
> When drawing the graph of $y = a \sin bx + c$ using a series of transformations of the graph $y = \sin x$, why is it necessary to do the translation last?

All the transformations in this section have been applied to the graph of $y = \sin x$. The same rules can be applied to the graphs of all trigonometric functions, and to other graphs as well as those shown in these examples.

Worked example

The diagram shows the graph of a function $y = f(x)$.
Sketch the graph of each of these functions.

a) $y = f(2x)$

b) $y = 3f(x)$

c) $y = 3f(2x)$

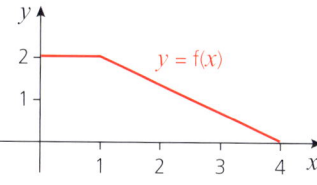

Solution

a) $y = f(2x)$ is obtained from $y = f(x)$ by applying a stretch of scale factor $\frac{1}{2}$ parallel to the x-axis.

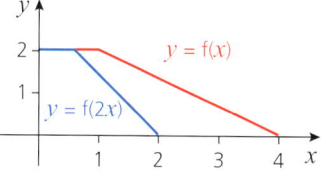

b) $y = 3f(x)$ is obtained from $y = f(x)$ by applying a stretch of scale factor 3 parallel to the y-axis.

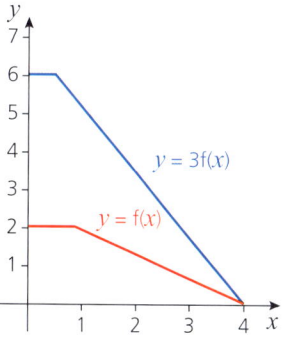

10 TRIGONOMETRY

The order of the transformations is not important in this example because the two directions are independent.

c) $y = 3f(2x)$ is obtained from $y = f(x)$ by applying a stretch of scale factor $\frac{1}{2}$ parallel to the x-axis and a stretch of scale factor 3 parallel to the y-axis.

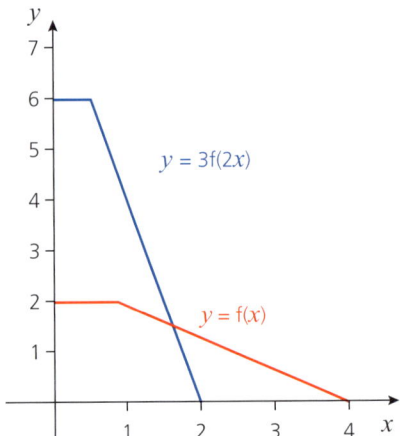

→ Worked example

Starting with the graph of $y = \cos x$
  i   State the transformations that can be used to sketch each curve.
  ii Sketch each curve for $0° \leqslant x \leqslant 360°$.

a) $y = \cos 2x$

b) $y = 3\cos 2x$

c) $y = 3\cos 2x - 1$

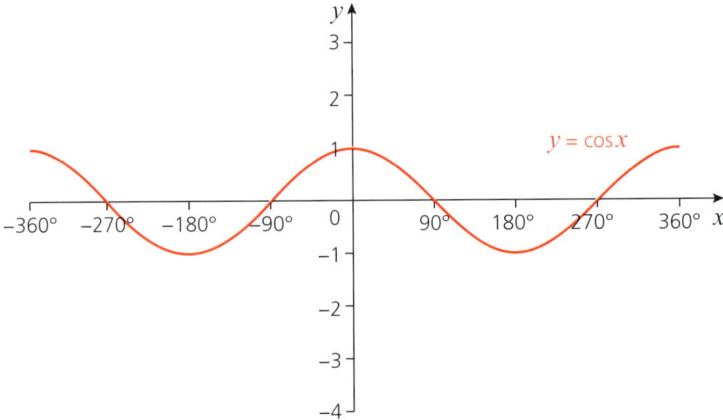

Solution

a) i The graph of $y = \cos 2x$ is a stretch of $y = \cos x$ by scale factor $\frac{1}{2}$ in the x-direction.

ii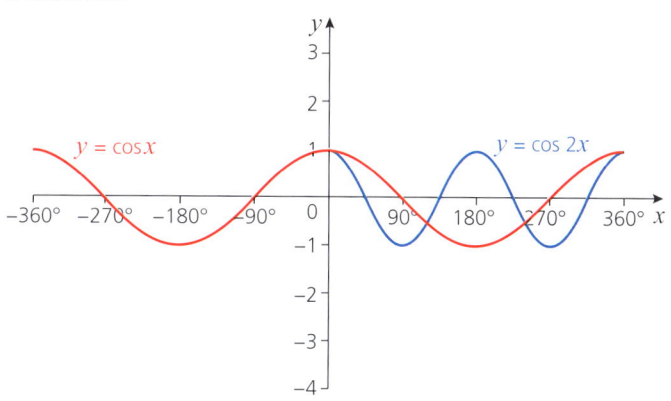

b) i The graph of $y = 3\cos 2x$ is a stretch of $y = \cos x$ by scale factor $\frac{1}{2}$ in the x-direction and by scale factor 3 in the y-direction.

ii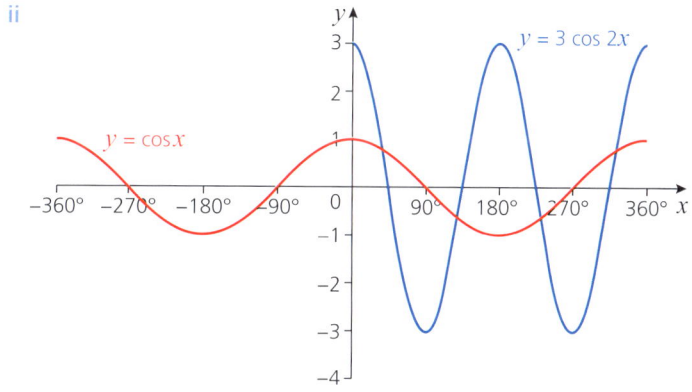

c) i The graph of $y = 3\cos 2x - 1$ is a stretch of $y = \cos x$ by scale factor $\frac{1}{2}$ in the x-direction and scale factor 3 in the y-direction, followed by a translation of 1 unit vertically downwards.

ii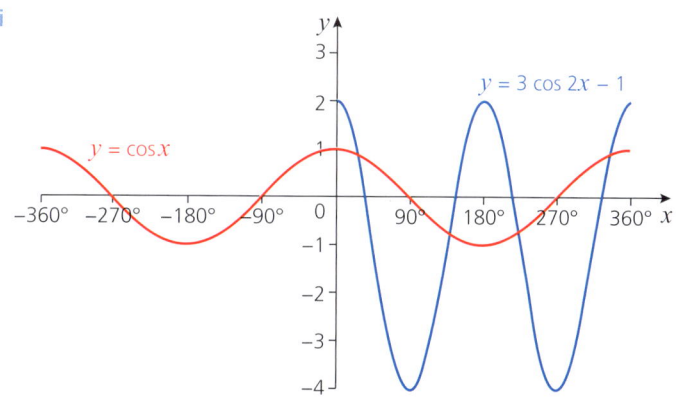

10 TRIGONOMETRY

> **Worked example**

a) State the transformations needed to transform the graph of $y = \sin x$ to the graph of $y = 2\sin 3x$. Sketch the graph of $y = 2\sin 3x$ for $0° \leq x \leq 360°$.

b) Sketch each of the following graphs for $0° \leq x \leq 360°$:
   i $y = |2\sin 3x|$
   ii $y = |2\sin 3x| + 1$
   iii $y = 2\sin 3x + 1$
   iv $y = |2\sin 3x + 1|$

Solution

a) The graph of $y = 2\sin 3x$ is a stretch of $y = \sin x$ by scale factor $\frac{1}{3}$ in the x-direction and by scale factor 2 in the y-direction.

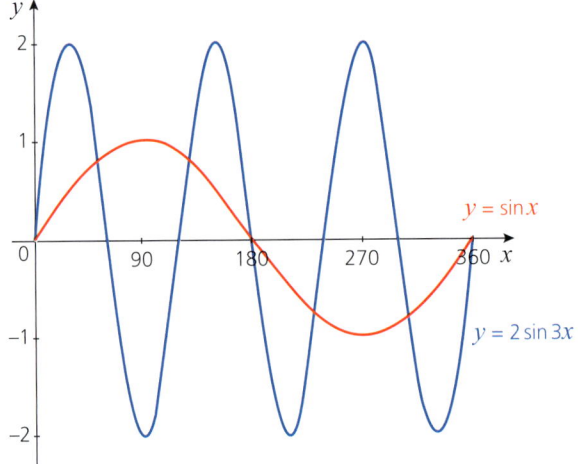

b) i
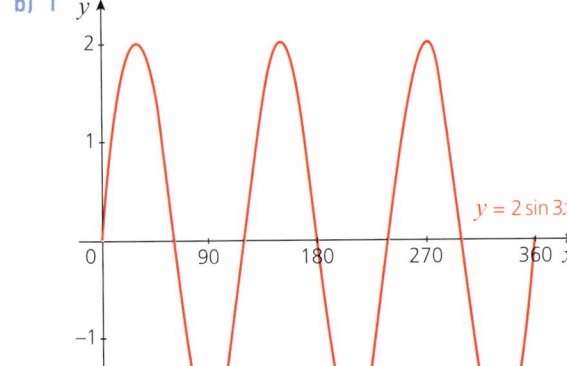
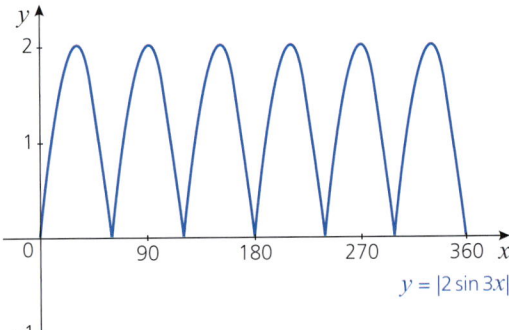

For the graph of the modulus function $y = |f(x)|$, any part of the corresponding graph of $y = f(x)$ where $y < 0$ is reflected in the x-axis.

Transformations of trigonometrical graphs

ii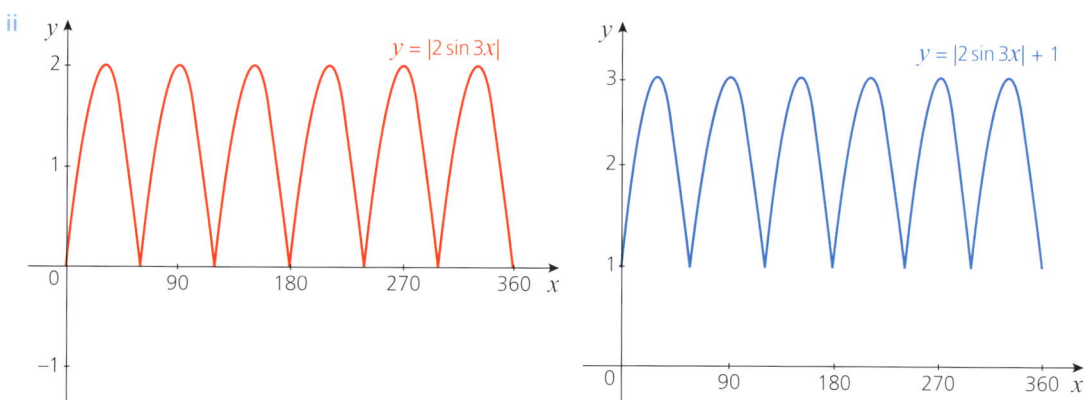

Translation of the graph $y = |2\sin 3x|$ by 1 unit vertically upwards

iii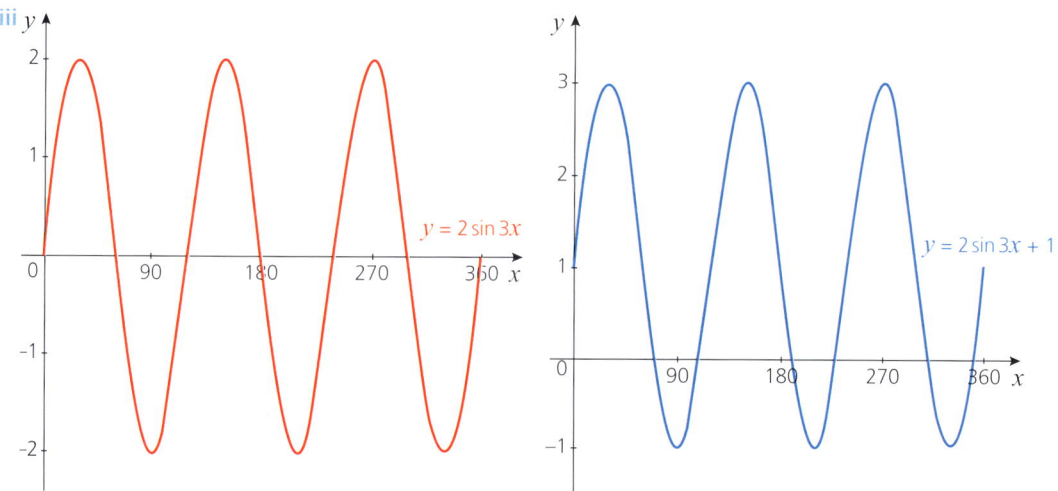

Translation of the graph $y = 2\sin 3x$ by 1 unit vertically upwards

iv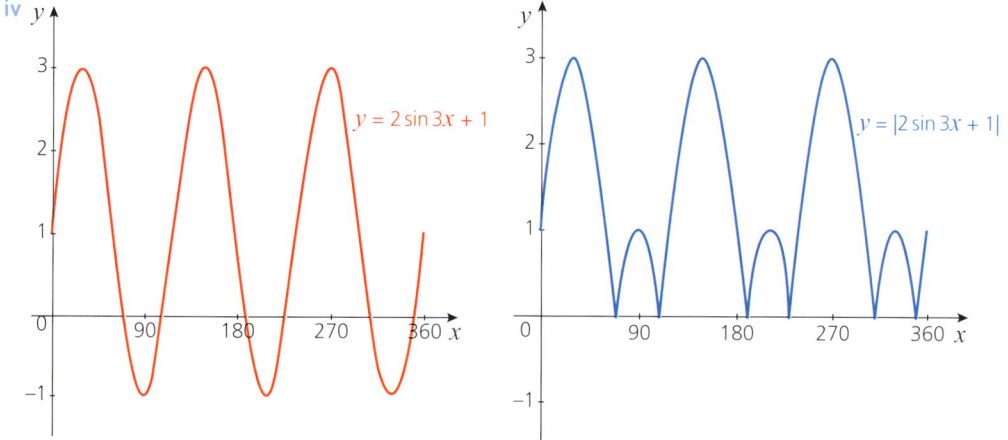

Any part of the graph of $y = 2\sin 3x + 1$ where $y < 0$ is reflected in the x-axis.

10 TRIGONOMETRY

The curve $y = \tan x$ can also be translated and stretched. However, because $y = \tan x$ has no finite boundary in the y-direction, it is not as straightforward to show such stretches when the graphs approach the asymptotes.

→ Worked example

a) Sketch the curve $y = \tan x$ for $0° \leq x \leq 180°$.

b) On the same axes, sketch the curve $y = \tan x + 2$.

Solution

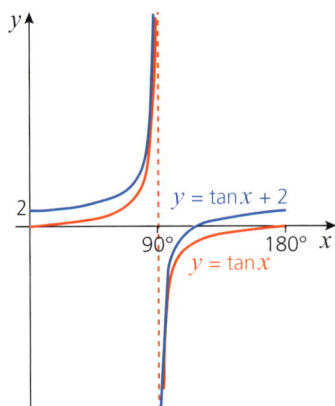

Exercise 10.4

1 For each transformation **i** to **iv**:
   a) Sketch the graph of $y = \sin x$ and on the same axes sketch its image under the transformation.
   b) State the amplitude and period of the transformed graph.
   c) What do you notice about your answers for **iii** and **iv**?
      i) a stretch, scale factor 2, parallel to the y-axis
      ii) a translation of 1 unit vertically downwards
      iii) a stretch, scale factor 2, parallel to the y-axis followed by a translation of 1 unit vertically downwards.
      iv) a translation of 1 unit vertically downwards followed by a stretch of scale factor 2 parallel to the y-axis

2 a) Apply each set of transformations to the graph of $y = \cos x$.
   b) Sketch the graph of $y = \cos x$ and the transformed curve on the same axes.
   c) State the amplitude and period of the transformed graph.
   d) What do you notice about your answers for **iii** and **iv**?
      i) a stretch, scale factor 2, parallel to the x-axis.
      ii) a translation of 180° in the negative x-direction
      iii) a stretch, scale factor 2, parallel to the x-axis followed by a translation of 180° in the negative x-direction
      iv) a translation of 180° in the negative x direction followed by a stretch of scale factor 2 parallel to the x-axis

Questions 1.c, 2.d and 3.d go beyond the syllabus but will help you understand curve sketching.

3 a) Apply these transformations to the graph of $y = \sin x$.
   b) Sketch the graph of $y = \sin x$ and the transformed curve on the same axes.
   c) State the amplitude and period of the transformed graph.
   d) What do you notice about your answers for **iii** and **iv**?
      i) a stretch, scale factor 2, parallel to the x-axis
      ii) a translation of 1 unit vertically upwards
      iii) a stretch, scale factor 2, parallel to the x-axis followed by a translation of 1 unit vertically upwards
      iv) a translation of 1 unit vertically upwards followed by a stretch, scale factor 2, parallel to the x-axis

4 State the transformations needed, in the correct order, to transform the first graph to the second graph.
   a) $y = \tan x$, $\quad y = 3 \tan 2x$
   b) $y = \tan x$, $\quad y = 2 \tan x + 1$
   c) $y = \tan x$, $\quad y = 2 \tan (x - 180°)$
   d) $y = \tan x$, $\quad y = 3 \tan (x + \frac{\pi}{2}) + 3$

5 State the transformations required, in the correct order, to obtain this graph from the graph of $y = \sin x$.

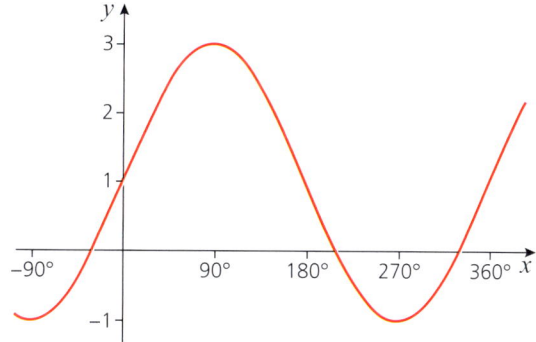

6 State the transformations required, in the correct order, to obtain this graph from the graph of $y = \tan x$.

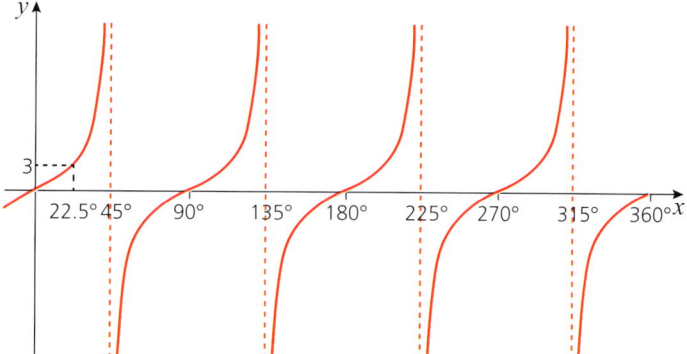

Exercise 10.4 (cont)

7 State the transformations required, in the correct order, to obtain this graph from the graph of $y = \cos x$.

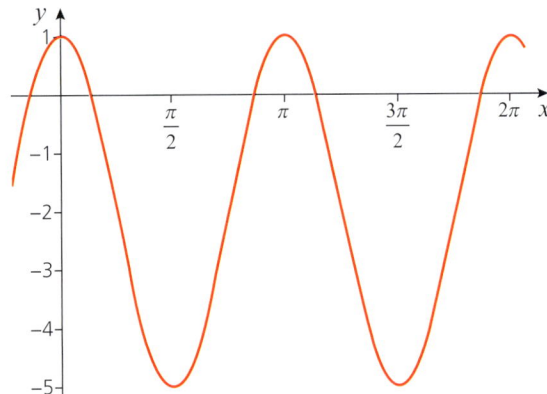

8 a) State the transformations needed to transform the graph of $y = \sin x$ to the graph of $y = 4\sin 2x$. Sketch the graph of $y = 4\sin 2x$ for $0° \leqslant x \leqslant 360°$.
   b) Sketch each of the following graphs for $0° \leqslant x \leqslant 360°$:
      i) $y = |4\sin 2x|$
      ii) $y = |4\sin 2x| - 1$
      iii) $y = 4\sin 2x - 1$
      iv) $y = |4\sin 2x - 1|$
9 Sketch each of the following graphs for $0 \leqslant x \leqslant 2\pi$:
   i) $y = |2\cos 3x|$
   ii) $y = |2\cos 3x| - 2$
   iii) $y = |2\cos 3x + 1|$

Identities and equations
sin, cos and tan

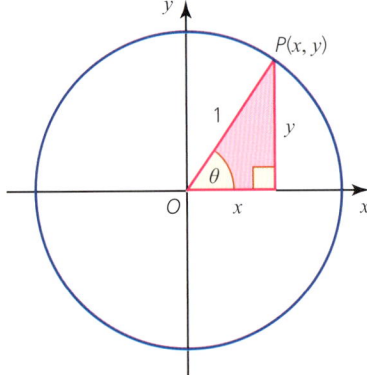

Look at the diagram of the unit circle. It shows you that

$$x = \cos\theta, \; y = \sin\theta \text{ and } \frac{y}{x} = \tan\theta.$$

It follows that

$$\tan\theta = \frac{\sin\theta}{\cos\theta}.$$

Identities and equations

However, it more accurate to use the identity sign here because the relationship is true for all values of θ, so

$$\tan\theta \equiv \frac{\sin\theta}{\cos\theta}$$

Remember that an equation is only true for certain values of the variable, called the **solution** of the equation.

For example, $\tan\theta = 1$ is an equation: in the range $0° \leq \theta \leq 360°$, it is only true when $\theta = 45°$ or $225°$.

By contrast, an **identity** is true for *all* values of the variable. For example,

$$\tan 45° \equiv \frac{\sin 45°}{\cos 45°}, \quad \tan 75° \equiv \frac{\sin 75°}{\cos 75°}, \quad \tan(-300°) \equiv \frac{\sin(-300°)}{\cos(-300°)}, \quad \tan\frac{\pi}{6} \equiv \frac{\sin\frac{\pi}{6}}{\cos\frac{\pi}{6}}$$

and so on for all values of the angle.

The identity below is found by applying Pythagoras' theorem to any point $P(x, y)$ on the unit circle.

$$y^2 + x^2 \equiv OP^2$$
$$(\sin\theta)^2 + (\cos\theta)^2 \equiv 1.$$

This is written as:

$$\sin^2\theta + \cos^2\theta \equiv 1.$$

You can use the identities $\tan\theta \equiv \frac{\sin\theta}{\cos\theta}$ and $\sin^2\theta + \cos^2\theta \equiv 1$ to prove other identities are true.

There are two methods you can use to prove an identity; you can use either method or a mixture of both.

Method 1

When both sides of the identity look equally complicated you can work with both the left-hand side (LHS) and the right-hand side (RHS) and show that LHS − RHS = 0 (as shown in the example below).

➜ Worked example

Prove the identity $\cos^2\theta - \sin^2\theta \equiv 1 - 2\sin^2\theta$.

Solution

You need to show that $\cos^2\theta - \sin^2\theta - 1 + 2\sin^2\theta \equiv 0.$

Both sides look equally complicated, so show
LHS − RHS $= \cos^2\theta - \sin^2\theta - 1 + 2\sin^2\theta$

Simplifying:
$\equiv \cos^2\theta + \sin^2\theta - 1$

$\equiv 1 - 1$
$\equiv 0$ as required

In this book, as in mathematics generally, an equals sign is often used where it would be more correct to use an identity sign. The identity sign is kept for situations where it is particularly important to emphasise that the relationship is an identity and not an equation.

Using $\sin^2\theta + \cos^2\theta = 1$

10 TRIGONOMETRY

Method 2
When one side of the identity looks more complicated than the other side, you can work with this side until you end up with the same as the simpler side, as shown in the next example. In this case you show LHS = RHS.

➡ Worked example

Prove the identity $\dfrac{\sin\theta}{1-\cos\theta} - \dfrac{1}{\sin\theta} \equiv \dfrac{1}{\tan\theta}$.

Solution

$\text{LHS} = \dfrac{\sin\theta}{1-\cos\theta} - \dfrac{1}{\sin\theta}$

$\equiv \dfrac{\sin^2\theta - (1-\cos\theta)}{(1-\cos\theta)\sin\theta}$

Since $\sin^2\theta + \cos^2\theta = 1$, $\sin^2\theta = 1 - \cos^2\theta$

$\equiv \dfrac{(1-\cos^2\theta) + \cos\theta - 1}{(1-\cos\theta)\sin\theta}$

$\equiv \dfrac{\cos\theta - \cos^2\theta}{(1-\cos\theta)\sin\theta}$

$\equiv \dfrac{\cos\theta(1-\cos\theta)}{(1-\cos\theta)\sin\theta}$

$\equiv \dfrac{\cos\theta}{\sin\theta} = \text{RHS}$

$\equiv \dfrac{1}{\tan\theta}$ as required

➡ Worked example

Solve the equation $4\sin\dfrac{\theta}{2} - 3\cos\dfrac{\theta}{2} = 0$ for $0° \leq \theta \leq 360°$.

Solution

$4\sin\dfrac{\theta}{2} - 3\cos\dfrac{\theta}{2} = 0$

$\Rightarrow 4\sin\dfrac{\theta}{2} = 3\cos\dfrac{\theta}{2}$

Use the identity $\tan\theta \equiv \dfrac{\sin\theta}{\cos\theta}$

$\Rightarrow 4\dfrac{\sin\dfrac{\theta}{2}}{\cos\dfrac{\theta}{2}} = 3$

$\Rightarrow 4\tan\dfrac{\theta}{2} = 3$

$\Rightarrow \tan\dfrac{\theta}{2} = \dfrac{3}{4}$

Using the substitution $u = \dfrac{\theta}{2}$

$\Rightarrow \tan u = \dfrac{3}{4}$

Principal value $\Rightarrow u = 36.87°$

Identities and equations

To find any other values you should draw a sketch graph.

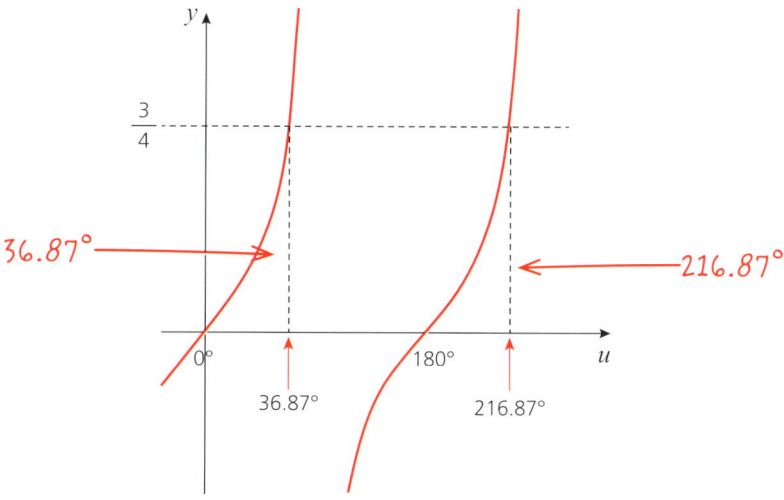

36.87°

216.87°

The substitution $u = \dfrac{\theta}{2} \Rightarrow \theta = 2u$, hence

$\theta = 433.74°$ is outside the interval $0° \leq \theta \leq 360°$

$u = 36.87° \Rightarrow \theta = 73.74°$
$u = 216.87° \Rightarrow \theta = 433.74°$

The solution is therefore $\theta = 73.74°$.

cosec, sec and cot

The reciprocal of the relationship $\tan\theta \equiv \dfrac{\sin\theta}{\cos\theta}$ is

$$\dfrac{1}{\tan\theta} \equiv 1 \div \dfrac{\sin\theta}{\cos\theta}$$

$$\Rightarrow \quad \cot\theta \equiv \dfrac{\cos\theta}{\sin\theta}.$$

Similarly, dividing the relationship $\sin^2\theta + \cos^2\theta \equiv 1$ through by $\sin^2\theta$ gives

$$\dfrac{\sin^2\theta}{\sin^2\theta} + \dfrac{\cos^2\theta}{\sin^2\theta} \equiv \dfrac{1}{\sin^2\theta}$$

$$\Rightarrow \quad 1 + \cot^2\theta \equiv \operatorname{cosec}^2\theta$$

This is usually presented as $\quad \operatorname{cosec}^2\theta \equiv 1 + \cot^2\theta$

If instead you divide $\sin^2\theta + \cos^2\theta \equiv 1$ by $\cos^2\theta$ you get

$$\dfrac{\sin^2\theta}{\cos^2\theta} + \dfrac{\cos^2\theta}{\cos^2\theta} \equiv \dfrac{1}{\cos^2\theta}$$

$$\Rightarrow \quad \tan^2\theta + 1 \equiv \sec^2\theta$$

This is usually presented as $\quad \sec^2\theta \equiv 1 + \tan^2\theta$

10 TRIGONOMETRY

> **Worked example**
>
> a) Show that $\sec^2 x - \text{cosec}^2 x \equiv \tan^2 x - \cot^2 x$.
>
> b) Prove that $\dfrac{\sec\theta}{\tan\theta} \equiv \text{cosec}\,\theta$.

Solution

a) Start with the left-hand side since this looks more complicated.

$$\text{LHS} \equiv \sec^2 x - \text{cosec}^2 x \equiv (1 + \tan^2 x) - (1 + \cot^2 x)$$
$$\equiv 1 + \tan^2 x - 1 - \cot^2 x$$
$$\equiv \tan^2 x - \cot^2 x \equiv \text{RHS}$$

Using $\tan^2 x + 1 = \sec^2 x$ and $1 + \cot^2 x = \text{cosec}^2 x$

b) Again, start with the left-hand side since this looks more complicated.

$$\text{LHS} \equiv \frac{\sec\theta}{\tan\theta} \equiv \sec\theta \div \tan\theta$$
$$\equiv \frac{1}{\cos\theta} \div \frac{\sin\theta}{\cos\theta}$$
$$\equiv \frac{1}{\cos\theta} \times \frac{\cos\theta}{\sin\theta}$$
$$\equiv \frac{1}{\sin\theta}$$
$$\equiv \text{cosec}\,\theta \equiv \text{RHS}$$

It is often more straightforward to go back to the basic trigonometric functions $\sin\theta$ and $\cos\theta$.

You can also use this approach to solve equations involving the reciprocal functions. This involves using the definitions of the functions to find equivalent equations using sin, cos and tan. You will usually be given a range of values within which your solution must lie, so a sketch graph is useful to ensure that you find all possible values.

> **Worked example**
>
> Solve the following equations for
>
> a) $\text{cosec}\,\theta = 2$ for $0° \leqslant \theta \leqslant 360°$
>
> b) $\sec^2\theta + 2\tan^2\theta = 4$ for $0 \leqslant \theta \leqslant 2\pi$

Solution

a) $\text{cosec}\,\theta = 2 \Rightarrow \dfrac{1}{\sin\theta} = 2$
$\Rightarrow 1 = 2\sin\theta$
$\Rightarrow \sin\theta = \dfrac{1}{2}$
$\Rightarrow \theta = 30°$

*This is called the **principal value**.*

Identities and equations

To find any other values in the interval $0° < \theta < 360°$, sketch the graph.

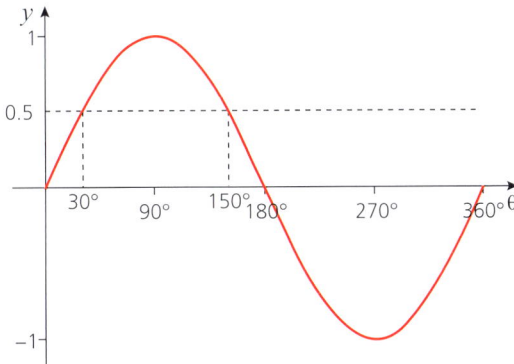

The graph shows that $\sin \theta$ is also $\frac{1}{2}$ when $\theta = 150°$.

The solution is therefore $\theta = 30°$ or $\theta = 150°$.

b) $\sec^2 \theta + 2\tan^2 \theta = 4 \Rightarrow (1 + \tan^2 \theta) + 2\tan^2 \theta = 4$

$\Rightarrow \quad 1 + 3\tan^2 \theta = 4$

$\Rightarrow \quad 3\tan^2 \theta = 3$

$\Rightarrow \quad \tan^2 \theta = 1$

$\Rightarrow \quad \tan \theta = \pm 1$

$\tan^{-1} 1 = \frac{\pi}{4}$ and $\tan^{-1}(-1) = -\frac{\pi}{4}$

To find the values in the interval $0 \leq \theta \leq 2\pi$, sketch the graph.

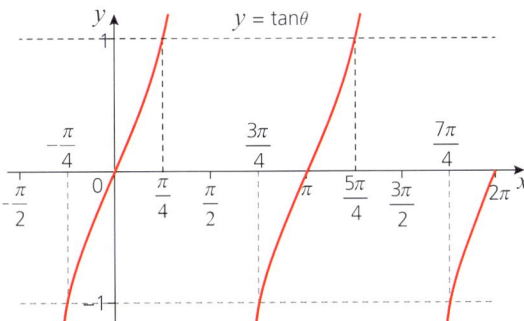

The graph shows the value of $\tan \theta$ is also 1 when $\theta = \frac{5\pi}{4}$.

The principal value for $\tan \theta = -1$ is outside the required range. The graph shows that the values in the required range are $\frac{3\pi}{4}$ and $\frac{7\pi}{4}$.

The solution is therefore $\theta = \frac{\pi}{4}, \theta = \frac{3\pi}{4}, \theta = \frac{5\pi}{4}$ and $\theta = \frac{7\pi}{4}$.

10 TRIGONOMETRY

→ Worked example

Solve the equation $3\sec\left(2\theta - \frac{\pi}{6}\right) = 6$ for $0 \leq \theta \leq 2\pi$.

Solution

$$3\sec\left(2\theta - \frac{\pi}{6}\right) = 6$$

$$\Rightarrow \sec\left(2\theta - \frac{\pi}{6}\right) = 2$$

$\sec\theta = \frac{1}{\cos\theta}$ ⟶ $\Rightarrow \dfrac{1}{\cos\left(2\theta - \frac{\pi}{6}\right)} = 2$

$$\Rightarrow \cos\left(2\theta - \frac{\pi}{6}\right) = \frac{1}{2}$$

Using the substitution $u = 2\theta - \frac{\pi}{6}$

$$\Rightarrow \cos u = \frac{1}{2}$$

Principal value ⟶ $\Rightarrow u = \frac{\pi}{3}$

To find any other values you should draw a sketch graph.

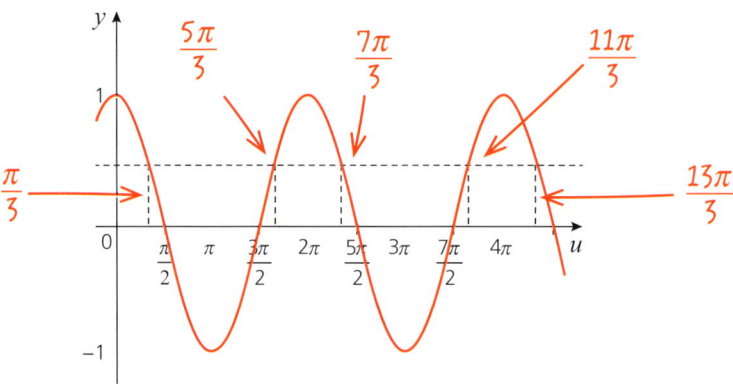

So, from the sketch graph you get the values

$$u = \frac{\pi}{3}, \frac{5\pi}{3}, \frac{7\pi}{3}, \frac{11\pi}{3} \text{ and } \frac{13\pi}{3}$$

The substitution $u = 2\theta - \frac{\pi}{6} \Rightarrow \theta = \dfrac{u + \frac{\pi}{6}}{2}$, hence the corresponding values for θ are

$$\theta = \frac{\pi}{4}, \frac{11\pi}{12}, \frac{5\pi}{4}, \frac{23\pi}{12} \text{ and } \frac{9\pi}{4}$$

However, the value $\frac{9\pi}{4}$ is outside the interval $0 \leq \theta \leq 2\pi$.

The solution is therefore $\theta = \frac{\pi}{4}, \frac{11\pi}{12}, \frac{5\pi}{4}$ or $\frac{23\pi}{12}$.

Identities and equations

Exercise 10.5

1. Prove that $\sin^2\theta - \cos^2\theta = 3 - 2\sin^2\theta - 4\cos^2\theta$.
2. Prove that $1 + \dfrac{1}{\tan^2\theta} = \dfrac{1}{\sin^2\theta}$.
3. Prove that $4\cos^2\theta + 5\sin^2\theta = \sin^2\theta + 4$.
4. Prove that $\dfrac{1 - (\sin\theta - \cos\theta)^2}{\sin\theta\cos\theta} = 2$.
5. Solve the equation $\operatorname{cosec}^2\theta + \cot^2\theta = 2$ for $0° < x < 180°$.
6. Show that $\dfrac{\operatorname{cosec} A}{\operatorname{cosec} A - \sin A} = \sec^2 A$.
7. Solve the equation $\sec^2\theta = 4$ for $0 \leqslant \theta \leqslant 2\pi$.
8. Prove the identity $\cot\theta + \tan\theta \equiv \sec\theta\operatorname{cosec}\theta$.
9. This is the graph of $y = \sec\theta$.

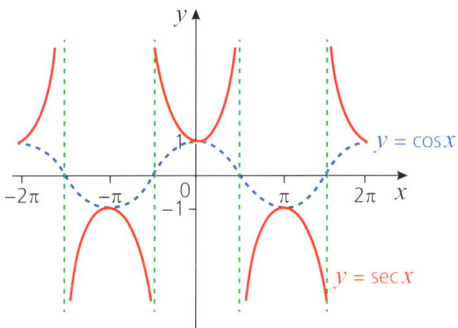

   a) Solve the equation $\sec\theta = 1$ for $-2\pi \leqslant \theta \leqslant 2\pi$.
   b) What happens if you try to solve $\sec\theta = 0.5$?
10. a) Solve the equation $\sin\theta = \operatorname{cosec}\theta$ for $-360° \leqslant \theta \leqslant 360°$.
    b) Solve the equation $5\sin\dfrac{\theta}{3} + 9\cos\dfrac{\theta}{3} = 0$ for $-360° \leqslant \theta \leqslant 360°$.
    c) Solve the equation $4\operatorname{cosec}\left(2\theta - \dfrac{\pi}{3}\right) = 5$ for $0 \leqslant \theta \leqslant 2\pi$.
11. a) Show that $12\sin^2 x + \cos x - 1 = 11 + \cos x - 12\cos^2 x$.
    b) Solve $12\sin^2 x + \cos x - 1 = 0$ for $0° \leqslant x \leqslant 360°$.
12. a) Show that $\sin^2\theta + 1 - \sin\theta - \cos^2\theta = 2\sin^2\theta - \sin\theta$.
    b) Solve $\sin^2\theta + 1 - \sin\theta - \cos^2\theta = 0$ for $0 \leqslant \theta \leqslant 2\pi$.

Past-paper questions

1. (a) Solve $4\sin x = \operatorname{cosec} x$ for $0° \leqslant x \leqslant 360°$. [3]
   (b) Solve $\tan^2 3y - 2\sec 3y - 2 = 0$ for $0° \leqslant y \leqslant 180°$. [6]
   (c) Solve $\tan\left(z - \dfrac{\pi}{3}\right) = \sqrt{3}$ for $0 \leqslant z \leqslant 2\pi$ radians. [3]

Cambridge O Level Additional Mathematics (4037)
Paper 11 Q10, June 2015
Cambridge IGCSE Additional Mathematics (0606)
Paper 11 Q10, June 2015

10 TRIGONOMETRY

2

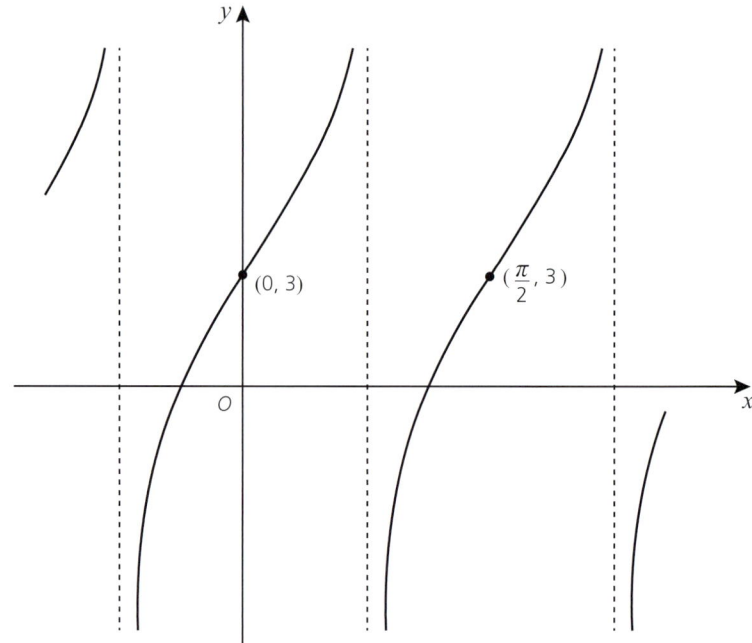

(a) (i) The diagram shows the graph of $y = A + C \tan(Bx)$ passing through the points $(0, 3)$ and $\left(\frac{\pi}{2}, 3\right)$. Find the value of A and of B. [2]

(ii) Given that the point $\left(\frac{\pi}{8}, 7\right)$ also lies on the graph, find the value of C. [1]

(b) Given that $f(x) = 8 - 5\cos 3x$, state the period and the amplitude of f. [2]

Cambridge O Level Additional Mathematics (4037)
Paper 23 Q4, November 2013
Cambridge IGCSE Additional Mathematics (0606)
Paper 23 Q4, November 2013

3 Show that $\dfrac{1}{1-\cos\theta} + \dfrac{1}{1+\cos\theta} = 2\operatorname{cosec}^2\theta$. [3]

Cambridge O Level Additional Mathematics (4037)
Paper 11 Q1, June 2011
Cambridge IGCSE Additional Mathematics (0606)
Paper 11 Q1, June 2011

Identities and equations

Now you should be able to:
- ★ know and use the six trigonometric functions of angles of any magnitude
- ★ understand and use the amplitude and period of a trigonometric function, including the relationship between graphs of related trigonometric functions
- ★ draw and use the graphs of
  - $y = a \sin bx + c$
  - $y = a \cos bx + c$
  - $y = a \tan bx + c$

  where a is a positive integer, b is a simple fraction or integer, and c is an integer
- ★ use the relationships:
  - $\sin^2 A + \cos^2 A = 1$
  - $\sec^2 A = 1 + \tan^2 A$
  - $\text{cosec}^2 A = 1 + \cot^2 A$
  - $\dfrac{\sin A}{\cos A} = \tan A$
  - $\dfrac{\cos A}{\sin A} = \cot A$
- ★ solve, for a given domain, trigonometric equations involving the six trigonometric functions
- ★ prove trigonometric relationships involving the six trigonometric functions.

Key points

✔ In a right-angled triangle

$\sin \theta = \dfrac{\text{opposite}}{\text{hypotenuse}}$

$\cos \theta = \dfrac{\text{adjacent}}{\text{hypotenuse}}$

$\tan \theta = \dfrac{\text{opposite}}{\text{adjacent}}$.

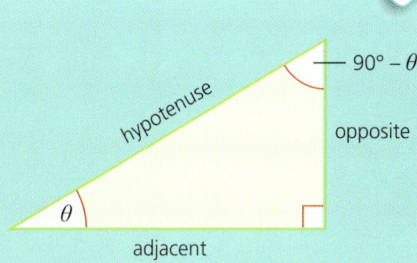

10 TRIGONOMETRY

✔ The graphs of the three main trigonometric functions have distinctive shapes as shown below.

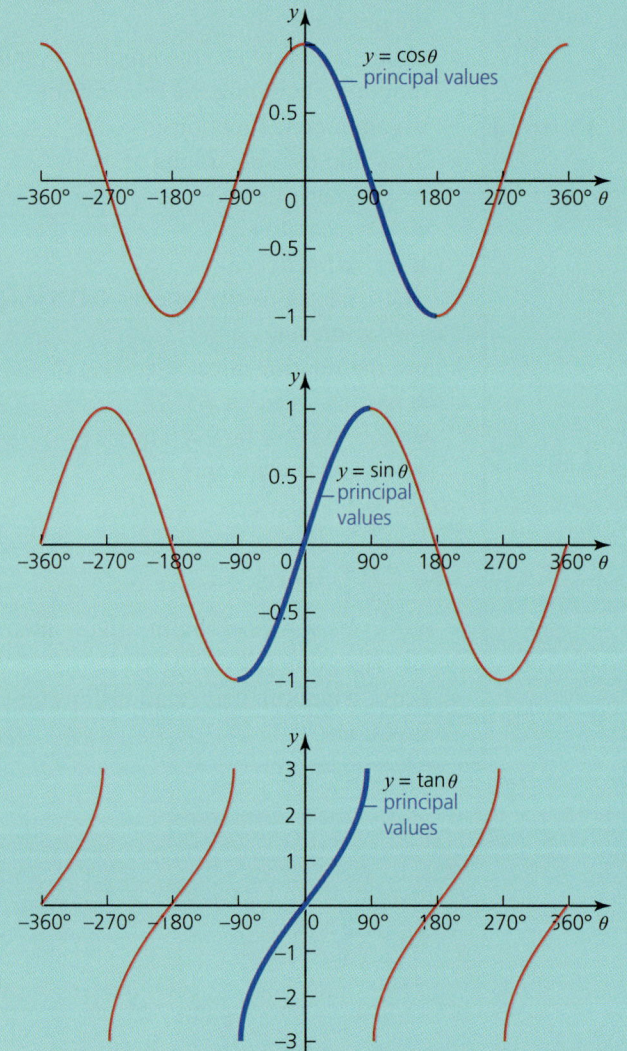

If θ is in radians, the shapes of these curves are exactly the same but the scale on the horizontal axis goes from -2π to 2π instead of from $-360°$ to $360°$.

✔ The reciprocal trigonometric functions are defined as:

$$\operatorname{cosec} \theta = \frac{1}{\sin \theta} \qquad \sec \theta = \frac{1}{\cos \theta} \qquad \cot \theta = \frac{1}{\tan \theta}$$

✔ The **amplitude** of an oscillating graph such as $y = \sin x$ or $y = \cos x$ is the largest displacement from the equilibrium position. For $y = \sin x$ or $y = \cos x$, the equilibrium position is the x-axis.

Identities and equations

- ✔ The **period** of the oscillations is the interval over which the graph does one complete oscillation.
- ✔ The graph of $y = a \sin bx + c$ is a transformation of the graph of $y = \sin x$ by:
  - **i** a stretch of scale factor a in the y-direction
  - **ii** a stretch of scale factor $\frac{1}{b}$ in the x-direction
  - **iii** a translation of c units in the y-direction

  Operation **(i)** must precede **(iii)** but otherwise the order of the transformations can be varied.

  The same rules apply if sin is replaced by cos or tan.
- ✔ The following relationships are referred to as **trigonometric identities.**

  $\sin^2 A + \cos^2 A \equiv 1$

  $\sec^2 A \equiv 1 + \tan^2 A$

  $\operatorname{cosec}^2 A \equiv 1 + \cot^2 A$

  $\dfrac{\sin A}{\cos A} \equiv \tan A$

  $\dfrac{\cos A}{\sin A} \equiv \cot A$

Review exercise 3

Ch 7 1 **Solutions to this question by accurate drawing will not be accepted.**
The points $A(3, 2)$, $B(7, -4)$, $C(2, -3)$ and $D(k, 3)$ are such that CD is perpendicular to AB.
Find the equation of the perpendicular bisector of CD. [6]

Cambridge O Level Additional Mathematics (4037)
Paper 22 Q5, February/March 2019
Cambridge IGCSE Additional Mathematics (0606)
Paper 22 Q5, February/March 2019

Ch 7 2 It is thought that the relationship $y = ax^n$, where a and n are constants, connects the variables x and y. An experiment was carried out recording the values of y for certain values of x.
a) Transform the relationship $y = ax^n$ into straight line form. [2]
The values of $\ln x$ and $\ln y$ were plotted and a line of best fit was drawn. It is given that the line of best fit crosses through the points with coordinates $(1.35, 4.81)$ and $(5.55, 2.29)$.
b) Calculate the constants a and n. [4]

Ch2, 5, 7, 8 3 The diagram shows the circle $x^2 + y^2 - 4x + 4y - 17 = 0$ and the lines l_1, $y = x + 1$, and l_2. The line l_1 intersects the circle at points P and Q and the line l_2 intersects the circle at points $R(5, 2)$ and $S(7, -2)$. The lines intersect at point T.

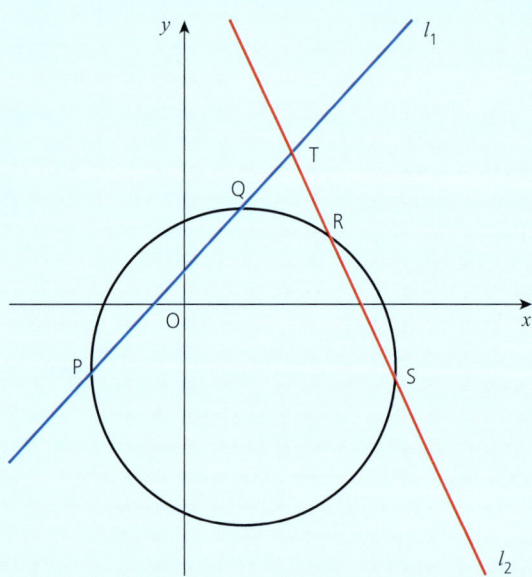

a) Find the coordinates of the point of intersection of l_1 and l_2. [5]
b) Give the coordinates of the points P and Q. [4]
c) Find the area of the triangle PST. [2]

Ch 8 4 Two circles with equations $x^2 + y^2 + 6x - 8y + 9 = 0$ and $x^2 + y^2 - 2x - 15 = 0$ intersect at points A and B.
a) Find the coordinates of the points A and B. [4]
b) State the equation of the line that passes through the points A and B. [1]

Ch 8 5 a) Show that the point $(2, 8)$ lies inside the circle $(x + 1)^2 + (y - 4)^2 = 100$. [2]
b) A second circle has equation $(x - 2)^2 + (y - 8)^2 = 25$. Deduce that the two circles touch at only one point. You are not required to find the point of intersection. [1]

Review exercise 3

Ch 9 6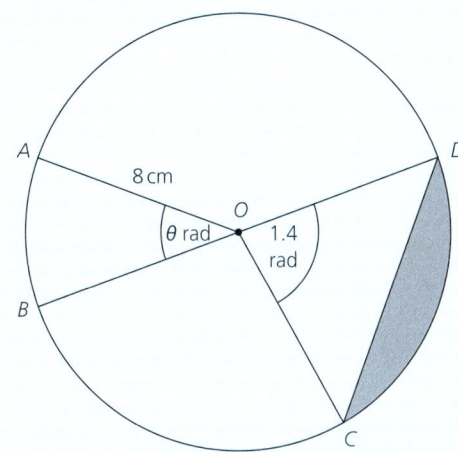
The diagram shows a circle with centre O and radius 8 cm. The points A, B, C and D lie on the circumference of the circle. Angle $AOB = \theta$ radians and angle $COD = 1.4$ radians. The area of sector AOB is 20 cm^2.
i   Find angle θ. [2]
ii Find the length of the arc AB. [2]
iii Find the area of the shaded segment. [3]

Cambridge O Level Additional Mathematics (4037)
Paper 22 Q7, February/March 2018
Cambridge IGCSE Additional Mathematics (0606)
Paper 22 Q7, February/March 2018

Ch 10 7 i The curve $y = a + b \sin cx$ has an amplitude of 4 and a period of $\frac{\pi}{3}$. Given that the curve passes through the point $\left(\frac{\pi}{12}, 2\right)$, find the value of each of the constants a, b and c. [4]

ii Using your values of a, b and c, sketch the graph of $y = a + b \sin cx$ for $0 \leq x \leq \pi$ radians. [3]

Cambridge O Level Additional Mathematics (4037)
Paper 11 Q4, May/June 2018
Cambridge IGCSE Additional Mathematics (0606)
Paper 11 Q4, May/June 2018

Ch 10 8 a) Show that $\dfrac{(1 - \cos A)(1 + \cos A)}{\sin A \cos A} = \tan A$. [2]

b) Hence solve $\dfrac{(1 - \cos 2x)(1 + \cos 2x)}{\sin 2x \cos 2x} = 1$ for $0° \leq x \leq 180°$. [4]

205

11 Permutations and combinations

It always seems impossible until it is done.

Nelson Mandela (1918–2013)

Discussion point

The combination lock has four numbers to be found and six choices for each number: 1, 2, 3, 4, 5 or 6. Suppose you have no idea what the code is, but you need to open the lock. It may seem like an impossible situation initially, but what if you try every possible combination of numbers systematically? How many possible combinations are there? Estimate how long it will take you to open the lock.

Factorials

 Worked example

Winni is tidying her bookshelf and wants to put her five maths books together. In how many different ways can she arrange them?

Solution

There are 5 possible books that can go in the first space on the shelf.

There are 4 possible books for the second space.

Factorials

There are 3 for the third space, 2 for the fourth and only 1 book left for the fifth space.

The total number of arrangements is therefore

$$5 \times 4 \times 3 \times 2 \times 1 = 120$$
Book 1   Book 2   Book 3   Book 4   Book 5

This number, $5 \times 4 \times 3 \times 2 \times 1$, is called **5 factorial** and is written 5!

n must be a positive integer.

This example illustrates a general result. The number of ways of placing n different objects in a line is $n!$, where
$n! = n \times (n-1) \times (n-2) \ldots \times 3 \times 2 \times 1$.
By convention, a special case is made for $n = 0$. The value of 0! is taken to be 1.

➡ Worked example

Find the value of each of the following:

a) 2!   b) 3!   c) 4!   d) 5!   e) 10!

Solution

a) $2! = 2 \times 1 = 2$

b) $3! = 3 \times 2 \times 1 = 6$

You can see that factorials go up very quickly in size.

c) $4! = 4 \times 3 \times 2 \times 1 = 24$

d) $5! = 5 \times 4 \times 3 \times 2 \times 1 = 120$

e) $10! = 10 \times 9 \times 8 \times 7 \times 6 \times 5 \times 4 \times 3 \times 2 \times 1 = 3\,628\,800$

➡ Worked example

a) Calculate $\dfrac{7!}{5!}$

b) Calculate $\dfrac{5! \times 4! \times 3!}{6! \times 2!}$

Solution

a) $7! = 7 \times 6 \times 5 \times 4 \times 3 \times 2 \times 1$ and $5! = 5 \times 4 \times 3 \times 2 \times 1$

So $\dfrac{7!}{5!} = \dfrac{7 \times 6 \times \cancel{5} \times \cancel{4} \times \cancel{3} \times \cancel{2} \times \cancel{1}}{\cancel{5} \times \cancel{4} \times \cancel{3} \times \cancel{2} \times \cancel{1}} = 42$

This result can be generalised as
$\dfrac{n!}{m!} = \dfrac{n \times (n-1) \times (n-2) \times \ldots \times (m+1) \times m!}{m!}$
$= n \times (n-1) \times (n-2) \times \ldots \times (m+1)$

You can also write 7! as $7 \times 6 \times 5!$

n > m

Using this, $\dfrac{7!}{5!} = \dfrac{7 \times 6 \times 5!}{5!} = 7 \times 6 = 42$

b) $\dfrac{5 \times 4 \times 3 \times 2 \times 1 \times 4 \times 3 \times 2 \times 1 \times 3 \times 2 \times 1}{6 \times 5 \times 4 \times 3 \times 2 \times 1 \times 2 \times 1} = 12$

11 PERMUTATIONS AND COMBINATIONS

> **Worked example**

a) Find the number of ways in which all six letters in the word FOURTH can be arranged.

b) In how many of these arrangements are the letters O and U next to each other?

Solution

a) There are six choices for the first letter (F, O, U, R, T, H). Then there are five choices for the next letter, then four for the third letter and so on. So the number of arrangements of the letters is

$6 \times 5 \times 4 \times 3 \times 2 \times 1 = 6! = 720$

b) The O and the U are to be together, so you can treat them as a single letter.

So there are five choices for the first letter (F, OU, R, T or H), four choices for the next letter and so on.

So the number of arrangements of these five 'letters' is

$5 \times 4 \times 3 \times 2 \times 1 = 5! = 120$

However

| OU | F | R | T | H |

is different from

| UO | F | R | T | H |

So each of the 120 arrangements can be arranged into two different orders.

The total number of arrangements with the O and U next to each other is

$2 \times 5! = 240$

Notice that the total number of ways of arranging the letters with the U and the O apart is $720 - 240 = 480$

Exercise 11.1

1 Calculate: a) $7!$ b) $\frac{9!}{7!}$ c) $\frac{4! \times 6!}{7! \times 2!}$

2 Simplify: a) $\frac{n!}{(n+1)!}$ b) $\frac{(n-2)!}{(n-3)!}$

3 Simplify: a) $\frac{(n+2)!}{n!}$ b) $\frac{(n+1)!}{(n-1)!}$

4 Write in factorial notation:

   a) $\frac{9 \times 8 \times 7}{6 \times 5 \times 4}$ b) $\frac{14 \times 15}{5 \times 4 \times 3 \times 2}$ c) $\frac{(n+2)(n+1)n}{4 \times 3 \times 2}$

5 Factorise: a) $6! + 7!$ b) $n! + (n-1)!$

6 Write the number 42 using factorials only.

7 How many different four-letter arrangements can be formed from the letters P, Q, R and S if letters cannot be repeated?

8 How many different ways can seven books be arranged in a row on a shelf?

9 There are five drivers in a motoring rally.
   How many different ways are there for the five drivers to finish?

10 There are five runners in a 60-metre hurdles race, one from each of the nations Japan, South Korea, Cambodia, Malaysia and Thailand.
   How many different finishing orders are there?

11 Toben listens to 15 songs from a playlist. If he selects 'shuffle' so the songs are played in a random order, in how many different orders could the songs be played?

12 How many different arrangements are there of the letters in each word?
   a) ASK
   b) QUESTION
   c) SINGAPORE
   d) GOVERN
   e) VIETNAM
   f) MAJORITY

13 How many arrangements of the letters in the word ARGUMENT are there if:
   a) there are no restrictions on the order of the letters
   b) the first letter is an A
   c) the letters A and R must be next to each other
   d) the letters G and M must not be next to each other.

Permutations

In some situations, such as a race, the finishing order matters. An ordered arrangement of a number of people, objects or operations is called a **permutation**.

Worked example

Joyeeta

I should be one of the judges! When I saw the 10 contestants in the cookery competition, I knew which ones I thought were the best three. Last night they announced the results and I had picked the same three contestants in the same order as the judges!

What is the probability of Joyeeta's result?

Solution

The winner can be chosen in 10 ways.

The second contestant can be chosen in 9 ways.

The third contestant can be chosen in 8 ways.

Thus the total number of ways of placing three contestants in the first three positions is $10 \times 9 \times 8 = 720$. So the probability that Joyeeta's selection is correct is $\frac{1}{720}$.

These examples will help you understand how to use permutations, but probability is not on the syllabus.

In this example attention is given to the order in which the contestants are placed. The solution required a permutation of three objects from ten.

In general the number of permutations, ^nP_r, of r objects from n is given by

$$^nP_r = n \times (n-1) \times (n-2) \times ... \times (n-r+1).$$

This can be written more compactly as

$$^nP_r = \frac{n!}{(n-r)!}$$

11 PERMUTATIONS AND COMBINATIONS

 Worked example

Five people go to the theatre. They sit in a row with eight seats. Find how many ways can this be done if:

a) they can sit anywhere

b) all the empty seats are next to each other.

Solution

a) The first person to sit down has a choice of eight seats.

The second person to sit down has a choice of seven seats.

The third person to sit down has a choice of six seats.

The fourth person to sit down has a choice of five seats.

The fifth person to sit down has a choice of four seats.

So the total number of arrangements is $8 \times 7 \times 6 \times 5 \times 4 = 6720$.

This is a permutation of five objects from eight, so a quicker way to work this out is:

number of arrangements $= {}^8P_5 = 6720$.

b) Since all three empty seats are to be together you can consider them to be a single 'empty seat', albeit a large one!

So there are six seats to seat five people.

So the number of arrangements is ${}^6P_5 = 720$.

Combinations

In other situations, order is not important, for example, choosing five of eight students to go to the theatre. You are not concerned with the order in which people or objects are chosen, only with which ones are picked. A selection where order is not important is called a **combination**.

A maths teacher is playing a game with her students. Each student selects six numbers out of a possible 19 (numbers 1, 2, ..., 19). The maths teacher then uses a random number machine to generate six numbers. If a student's numbers match the teacher's numbers then they win a prize.

Discussion point

You have the six winning numbers. Does it matter in which order the machine picked them?

The teacher says that the probability of an individual student picking the winning numbers is about 1 in 27 000. How can you work out this figure?

The key question is, how many ways are there of choosing six numbers out of 19?

These examples will help you understand how to use combinations, but probability is not on the syllabus.

Combinations

If the order mattered, the answer would be $^{19}P_6$, or $19 \times 18 \times 17 \times 16 \times 15 \times 14$.

However, the order does not matter. The selection 1, 3, 15, 19, 5 and 18 is the same as 15, 19, 1, 5, 3, 18 and as 18, 1, 19, 15, 3, 5, and lots more. For each set of six numbers there are 6! arrangements that all count as being the same.

So, the number of ways of selecting six numbers, given that the order does not matter, is

$$\frac{19 \times 18 \times 17 \times 16 \times 15 \times 14}{6!}.$$ ← This is $\frac{^{19}P_6}{6!}$

This is called the number of **combinations** of 6 objects from 19 and is denoted by $^{19}C_6$.

> **Discussion point**
>
> Show that $^{19}C_6$ can be written as $\frac{19!}{6!\,13!}$.

Returning to the maths teacher's game, it follows that the probability of a student winning is $\frac{1}{^{19}C_6}$. ← *This is about 27000.*

> **Discussion point**
>
> How does the probability change if there are 29, 39 and 49 numbers to choose from?

This example shows a general result, that the number of ways of selecting r objects from n, when the order does not matter, is given by

$$^nC_r = \frac{n!}{r!(n-r)!} = \frac{^nP_r}{r!}$$

> **Discussion point**
>
> How can you prove this general result?

Another common notation for ^nC_r is $\binom{n}{r}$. Both notations are used in this book to help you become familiar with them.

Caution: The notation $\binom{n}{r}$ looks exactly like a column vector and so there is the possibility of confusing the two. However, the context will usually make the meaning clear.

11 PERMUTATIONS AND COMBINATIONS

 Worked example

A student representative committee of five people is to be chosen from nine applicants. How many different selections are possible?

Solution

Number of selections $= \binom{9}{5} = \frac{9!}{5! \times 4!} = \frac{9 \times 8 \times 7 \times 6}{4 \times 3 \times 2 \times 1} = 126$

 Worked example

In how many ways can a committee of five people be selected from five applicants?

Solution

Common sense tells us that there is only one way to make the committee, that is by appointing all applicants. So $^5C_5 = 1$. However, if we work from the formula

$$^5C_5 = \frac{5!}{5!\,0!} = \frac{1}{0!}$$

For this to equal 1 requires the convention that 0! is taken to be 1.

Discussion point

Use the convention $0! = 1$ to show that $^nC_0 = {^nC_n} = 1$ for all values of n.

Exercise 11.2

1 a) Find the values of i) $^7P_3$ ii) $^9P_4$ iii) $^{10}P_8$
   b) Find the values of i) $^7C_3$ ii) $^9C_4$ iii) $^{10}C_8$
   c) Show that, for the values of n and r in parts **a** and **b**, $^nC_r = \frac{^nP_r}{r!}$.

2 There are 15 competitors in a camel race. How many ways are there of guessing the first three finishers?

3 A group of 6 computer programmers is to be chosen to work the night shift from a set of 14 programmers. In how many ways can the programmers be chosen if the 6 chosen must include the shift-leader who is one of the 14?

4 Zaid decides to form a band. He needs a bass player, a guitarist, a keyboard player and a drummer. He invites applications and gets 6 bass players, 8 guitarists, 4 keyboard players and 3 drummers. Assuming each person applies only once, in how many ways can Zaid put the band together?

5 A touring party of cricket players is made up of 6 players from each of India, Pakistan and Sri Lanka and 3 from Bangladesh.
   a) How many different selections of 11 players can be made for a team?
   b) In one match, it is decided to have 3 players from each of India, Pakistan and Sri Lanka and 2 from Bangladesh. How many different team selections can now be made?

6 A committee of four is to be selected from ten candidates, five men and five women.
  a) In how many distinct ways can the committee be chosen?
  b) Assuming that each candidate is equally likely to be selected, determine the probabilities that the chosen committee contains:
   i) no women
   ii) two men and two women.

7 A committee of four is to be selected from four boys and six girls. The members are selected at random.
  a) How many different selections are possible?
  b) What is the probability that the committee will be made up of:
   i) all girls
   ii) more boys than girls?

8 A factory advertises six positions. Nine men and seven women apply.
  a) How many different selections are possible?
  b) How many of these include equal numbers of men and women?
  c) How many of the selections include no men?
  d) How many of the selections include no women?

9 A small business has 14 staff; 6 men and 8 women. The business is struggling and needs to make four members of staff redundant.
  a) How many different selections are possible if the four staff are chosen at random?
  b) How many different selections are possible if equal numbers of men and women are chosen?
  c) How many different selections are possible if there are equal numbers of men and women remaining after the redundancies?

10 A football team consists of a goalkeeper, two defence players, four midfield players and four forwards. Three players are chosen to collect a medal at the closing ceremony of a competition.
How many selections are possible if one midfield player, one defence player and one forward must be chosen?

11 Nimisha is going to install 5 new game apps on her phone. She has shortlisted 2 word games, 5 quizzes and 16 saga games. Nimisha wants to have at least one of each type of game. How many different selections of apps could Nimisha possibly choose?

12 An MPV has seven passenger seats – one in the front, and three in each of the other two rows.

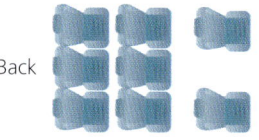

Back

Front

  a) In how many ways can all 8 seats be filled from a party of 12 people, assuming that they can all drive?
  b) In a party of 12 people, 3 are qualified drivers. They hire an MPV and a four-seater saloon car. In how many ways can the party fill the MPV given that one of the drivers must drive each vehicle?

11 PERMUTATIONS AND COMBINATIONS

Exercise 11.2 (cont)

13 Iram has 12 different DVDs of which 7 are films, 3 are music videos and 2 are documentaries.
   a) How many different arrangements of all 12 DVDs on a shelf are possible if the music videos are all next to each other?
   b) Iram makes a selection of 2 films, 2 music videos and 1 documentary. How many possible selections can be made?

14 A string orchestra consists of 15 violins, 8 violas, 7 cellos and 4 double basses. A chamber orchestra consisting of 8 violins, 4 violas, 2 cellos and 2 double basses is to be chosen from the string orchestra.
   a) In how many different ways can the chamber orchestra be chosen?
   b) Once the chamber orchestra is chosen, how many seating arrangements are possible if each instrument group has their own set of chairs?
   c) The violinists work in pairs. How many seating arrangements are possible for the violinists if they must sit with their partner?

15 An office car park has 12 parking spaces in a row. There are 9 cars to be parked.
   a) How many different arrangements are there for parking the 9 cars and leaving 3 empty spaces?
   b) How many different arrangements are there if the 3 empty spaces are next to each other?

Past-paper questions

1 A school council of 6 people is to be chosen from a group of 8 students and 6 teachers. Calculate the number of different ways that the council can be selected if

   (i) there are no restrictions, [2]

   (ii) there must be at least 1 teacher on the council and more students than teachers. [3]

   After the council is chosen, a chairperson and a secretary have to be selected from the 6 council members.

   (iii) Calculate the number of different ways in which a chairperson and a secretary can be selected. [1]

   Cambridge O Level Additional Mathematics (4037)
   Paper 23 Q5, November 2011
   Cambridge IGCSE Additional Mathematics (0606)
   Paper 23 Q5, November 2011

2 (a) (i) Find how many different 4-digit numbers can be formed from the digits 1, 3, 5, 6, 8 and 9 if each digit may be used only once. [1]

   (ii) Find how many of these 4-digit numbers are even. [1]

Combinations

(b) A team of 6 people is to be selected from 8 men and 4 women. Find the number of different teams that can be selected if

  (i) there are no restrictions, [1]

  (ii) the team contains all 4 women, [1]

  (iii) the team contains at least 4 men. [3]

Cambridge O Level Additional Mathematics (4037)
Paper 12 Q7, November 2013
Cambridge IGCSE Additional Mathematics (0606)
Paper 12 Q7, November 2013

3 Arrangements containing 5 different letters from the word AMPLITUDE are to be made. Find

  (a) (i) the number of 5-letter arrangements if there are no restrictions, [1]

   (ii) the number of 5-letter arrangements which start with the letter A and end with the letter E. [1]

Cambridge O Level Additional Mathematics (4037)
Paper 11 Q4 a, June 2012
Cambridge IGCSE Additional Mathematics (0606)
Paper 11 Q4 a, June 2012

Now you should be able to:
★ recognise the difference between permutations and combinations and know when each should be used
★ know and use the notation $n!$ and the expressions for permutations and combinations of n items taken r at a time
★ answer problems on arrangement and selection using permutations or combinations.

Key points

✔ The number of ways of arranging n different objects in a line is $n!$. This is read as n factorial.
✔ $n! = n \times (n-1) \times (n-2) \ldots \times 3 \times 2 \times 1$ where n is a positive integer.
✔ By convention, $0! = 1$.
✔ The number of permutations of r objects from n is $^nP_r = \frac{n!}{(n-r)!}$
✔ The number of combinations of r objects from n is $^nC_r = \frac{n!}{(n-r)!r!}$
✔ The order matters for permutations, but not for combinations.

12 Series

Math has the beauty of poetry, its abstractions are combined with perfect rigor.

Professor Raman Parimala (born 1948)

Discussion point

The origin of the game of chess is uncertain, both in time and place. According to one legend it was invented by Sissa ben Dahir, Vizier to Indian king Shirham. The king asked Sissa ben Dahir what he would like for a reward. This is what he replied:

'I would like one grain of wheat to be put on the first square of my board, two on the second square, four on the third square, eight on the fourth and so on.'

The king agreed without doing any calculations.

Given that one grain of wheat weighs about 50 mg, what mass of wheat would have been placed on the last square?

Definitions and notation

A **sequence** is a set of numbers in a given order, for example

$$\frac{1}{2}, \frac{1}{4}, \frac{1}{8}, \frac{1}{16}, \ldots$$

Each of these numbers is called a **term** of the sequence. When the terms of a sequence are written algebraically, the position of any term in the sequence is usually shown by a subscript, so that a general sequence is written:

$u_1, u_2, u_3, \ldots$, with general term u_k.

For the previous sequence, the first term is $u_1 = \frac{1}{2}$, the second term is $u_2 = \frac{1}{4}$, and so on.

When the terms of a sequence are added together, for example,

$$\frac{1}{2} + \frac{1}{4} + \frac{1}{8} + \frac{1}{16} + \ldots$$

the resulting sum is called a **series**. The process of adding the terms together is called **summation** and indicated by the symbol $\sum$ (the Greek letter sigma), with the position of the first and last terms involved given as **limits**.

So $u_1 + u_2 + u_3 + u_4 + u_5$ is written $\sum_{k=1}^{k=5} u_k$ or $\sum_{k=1}^{5} u_k$.

In cases like this one, where there is no possibility of confusion, the sum is normally written more simply as $\sum_{1}^{5} u_k$.

If all the terms are to be summed, it is usually denoted even more simply as $\sum_{k} u_k$, or even $\sum u_k$.

A sequence may have an infinite number of terms, in which case it is called an **infinite sequence**. The corresponding series is called an **infinite series**.

The phrase 'sum of a sequence' is often used to mean the sum of the terms of a sequence (i.e. the series).

Although the word **series** can describe the sum of the terms of any sequence in mathematics, it is usually used only when summing the sequence provides a useful or interesting overall result.

For example:

$(1 + t)^4 = 1 + 4t + 6t^2 + 4t^3 + t^4$ ← *This series has a finite number of terms (5).*

$\sqrt{11} = \frac{10}{3}\left[1 - \frac{1}{2}(0.01) - \frac{1}{8}(0.01)^2 - \frac{1}{16}(0.01)^3 \ldots\right]$

↑ *This series has an infinite number of terms.*

The binomial theorem

A special type of series is produced when a binomial (i.e. two-part) expression such as $(x + 1)$ is raised to a power. The resulting expression is often called a **binomial expansion**.

12 SERIES

*Expressions like these, consisting of integer powers of x and constants are called **polynomials**.*

The simplest binomial expansion is $(x + 1)$ itself. This and other powers of $(x + 1)$ are given below.

$$
\begin{aligned}
(x+1)^1 &= & & & & & 1x & + & 1 \\
(x+1)^2 &= & & & 1x^2 & + & 2x & + & 1 \\
(x+1)^3 &= & 1x^3 & + & 3x^2 & + & 3x & + & 1 \\
(x+1)^4 &= 1x^4 + 4x^3 + 6x^2 + 4x + 1 \\
(x+1)^5 &= 1x^5 + 5x^4 + 10x^3 + 10x^2 + 5x + 1
\end{aligned}
$$

If you look at the coefficients on the right-hand side you will see that they form a pattern.

(1)

*These numbers are called **binomial coefficients**.*

```
              1         1
           1      2        1
         1     3      3       1
      1     4      6      4      1
   1     5     10     10     5      1
```

This is called **Pascal's triangle**, or the **Chinese triangle**. Each number is obtained by adding the two above it, for example

$$4 + 6$$

gives 10

This pattern of coefficients is very useful when you need to write down the expansions of other binomial expressions. For example,

Notice how in each term the sum of the powers of x and y is the same as the power of $(x + y)$.

$$
\begin{aligned}
(x+y) &= & & 1x + 1y \\
(x+y)^2 &= & & 1x^2 + 2xy + 1y^2 \\
(x+y)^3 &= & 1x^3 + 3x^2y + 3xy^2 + 1y^3
\end{aligned}
$$

This is a binomial expression.

These numbers are binomial coefficients.

➡ Worked example

Write out the binomial expansion of $(a + 3)^5$.

Solution

The binomial coefficients for power 5 are   1 5 10 10 5 1.
In each term, the sum of the powers of a and 3 must equal 5.
So the expansion is:

$$1 \times a^5 + 5 \times a^4 \times 3 + 10 \times a^3 \times 3^2 + 10 \times a^2 \times 3^3 + 5 \times a \times 3^4 + 1 \times 3^5$$

i.e. $a^5 + 15a^4 + 90a^3 + 270a^2 + 405a + 243$.

Worked example

Write out the binomial expansion of $(3x - 2y)^4$.

Solution
The binomial coefficients for power 4 are   1   4   6   4   1.
The expression $(3x - 2y)$ is treated as $(3x + (-2y))$.
So the expansion is

$1 \times (3x)^4 + 4 \times (3x)^3 \times (-2y) + 6 \times (3x)^2 \times (-2y)^2 + 4 \times (3x) \times (-2y)^3 + 1 \times (-2y)^4$

i.e. $81x^4 - 216x^3y + 216x^2y^2 - 96xy^3 + 16y^4$

Pascal's triangle (and the binomial theorem) had actually been discovered by Chinese mathematicians several centuries earlier, and can be found in the works of Yang Hui (around AD1270) and Chu Shi-kie (in AD1303). However, Pascal is remembered for his application of the triangle to elementary probability, and for his study of the relationships between binomial coefficients.

Tables of binomial coefficients

Values of binomial coefficients can be found in books of tables. It can be helpful to use these when the power becomes large, since writing out Pascal's triangle becomes progressively longer and more tedious, row by row. Note that since the numbers are symmetrical about the middle number, tables do not always give the complete row of numbers.

Worked example

Write out the full expansion of $(a + b)^8$.

Solution
The binomial coefficients for the power 8 are

1   8   28   56   70   56   28   8   1

and so the expansion is

$a^8 + 8a^7b + 28a^6b^2 + 56a^5b^3 + 70a^4b^4 + 56a^3b^5 + 28a^2b^6 + 8ab^7 + b^8$.

The formula for a binomial coefficient

You may need to find binomial coefficients that are outside the range of your tables. The tables may, for example, list the binomial coefficients for powers up to 20. What happens if you need to find the coefficient of x^{17} in the expansion of $(x + 2)^{25}$? Clearly you need a formula that gives binomial coefficients.

The first thing you need is a notation for identifying binomial coefficients. It is usual to denote the power of the binomial expression by n, and the position in the row of binomial coefficients by r, where r can take any value from 0 to n. So, for row 5 of Pascal's triangle

$n = 5$:   1   5   10   10   5   1
         $r=0$ $r=1$ $r=2$ $r=3$ $r=4$ $r=5$

The general binomial coefficient corresponding to values of n and r is written as $\binom{n}{r}$. An alternative notation is ^nC_r, which is said as 'N C R'.

Thus $\binom{5}{3} = {}^5C_3 = 10$.

The next step is to find a formula for the general binomial coefficient $\binom{n}{r}$.

Real-world activity

The table shows an alternative way of laying out Pascal's triangle.

		Column (r)							...	r
		0	1	2	3	4	5	6	...	r
Row (n)	1	1	1							
	2	1	2	1						
	3	1	3	3	1					
	4	1	4	6	4	1				
	5	1	5	10	10	5	1			
	6	1	6	15	20	15	6	1		
	...	...	...	...	...	...	...	...	...	
	...	...	...	...	...	...	...	...	...	...
	n	1	n	?	?	?	?	?	?	?

Note that 0! is defined to be 1. You will see the need for this when you use the formula for $\binom{n}{r}$.

Show that $\binom{n}{r} = \dfrac{n!}{r!(n-r)!}$, by following the procedure below.

The numbers in column 0 are all 1.

To find each number in column 1 you multiply the 1 in column 0 by the row number, n.

1. Find, in terms of n, what you must multiply each number in column 1 by to find the corresponding number in column 2.
2. Repeat the process to find the relationship between each number in column 2 and the corresponding number in column 3.
3. Show that repeating the process leads to
$$\binom{n}{r} = \frac{n(n-1)(n-2)\ldots(n-r+1)}{1 \times 2 \times 3 \times \ldots \times r} \text{ for } r \geq 1.$$
4. Show that this can also be written as
$$\binom{n}{r} = \frac{n!}{r!(n-r)!}$$
and that it is also true for $r = 0$.

The binomial theorem

→ Worked example

Use the formula $\binom{n}{r} = \dfrac{n!}{r!(n-r)!}$ to calculate these.

a) $\binom{7}{0}$ b) $\binom{7}{1}$ c) $\binom{7}{2}$ d) $\binom{7}{3}$

e) $\binom{7}{4}$ f) $\binom{7}{5}$ g) $\binom{7}{6}$ h) $\binom{7}{7}$

Solution

a) $\binom{7}{0} = \dfrac{7!}{0!(7-0)!} = \dfrac{5040}{1 \times 5040} = 1$

b) $\binom{7}{1} = \dfrac{7!}{1!6!} = \dfrac{5040}{1 \times 720} = 7$

c) $\binom{7}{2} = \dfrac{7!}{2!5!} = \dfrac{5040}{2 \times 120} = 21$

d) $\binom{7}{3} = \dfrac{7!}{3!4!} = \dfrac{5040}{6 \times 24} = 35$

e) $\binom{7}{4} = \dfrac{7!}{4!3!} = \dfrac{5040}{24 \times 6} = 35$

f) $\binom{7}{5} = \dfrac{7!}{5!2!} = \dfrac{5040}{120 \times 2} = 21$

g) $\binom{7}{6} = \dfrac{7!}{6!1!} = \dfrac{5040}{720 \times 1} = 7$

h) $\binom{7}{7} = \dfrac{7!}{7!0!} = \dfrac{5040}{5040 \times 1} = 1$

Note

Most scientific calculators have factorial buttons, e.g. $\boxed{x!}$. Many also have $\boxed{^nC_r}$ buttons. Find out how best to use your calculator to find binomial coefficients, as well as practising non-calculator methods.

→ Worked example

Find the coefficient of x^{19} in the expansion of $(x + 3)^{25}$.

Solution

$(x+3)^{25} = \binom{25}{0}x^{25} + \binom{25}{1}x^{24}3^1 + \binom{25}{2}x^{23}3^2 + \ldots + \binom{25}{6}x^{19}3^6 + \ldots \binom{25}{25}3^{25}$

So the required term is $\binom{25}{6} \times x^{19} 3^6$

$\binom{25}{6} = \dfrac{25!}{6!19!} = \dfrac{25 \times 24 \times 23 \times 22 \times 21 \times 20 \times \cancel{19!}}{6! \times \cancel{19!}}$

$= 177\,100$

So the coefficient of x^{19} is $177\,100 \times 3^6 = 129\,105\,900$.

Notice how 19! was cancelled in working out $\binom{25}{6}$. Factorials become large numbers very quickly and you should keep a look-out for such opportunities to simplify calculations.

12 SERIES

> **Worked example**

Find the value of the term that is independent of x in the expansion of $\left(3x + \dfrac{1}{x}\right)^8$.

Solution

$$\left(3x + \frac{1}{x}\right)^8 = \binom{8}{0}(3x)^8 + \binom{8}{1}(3x)^7\left(\frac{1}{x}\right)^1 + \ldots + \binom{8}{4}(3x)^4\left(\frac{1}{x}\right)^4 + \ldots$$

The required term is $\binom{8}{4}(3x)^4\left(\dfrac{1}{x}\right)^4$.

Notice that the powers are the same; the x terms will cancel out when the term is simplified.

$$\binom{8}{4} = \frac{8!}{4!4!} = \frac{8 \times 7 \times 6 \times 5 \times 4!}{4 \times 3 \times 2 \times 1 \times 4!} = 70$$

$$\Rightarrow \binom{8}{4}(3x)^4\left(\frac{1}{x}\right)^4 = 70 \times 81x^4 \times \frac{1}{x^4} = 5670$$

So the term that is independent of x is 5670.

The expansion of $(1 + x)^n$

When deriving the result for $\binom{n}{r}$ you found the binomial coefficients in the form

$$1 \quad n \quad \frac{n(n-1)}{2!} \quad \frac{n(n-1)(n-2)}{3!} \quad \frac{n(n-1)(n-2)(n-3)}{4!} \quad \ldots$$

This form is commonly used in the expansion of expressions of the type $(1 + x)^n$.

The first few terms →
$$(1+x)^n = 1 + nx + \frac{n(n-1)x^2}{1 \times 2} + \frac{n(n-1)(n-2)x^3}{1 \times 2 \times 3} + \frac{n(n-1)(n-2)(n-3)x^4}{1 \times 2 \times 3 \times 4} + \ldots$$

The last few terms → $+ \dfrac{n(n-1)}{1 \times 2}x^{n-2} + nx^{n-1} + 1x^n$

> **Worked example**

Use the binomial expansion to write down the first four terms, in ascending powers of x, of $(1 + x)^8$.

Solution

$$(1+x)^8 = 1 + 8x + \frac{8 \times 7}{1 \times 2}x^2 + \frac{8 \times 7 \times 6}{1 \times 2 \times 3}x^3 + \ldots$$

The power of x is the same as the largest number underneath.

Two numbers on top, two underneath.

Three numbers on top, three underneath.

$$= 1 + 8x + 28x^2 + 56x^3 + \ldots$$

The binomial theorem

An expression like $1 + 8x + 28x^2 + 56x^3 \ldots$ is said to be in **ascending** powers of x, because the powers of x are increasing from one term to the next.

An expression like $x^8 + 8x^7 + 28x^6 + 56x^5 \ldots$ is in **descending** powers of x, because the powers of x are decreasing from one term to the next.

→ Worked example

Use the binomial expansion to write down the first four terms, in ascending powers of x, of $(1 - 2x)^6$. Simplify the terms.

Solution

Think of $(1 - 2x)^6$ as $(1 + (-2x))^6$. Keep the brackets while you write out the terms.

$$(1+(-2x))^6 = 1 + 6(-2x) + \frac{6 \times 5}{1 \times 2}(-2x)^2 + \frac{6 \times 5 \times 4}{1 \times 2 \times 3}(-2x)^3 + \ldots$$
$$= 1 - 12x + 60x^2 - 160x^3 + \ldots$$

← Notice how the signs alternate.

Exercise 12.1

1. Write out the following binomial expressions:
   a) $(1+x)^4$
   b) $(1+2x)^4$
   c) $(1+3x)^4$

2. Write out the following binomial expressions:
   a) $(2+x)^4$
   b) $(3+x)^4$
   c) $(4+x)^4$

3. Write out the following binomial expressions:
   a) $(x+y)^4$
   b) $(x+2y)^4$
   c) $(x+3y)^4$

4. Use a non-calculator method to calculate the following binomial coefficients. Check your answers using your calculator's shortest method.
   a) $\binom{5}{3}$
   b) $\binom{7}{2}$
   c) $\binom{7}{4}$
   d) $\binom{7}{5}$
   e) $\binom{5}{0}$
   f) $\binom{13}{3}$

5. Find the coefficients of the term shown for each expansion:
   a) x^4 in $(1+x)^6$
   b) x^5 in $(1+x)^7$
   c) x^6 in $(1+x)^8$

6. Find the first three terms, in ascending powers of x, in the expansion of $(3+kx)^5$.

7. Find the first three terms, in descending powers of x, in the expansion of $\left(3x - \frac{3}{x}\right)^6$.

8. a) Simplify $(1+t)^3 - (1-t)^3$.
   b) Show that $x^3 - y^3 = (x-y)(x^2 + xy + y^2)$.
   c) Substitute $x = 1+t$ and $y = 1-t$ in the result in part **b** and show that your answer is the same as that for part **a**.

12 SERIES

Exercise 12.1 (cont)

9 Find the coefficients of x^3 and x^4 for each of the following:
  a) $(1+x)(1-x)^6$
  b) $(1-x)(1+x)^6$

10 Write down the first four terms, in ascending powers of x, of the following binomial expressions:
  a) $(1-2x)^6$
  b) $(2-3x)^6$
  c) $(3-4x)^6$

11 Find the first four terms, in descending powers of x, of the following binomial expressions:
  a) $\left(x^2 + \frac{1}{x}\right)^5$
  b) $\left(x^2 - \frac{1}{x}\right)^5$
  c) $\left(x^3 + \frac{1}{x}\right)^5$
  d) $\left(x^3 - \frac{1}{x}\right)^5$

12 Find the coefficients of the term shown for each expansion:
  a) x^6 in $\left(2x + \frac{1}{x}\right)^{10}$
  b) x^3 in $\left(x^2 + \frac{1}{x}\right)^{12}$

13 Find the term that is independent of x in the following expansions:
  a) $\left(3x + \frac{1}{x}\right)^{14}$
  b) $\left(5x - \frac{2}{x}\right)^{10}$

14 The first three terms in the expansion of $(2 - ax)^n$ in ascending powers of x are 32, −240 and 720. Find the values of a and n.

Arithmetic progressions

The smallest square shape in this toy has sides 1 cm long, and the lengths of the sides increase in steps of 1 cm.

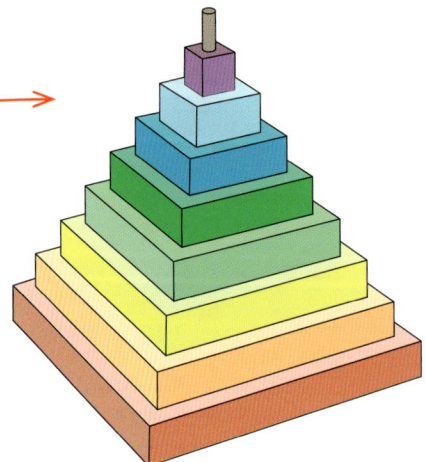

Any ordered set of numbers, like the areas of the squares in this toy, form a sequence. In mathematics, we are particularly interested in sequences with a well-defined pattern, often in the form of an algebraic formula linking the terms. The area of the squares in the toy, in cm², are $1^2, 2^2, 3^2, 4^2, \ldots$ or 1, 4, 9, 16....

A sequence in which the terms increase by the addition of a fixed amount (or decrease by the subtraction of a fixed amount) is described as an **arithmetic sequence** or **arithmetic progression (A. P.)**. The increase from one term to the next is called the **common difference**.

Arithmetic progressions

Thus the sequence 8 11 14 17... is arithmetic with
common difference 3. This sequence can be written algebraically as

$$u_k = 5 + 3k \text{ for } k = 1, 2, 3, \ldots$$

When $k = 1, u_1 = 5 + 3 = 8$
$k = 2, u_2 = 5 + 6 = 11$
$k = 3, u_3 = 5 + 9 = 14$ and so on.

This version has the advantage that the right-hand side begins with the first term of the sequence.

(You can also write this as $u_k = 8 + 3(k-1)$ for $k = 1, 2, 3, \ldots$.)

As successive terms of an arithmetic progression increase (or decrease) by a fixed amount called the common difference, d, you can define each term in the sequence in relation to the previous term:

$$u_{k+1} = u_k + d.$$

When the terms of an arithmetic progression are added together, the sum is called an **arithmetic series**.

Notation

The following conventions are used in this book to describe arithmetic progressions and sequences:
» first term, $u_1 = a$
» number of terms $= n$
» last term, $u_n = l$
» common difference $= d$
» the general term, u_k, is that in position k (i.e. the kth term).

Thus in the arithmetic progression 7, 9, 11, 13, 15, 17, 19

$$a = 7, l = 19, d = 2 \text{ and } n = 7.$$

The terms are formed as follows:

$u_1 = a \qquad\qquad = 7$
$u_2 = a + d \qquad = 7 + 2 \qquad = 9$
$u_3 = a + 2d \qquad = 7 + 2 \times 2 = 11$
$u_4 = a + 3d \qquad = 7 + 3 \times 2 = 13$
$u_5 = a + 4d \qquad = 7 + 4 \times 2 = 15$
$u_6 = a + 5d \qquad = 7 + 5 \times 2 = 17$
$u_7 = a + 6d \qquad = 7 + 6 \times 2 = 19$

The 7th term is the 1st term (7) plus six times the common difference (2).

This shows that any term is given by the first term plus a number of differences. The number of differences is, in each case, one less than the number of the term. You can express this mathematically as

$$u_k = a + (k-1)d.$$

For the last term, this becomes

$$l = a + (n-1)d.$$

These are both general formulae so apply to any arithmetic progression.

12 SERIES

Worked example

Find the 19th term in the arithmetic progression 20, 16, 12, …

Solution

In this case $a = 20$ and $d = -4$.

Using $u_k = a + (k - 1)d$, you obtain

$$u_{19} = 20 + (19 - 1) \times (-4)$$
$$= 20 - 72$$
$$= -52.$$

The 19th term is −52.

> **Note**
>
> The relationship $l = a + (n - 1)d$ may be rearranged to give
>
> $$n = \frac{l - a}{d} + 1$$
>
> This gives the number of terms in an A.P. directly if you know the first term, the last term and the common difference.

Worked example

How many terms are there in the sequence 12, 16, 20, …, 556?

Solution

This is an arithmetic sequence with first term $a = 12$, last term $l = 556$ and common difference $d = 4$.

Using the result $l = a + (n - 1)d$, you have

$$556 = 12 + 4(n - 1)$$
$$\Rightarrow \quad 4n = 556 - 12 + 4$$
$$\Rightarrow \quad n = 137$$

There are 137 terms.

The sum of the terms of an arithmetic progression

When Carl Friederich Gauss (1777–1855) was at school he was always quick to answer mathematics questions. One day his teacher, hoping for half an hour of peace and quiet, told his class to add up all the whole numbers from 1 to 100. Almost at once the 10-year-old Gauss announced that he had done it and that the answer was 5050.

Gauss had not of course added the terms one by one. Instead he wrote the series down twice, once in the given order and once backwards, and added the two together:

$$S = 1 + 2 + 3 + \ldots + 98 + 99 + 100$$
$$S = 100 + 99 + 98 + \ldots + 3 + 2 + 1.$$

Arithmetic progressions

Adding, $2S = 101 + 101 + 101 + \ldots + 101 + 101 + 101$.

Since there are 100 terms in the series,

$$2S = 101 \times 100$$
$$S = 5050.$$

The numbers 1, 2, 3, ... , 100 form an arithmetic sequence with common difference 1. Gauss' method can be used for finding the sum of any arithmetic series.

It is common to use the letter S to denote the sum of a series. When there is any doubt as to the number of terms that are being summed, this is indicated by a subscript: S_5 indicates five terms, S_n indicates n terms.

→ Worked example

Find the value of $6 + 4 + 2 + \ldots + (-32)$.

Solution

This is an arithmetic progression, with common difference -2. The number of terms, n, can be calculated using

$$n = \frac{l-a}{d} + 1$$

$$n = \frac{-32-6}{-2} + 1$$

$$= 20$$

The sum S of the progression is then found as follows:

$$S = 6 + 4 + \ldots - 30 - 32$$
$$\underline{S = -32 + (-30) - \ldots + 4 + 6}$$
$$2S = -26 + (-26) + \ldots + (-26) + (-26).$$

Since there are 20 terms, this gives $2S = -26 \times 20$, so $S = -26 \times 10 = -260$.

Generalising this method by writing the series in the conventional notation gives:

$$S_n = [a] + [a+d] + \ldots + [a+(n-2)d] + [a+(n-1)d]$$
$$\underline{S_n = [a+(n-1)d] + [a+(n-2)d] + \ldots + [a+d] + [a]}$$
$$2S_n = [2a+(n-1)d] + [2a+(n-1)d] + \ldots + [2a+(n-1)d] + [2a+(n-1)d]$$

12 SERIES

Since there are n terms, it follows that

$$S_n = \tfrac{1}{2}n[2a+(n-1)d].$$

This result can also be written as

$$S_n = \tfrac{1}{2}n(a+l).$$

→ Worked example

Find the sum of the first 100 terms of the progression

$$3\tfrac{1}{3},\ 3\tfrac{2}{3},\ 4,\ \ldots$$

Solution

In this arithmetic progression

$$a = 3\tfrac{1}{3},\ d = \tfrac{1}{3} \text{ and } n = 100.$$

$$S_n = \tfrac{1}{2} \times 100\left(6\tfrac{2}{3} + 99 \times \tfrac{1}{3}\right)$$

$$= 1983\tfrac{1}{3}$$

Using
$$S_n = \tfrac{1}{2}n[2a+(n-1)d]$$

→ Worked example

Tatjana starts a part-time job on a salary of $10 000 per year, and this increases by $500 each year. Assuming that, apart from the annual increment, Tatjana's salary does not increase, find

a) her salary in the 5th year

b) the length of time she has been working to receive total earnings of $122 500.

Solution

Tatjana's annual salaries (in dollars) form the arithmetic sequence

10 000, 10 500, 11 000, ...

with first term $a = 10 000$, and common difference $d = 500$.

a) Her salary in the 5th year is calculated using:

$$u_k = a + (k-1)d$$

 $u_5 = 10 000 + (5-1) \times 500$

$$= 12 000$$

Arithmetic progressions

b) The number of years that have elapsed when her total earnings are $122 500 is given by:

$$S = \frac{1}{2}n[2a + (n-1)d]$$

where $S = 122\,500$, $a = 10\,000$ and $d = 500$.

This gives $122\,500 = \frac{1}{2}n[2 \times 10\,000 + 500(n-1)]$.

This simplifies to the quadratic equation:

$$n^2 + 39n - 490 = 0.$$

Factorising,

$$(n - 10)(n + 49) = 0$$
$$\Rightarrow n = 10 \text{ or } n = -49.$$

The root $n = -49$ is irrelevant, so the answer is $n = 10$.

Tatjana has earned a total of $122 500 after 10 years.

Exercise 12.2

1. Are the following sequences arithmetic?
   If so, state the common difference and the seventh term.
   a) 28, 30, 32, 34, …
   b) 1, 1, 2, 3, 5, 8, …
   c) 3, 9, 27, 81, …
   d) 5, 9, 13, 17, …
   e) 12, 8, 4, 0, …

2. The first term of an arithmetic sequence is −7 and the common difference is 4.
   a) Find the eighth term of the sequence.
   b) The last term of the sequence is 65. How many terms are there in the sequence?

3. The first term of an arithmetic sequence is 10, the seventh term is 46 and the last term is 100.
   a) Find the common difference.
   b) Find how many terms there are in the sequence.

4. There are 30 terms in an arithmetic progression.
   The first term is −4 and the last term is 141.
   a) Find the common difference.
   b) Find the sum of the terms in the progression.

5. The kth term of an arithmetic progression is given by
   $u_k = 12 + 4k$.
   a) Write down the first three terms of the progression.
   b) Calculate the sum of the first 12 terms of this progression.

6. Below is an arithmetic progression.
   $118 + 112 + \ldots + 34$
   a) How many terms are there in the progression?
   b) What is the sum of the terms in the progression?

7. The fifth term of an arithmetic progression is 32 and the tenth term is 62.
   a) Find the first term and the common difference.
   b) The sum of all the terms in this progression is 350. How many terms are there?

12 SERIES

Exercise 12.2 (cont)

8 The ninth term of an arithmetic progression is three times the second term, and the first term is 5. The sequence has 20 terms.
   a) Find the common difference.
   b) Find the sum of all the terms in the progression.
9 a) Find the sum of all the odd numbers between 150 and 250.
   b) Find the sum of all the even numbers from 150 to 250 inclusive.
   c) Find the sum of the terms of the arithmetic sequence with first term 150, common difference 1 and 101 terms.
   d) Explain the relationship between your answers to parts **a**, **b** and **c**.
10 The first term of an arithmetic progression is 9000 and the tenth term is 3600.
   a) Find the sum of the first 20 terms of the progression.
   b) After how many terms does the sum of the progression become negative?
11 An arithmetic progression has first term −2 and common difference 7.
   a) Write down a formula for the nth term of the progression. Which term of the progression equals 110?
   b) Write down a formula for the sum of the first n terms of the progression. How many terms of the progression are required to give a sum equal to 2050?
12 Luca's starting salary in a company is $45 000. During the time he stays with the company, it increases by $1800 each year.
   a) What is his salary in his sixth year?
   b) How many years has Luca been working for the company when his total earnings for all his years there are $531 000?
13 A jogger is training for a 5 km charity run. He starts with a run of 400 m, then increases the distance he runs in training by 100 m each day.
   a) How many days does it take the jogger to reach a distance of 5 km in training?
   b) What total distance will he have run in training by then?
14 A piece of string 20 m long is to be cut into pieces such that the lengths of the pieces form an arithmetic sequence.
   a) If the lengths of the longest and shortest pieces are 2 m and 50 cm respectively, how many pieces are there?
   b) The string is cut into 20 pieces. If the length of the longest piece is 185 cm, how long is the shortest piece?
15 The ninth term of an arithmetic progression is 95 and the sum of the first four terms is −10.
   a) Find the first term of the progression and the common difference.
   The nth term of the progression is 200.
   b) Find the value of n.
16 Following knee surgery, Adankwo has to do squats as part of her physiotherapy programme. Each day she must do 4 more squats than the day before. On the eighth day she did 31 squats. Calculate how many squats Adankwo completed:
   a) on the first day
   b) in total by the end of the seventh day
   c) in total by the end of the nth day
   d) in total from the end of the nth day to the end of the $(2n)$th day. Simplify your answer.

Geometric progressions

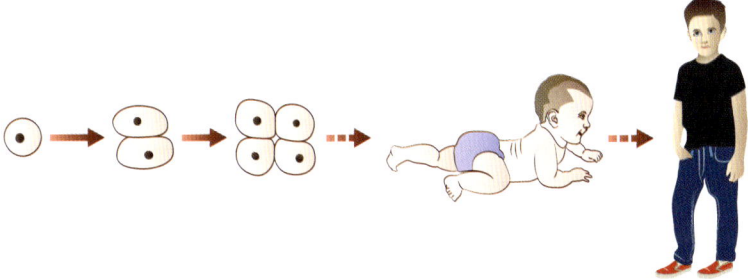

A human being begins life as one cell, which divides into two, then four…

The terms of a **geometric sequence** or **geometric progression (G.P.)** are formed by multiplying one term by a fixed number, the **common ratio**, to obtain the next. This can be written inductively as:

$u_{k+1} = ru_k$ with first term u_1.

The sum of the terms of a geometric sequence is called a **geometric series**.

Notation

The following conventions are used in this book to describe geometric progressions:
- first term $u_1 = a$
- common ratio $= r$
- number of terms $= n$
- the general term, u_k, is that in position k (i.e. the kth term).

Thus in the geometric progression 2, 6, 18, 54, 162

$a = 2, r = 3$ and $n = 5$.

The terms of this sequence are formed as follows:

$$u_1 = a = 2$$
$$u_2 = a \times r = 2 \times 3 = 6$$
$$u_3 = a \times r^2 = 2 \times 3^2 = 18$$
$$u_4 = a \times r^3 = 2 \times 3^3 = 54$$
$$u_5 = a \times r^4 = 2 \times 3^4 = 162.$$

This shows that in each case the power of r is one less than the number of the term: $u_5 = ar^4$ and 4 is one less than 5. This can be written deductively as

$u_k = ar^{k-1}$.

For the last term this becomes

$u_n = ar^{n-1}$.

These are both general formulae so apply to any geometric sequence.

12 SERIES

Given two consecutive terms of a geometric sequence, you can always find the common ratio by dividing the later term by the earlier term. For example, the geometric sequence ... 7, 9, ... has common ratio $r = \frac{9}{7}$.

➡ Worked example

Find the ninth term in the geometric sequence 7, 28, 112, 448, ...

Solution

In the sequence, the first term $a = 7$ and the common ratio $r = 4$.

Using $u_k = ar^{k-1}$

$$u_9 = 7 \times 4^8$$
$$= 458\,752$$

➡ Worked example

How many terms are there in the geometric sequence 3, 15, 75, ... , 29 296 875?

Solution

Since it is a geometric sequence and the first two terms are 3 and 15, you can immediately write down

First term: $\quad a = 3$

Common ratio: $\quad r = 5$

The third term allows you to check you are right.

$15 \times 5 = 75 \quad \checkmark$

The nth term of a geometric sequence is ar^{n-1}, so in this case

$$3 \times 5^{n-1} = 29\,296\,875$$

Dividing by 3 gives

$$5^{n-1} = 9\,765\,625$$

Using logarithms, $\quad \lg(5^{n-1}) = \lg 9\,765\,625$

$\Rightarrow \quad (n-1)\lg 5 = \lg 9\,765\,625$

$\Rightarrow \quad n - 1 = \dfrac{\lg 9\,765\,625}{\lg 5} = 10$

So $n = 11$ and there are 11 terms in the sequence.

Alternatively, you could find the solution by using trial and improvement and a calculator, since you know n must be a whole number.

Discussion point

How would you use a spreadsheet to solve the equation $5^{n-1} = 9\,765\,625$?

The sum of the terms of a geometric progression

This chapter began with the story of Sissa ben Dahir's reward for inventing chess. In the discussion point you were asked how much grain would have been placed on the last square. This situation also gives rise to another question:

How many grains of wheat was the inventor actually asking for?

The answer is the geometric series with 64 terms and common ratio 2:

$$1 + 2 + 4 + 8 + \ldots + 2^{63}.$$

This can be summed as follows.

Call the series S:

$$S = 1 + 2 + 4 + 8 + \ldots + 2^{63}. \quad \text{①}$$

Now multiply it by the common ratio, 2:

$$2S = 2 + 4 + 8 + 16 + \ldots + 2^{64}. \quad \text{②}$$

Then subtract ① from ②:

② $2S = 2 + 4 + 8 + 16 + \ldots + 2^{63} + 2^{64}$

① $S = 1 + 2 + 4 + 8 + \ldots + 2^{63}$

Subtracting: $S = -1 + 0 + 0 + 0 + \ldots + 2^{64}.$

The total number of wheat grains requested was therefore $2^{64} - 1$ (which is about 1.84×10^{19}).

> **Note**
>
> The method shown here can be used to sum any geometric progression.

Discussion point

How many tonnes of wheat is this, and how many tonnes would you expect there to be in China at any time?

(One hundred grains of wheat weigh about 2 grams. The world annual production of all cereals is about 1.8×10^9 tonnes.)

➡ Worked example

Find the sum of $0.04 + 0.2 + 1 + \ldots + 78\,125$.

Solution

This is a geometric progression with common ratio 5.

Let $S = 0.04 + 0.2 + 1 + \ldots + 78\,125.$ ①

Multiplying by the common ratio, 5, gives:

$$5S = 0.2 + 1 + 5 + \ldots + 78\,125 + 390\,625. \quad \text{②}$$

12 SERIES

Subtracting ① from ②:

$$5S = 0.2 + 1 + 5 + \ldots + 78\,125 + 390\,625$$
$$S = 0.04 + 0.2 + 1 + 5 + \ldots + 78\,125 $$
$$4S = -0.04 + 0 + \ldots + 0 + 390\,625$$

This gives $\quad 4S = 390\,624.96$

$\Rightarrow \quad S = 97\,656.24$

The same method can be applied to the general geometric progression to give a formula for its value:

$$S_n = a + ar + ar^2 + \ldots + ar^{n-1}. \qquad ①$$

Multiplying by the common ratio, r, gives:

$$rS_n = ar + ar^2 + ar^3 + \ldots + ar^n. \qquad ②$$

Subtracting ① from ②, as before, gives:

$$rS_n - S_n = ar^n - a$$
$$S_n(r-1) = a(r^n - 1)$$

so $\quad S_n = \dfrac{a(r^n - 1)}{(r - 1)}.$

This can also be written as:

$$S_n = \dfrac{a(1 - r^n)}{(1 - r)}.$$

→ Worked example

a) Solve the simultaneous equations $ar^2 = 6$
$ar^4 = 54$

b) Find in each case the sum of the first five terms of the geometric progression.

Solution

a) $ar^2 = 6 \Rightarrow a = \dfrac{6}{r^2}$

Substituting into $ar^4 = 54$ gives $\dfrac{6}{r^2} \times r^4 = 54$

$\Rightarrow r^2 = 9$
$\Rightarrow r = \pm 3$

Substituting in $ar^2 = 6$ gives $a = \dfrac{2}{3}$ in both cases.

b) When $r = +3$ terms are $\dfrac{2}{3}$, 2, 6, 18, 54 $\quad$ Sum $= 80\dfrac{2}{3}$

When $r = -3$ terms are $\dfrac{2}{3}$, -2, 6, -18, 54 $\quad$ Sum $= 40\dfrac{2}{3}$

Infinite geometric progressions

The progression $1 + \frac{1}{2} + \frac{1}{4} + \frac{1}{8} + \frac{1}{16} + \ldots$ is geometric, with common ratio $\frac{1}{2}$.

Summing the terms one by one gives $1, 1\frac{1}{2}, 1\frac{3}{4}, 1\frac{7}{8}, 1\frac{15}{16} \ldots$

Clearly the more terms you add, the nearer the sum gets to 2. In the limit, as the number of terms tends to infinity, the sum tends to 2.

$$\text{As } n \to \infty, S_n \to 2.$$

This is an example of a **convergent** series. The sum to infinity is a finite number.

You can see this by substituting $a = 1$ and $r = \frac{1}{2}$ in the formula for the sum of the series:

$$S_n = \frac{a(1-r^n)}{1-r}$$

giving
$$S_n = \frac{1 \times \left(1 - \left(\frac{1}{2}\right)^n\right)}{\left(1 - \frac{1}{2}\right)}$$

$$= 2 \times \left(1 - \left(\frac{1}{2}\right)^n\right).$$

The larger the number of terms, n, the smaller $\left(\frac{1}{2}\right)^n$ becomes and so the nearer S_n is to the limiting value of 2, as shown on the left. Notice that $\left(\frac{1}{2}\right)^n$ can never be negative, however large n becomes; so S_n can never exceed 2.

Notice how representing all of the terms of the geometric progression as in these diagrams shows that the sum can never exceed 2.

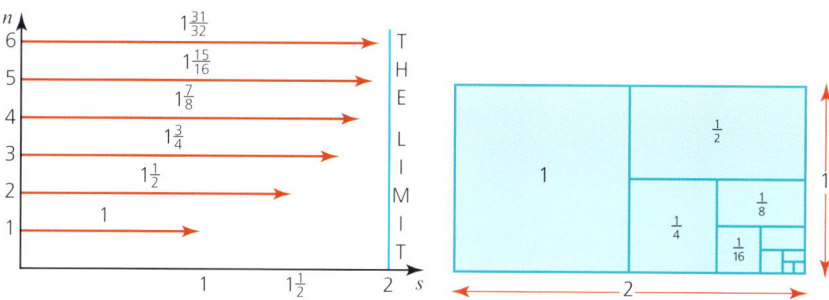

In the general geometric series $a + ar + ar^2 + \ldots$ the terms become progressively smaller in size if the common ratio r is between -1 and 1. In such cases, the geometric series is **convergent**.

If, on the other hand, the value of r is greater than 1 (or less than -1), the terms in the series become larger and larger in size and so the series is described as **divergent**.

12 SERIES

A series corresponding to a value of r of exactly +1 consists of the first term a repeated over and over again. A sequence corresponding to a value of r of exactly −1 oscillates between $+a$ and $-a$. Neither of these is convergent.

It only makes sense to talk about the sum of an infinite series if it is convergent. Otherwise the sum is undefined.

The condition for a geometric series to converge, $-1 < r < 1$, ensures that as $n \to \infty$, $r^n \to 0$, and so the formula for the sum of a geometric series:

$$S_n = \frac{a(1-r^n)}{(1-r)}$$

can be rewritten for an infinite series as:

$$S_\infty = \frac{a}{1-r}.$$

Note

You may have noticed that the sum of the series $0.4 + 0.04 + 0.004 + \ldots$ is $0.\dot{4}$, and that this recurring decimal is the same as $\frac{4}{9}$.

➡ Worked example

Find the sum of the terms of the infinite progression $0.4, 0.04, 0.004, \ldots$

Solution

This is a geometric progression with $a = 0.4$ and $r = 0.1$.

Its sum is given by:

$$S_\infty = \frac{a}{1-r}$$
$$= \frac{0.4}{1-0.1}$$
$$= \frac{0.4}{0.9}$$
$$= \frac{4}{9}$$

➡ Worked example

The first three terms of an infinite geometric progression are 75, 45 and 27.

a) Write down the common ratio.

b) Find the sum of the terms of the progression.

Solution

a) The common ratio is $\frac{45}{75} = \frac{3}{5}$.

using $S_\infty = \frac{a}{1-r}$ ⟶ b) $S_\infty = \dfrac{75}{1-\frac{3}{5}} = 187.5$

The sum of the terms of a geometric progression

> **Discussion point**
>
> **A paradox**
> Consider the following arguments.
>
> i $\quad S = 1 - 2 + 4 - 8 + 16 - 32 + 64 - \ldots$
> $\Rightarrow \quad S = 1 - 2(1 - 2 + 4 - 8 + 16 - 32 + \ldots)$
> $\qquad = 1 - 2S$
> $\Rightarrow \quad 3S = 1$
> $\Rightarrow \quad S = \frac{1}{3}$
>
> ii $\quad S = 1 + (-2 + 4) + (-8 + 16) + (-32 + 64) + \ldots$
> $\Rightarrow \quad S = 1 + 2 + 8 + 32 + \ldots$
> So S diverges towards $+\infty$.
>
> iii $\quad S = (1 - 2) + (4 - 8) + (16 - 32) + \ldots$
> $\Rightarrow \quad S = -1 - 4 - 8 - 16 \ldots$
> So S diverges towards $-\infty$.
>
> What is the sum of the series: $\frac{1}{3}$, $+\infty$, $-\infty$, or something else?

Exercise 12.3

1 Are the following sequences geometric?
   If so, state the common ratio and calculate the seventh term.
   a) 3, 6, 12, 24, …
   b) 3, 6, 9, 12, …
   c) 10, −10, 10, −10, 10, …
   d) 1, 1, 1, 1, 1, 1, …
   e) 15, 10, 5, 0, −5, …
   f) 10, 5, $\frac{5}{2}$, $\frac{5}{4}$, $\frac{5}{8}$, …
   g) 2, 2.2, 2.22, 2.222, …

2 A geometric sequence has first term 5 and common ratio 2.
   The sequence has seven terms.
   a) Find the last term.
   b) Find the sum of the terms in the sequence.

3 The first term of a geometric sequence of positive terms is 3 and the fifth term is 768.
   a) Find the common ratio of the sequence.
   b) Find the eighth term of the sequence.

4 A geometric sequence has first term $\frac{1}{16}$ and common ratio 4.
   a) Find the fifth term.
   b) Which is the first term of the sequence that exceeds 1000?

5 a) Find how many terms there are in the following geometric sequence:
      7, 14, …, 3584.
   b) Find the sum of the terms in this sequence.

6 a) Find how many terms there are in the following geometric sequence:
      100, 50, …, 0.390625.
   b) Find the sum of the terms in this sequence.

12 SERIES

Exercise 12.3 (cont)

7. The fourth term of a geometric progression is 36 and the eighth term is 576. All the terms are positive.
   a) Find the common ratio.
   b) Find the first term.
   c) Find the sum of the first ten terms.

8. The first three terms of an infinite geometric progression are 8, 4 and 2.
   a) State the common ratio of this progression.
   b) Calculate the sum to infinity of its terms.

9. The first three terms of an infinite geometric progression are 0.8, 0.08 and 0.008.
   a) Write down the common ratio for this progression.
   b) Find, as a fraction, the sum to infinity of the terms of this progression.
   c) Find the sum to infinity of the geometric progression
      $0.8 - 0.08 + 0.008 - \ldots$
      and hence show that $\frac{8}{11} = 0.\dot{7}\dot{2}$.

10. The first three terms of a geometric sequence are 100, 70 and 49.
    a) Write down the common ratio of the sequence.
    b) Which is the position of the first term in the sequence that has a value less than 1?
    c) Find the sum to infinity of the terms of this sequence.
    d) After how many terms is the sum of the sequence greater than 99% of the sum to infinity?

11. A geometric progression has first term 10 and its sum to infinity is 15.
    a) Find the common ratio.
    b) Find the sum to infinity if the first term is excluded from the progression.

12. The first four terms in an infinite geometric series are 216, 72, 24, 8.
    a) What is the common ratio r?
    b) Write down an expression for the nth term of the series.
    c) Find the sum of the first n terms of the series.
    d) Find the sum to infinity.
    e) How many terms are needed for the sum to be greater than 323.999?

13. A tank is filled with 10 litres of water. Half the water is removed and replaced with anti-freeze and then thoroughly mixed. Half this mixture is then removed and replaced with anti-freeze. The process continues.
    a) Find the first five terms in the sequence of amounts of water in the tank at each stage.
    b) Find the first five terms in the sequence of amounts of anti-freeze in the tank at each stage.
    c) Is either of these sequences geometric? Explain.

14. A pendulum is set swinging. Its first oscillation is through an angle of 20°, and each following oscillation is through 95% of the angle of the one before it.
    a) After how many swings is the angle through which it swings less than 1°?
    b) What is the total angle it has swung through at the end of its tenth oscillation?

The sum of the terms of a geometric progression

15 A ball is thrown vertically upwards from the ground. It rises to a height of 15 m and then falls and bounces. After each bounce it rises vertically to $\frac{5}{8}$ of the height from which it fell.
   a) Find the height to which the ball bounces after the nth impact with the ground.
   b) Find the total distance travelled by the ball from the first throw to the tenth impact with the ground.

16 The first three terms of an arithmetic sequence, a, $a + d$ and $a + 2d$, are the same as the first three terms, a, ar, ar^2, of a geometric sequence ($a \neq 0$). Show that this is only possible if $r = 1$ and $d = 0$.

17 a, b and c are three consecutive terms in a sequence.
   a) Prove that if the sequence is an arithmetic progression then $a + c = 2b$.
   b) Prove that if the sequence is a geometric progression then $ac = b^2$.

18 a) Solve the simultaneous equations $ar = 12$, $ar^5 = 3072$ (there are two possible answers).
   b) In each case, find the sum of the first ten terms of the geometric progression with first term a and common ratio r.

Past-paper questions

1 Find the values of the positive constants p and q such that, in the binomial expansion of $(p + qx)^{10}$, the coefficient of x^5 is 252 and the coefficient of x^3 is 6 times the coefficient of x^2. [8]

Cambridge O Level Additional Mathematics (4037)
Paper 11 Q9, June 2012
Cambridge IGCSE Additional Mathematics (0606)
Paper 11 Q9, June 2012

2 (i) Find the coefficient of x^3 in the expansion of $\left(1 - \frac{x}{2}\right)^{12}$. [2]

  (ii) Find the coefficient of x^3 in the expansion of $(1 + 4x)\left(1 - \frac{x}{2}\right)^{12}$. [3]

Cambridge O Level Additional Mathematics (4037)
Paper 21 Q2, June 2011
Cambridge IGCSE Additional Mathematics (0606)
Paper 21 Q2, June 2011

3 (i) Find the first four terms in the expansion of $(2 + x)^6$ in ascending powers of x. [3]

  (ii) Hence find the coefficient of x^3 in the expansion of $(1 + 3x)(1 - x)(2 + x)^6$. [4]

Cambridge O Level Additional Mathematics (4037)
Paper 21 Q7, June 2013
Cambridge IGCSE Additional Mathematics (0606)
Paper 21 Q7, June 2013

12 SERIES

> Now you should be able to:
> ★ use the binomial theorem for expansion of $(a + b)^n$ for positive integer n
> ★ use the general term $\binom{n}{r} a^{n-r} b^r, 0 \leqslant r \leqslant n$
> ★ recognise arithmetic and geometric progressions and understand the difference between them
> ★ use the formulae for the nth term and for the sum of the first n terms to solve problems involving arithmetic or geometric progressions
> ★ use the condition for the convergence of a geometric progression, and the formula for the sum to infinity of a convergent geometric progression.

Key points

✓ An expression of the form $(ax + b)^n$ where n is an integer is called a **binomial expression**.

✓ **Binomial coefficients**, denoted by $\binom{n}{r}$ or ^nC_r can be found:
  • using Pascal's triangle
  • using tables
  • using the formula $\binom{n}{r} = \dfrac{n!}{r!(n-r)!}$

✓ The binomial expansion of $(1+x)^n$ can also be written as
$$(1+x)^n = 1 + nx + \frac{n(n-1)}{2!}x^2 + \frac{n(n-1)(n-2)}{3!}x^3 + \ldots nx^{n-1} + x^n$$

✓ A sequence is an ordered set of numbers, $u_1, u_2, u_3, \ldots, u_k, \ldots u_n$, where u_k is the general term.

✓ In an arithmetic sequence, $u_{k+1} = u_k + d$ where d is a fixed number called the **common difference**.

✓ In a geometric sequence, $u_{k+1} = ru_k$ where r is a fixed number called the **common ratio**.

✓ For an arithmetic progression with first term a, common difference d and n terms
  • the kth term $u_k = a + (k-1)d$
  • the last term $l = a + (n-1)d$
  • the sum of the terms $= \frac{1}{2}n(a+l) = \frac{1}{2}n[2a + (n-1)d]$

✓ For a geometric progression with first term a, common ratio r and n terms
  • the kth term $a_k = ar^{k-1}$
  • the last term $a_n = ar^{n-1}$
  • the sum of the terms $= \dfrac{a(r^n - 1)}{(r-1)}$ for $r > 1$ or $\dfrac{a(1-r^n)}{(1-r)}$ for $r < 1$

✓ For an infinite geometric series to converge, $-1 < r < 1$. In this case the sum of all terms is given by $\dfrac{a}{1-r}$.

13 Vectors in two dimensions

Thought is an idea in transit.

Pythagoras (circa 569–475BC)

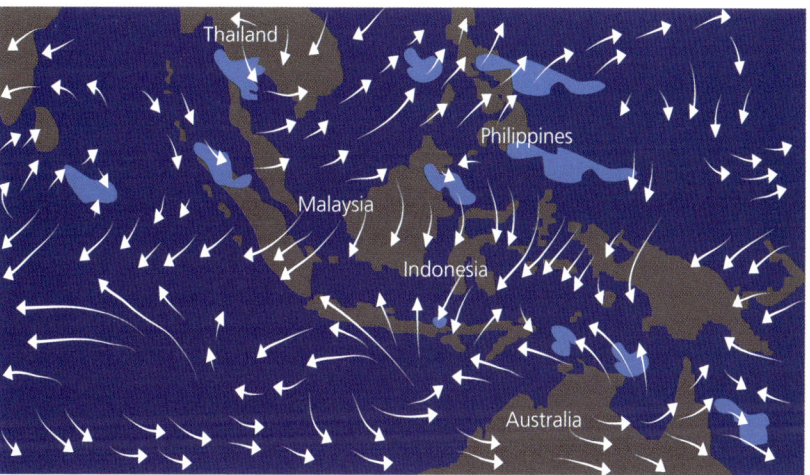

Discussion point

The lines on this weather map are examples of vectors. What do they tell you about the wind at any place?

Terminology and notation

The focus of this chapter is vectors in two dimensions. A **vector** is a quantity that has both magnitude and direction, for example, a **velocity** of $60\,\text{km}\,\text{h}^{-1}$ in a southerly direction. In contrast, a **scalar** quantity has a magnitude but no direction attached to it, for example, a **speed** of $60\,\text{km}\,\text{h}^{-1}$ where no direction is given.

A vector in two dimensions can be represented by drawing a straight line with an arrowhead to define the direction. The direction is often given as the angle the vector makes with the positive x-axis, with anticlockwise taken as positive.

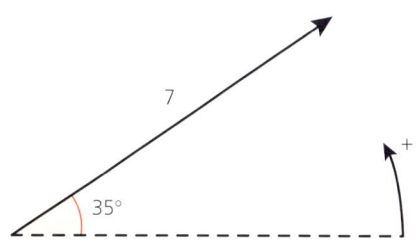

13 VECTORS IN TWO DIMENSIONS

You need to be able to recognise and use a number of different conventions for writing vectors, as outlined below.

A vector joining the points A and B can be written as $\overrightarrow{AB}$ or just given a single letter, e.g. **r**.

In print, vectors are usually in bold type, for example, **a**, but when you are writing them by hand it is usual to underline them, as in $\underline{a}$, or put an arrow above as in $\overrightarrow{OA}$.

A **unit vector** is any vector of length 1. **i** and **j** are principal unit vectors of length 1 in the positive x and y directions respectively.

Position vectors start at the origin and are the vector equivalent of coordinates. For example, the vector joining the origin to the point $(2, 5)$ is written as

$\begin{pmatrix} 2 \\ 5 \end{pmatrix}$ ← This is a **column vector**.

or $2\mathbf{i} + 5\mathbf{j}$.

*This vector is in **component form**: the two components represent distances in the x and y directions.*

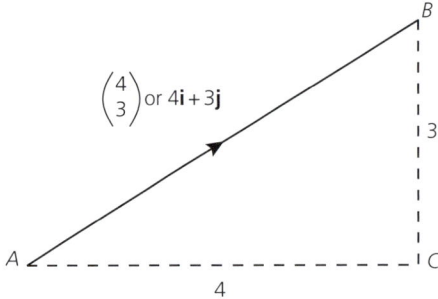

In the diagram, the vector $\overrightarrow{AB}$ is 4 units in the x direction and 3 units in the y direction and can be written either as the column vector $\begin{pmatrix} 4 \\ 3 \end{pmatrix}$ or as $4\mathbf{i} + 3\mathbf{j}$.

This can be found by using Pythagoras' theorem or, in this example, by recognising it as a 3, 4, 5 triangle.

The **magnitude** or **modulus** of a vector is its length and is denoted by vertical lines on either side of the vector, for example, $|\mathbf{a}|$ or $|\mathbf{OA}|$.

In the diagram above, $|\overrightarrow{AC}| = 4$, $|\overrightarrow{BC}| = 3$ and $|\overrightarrow{AB}| = 5$.

All of the vectors introduced so far have had both their x and y components in the positive directions, but this is not always the case.

The **negative** of a vector **a** is the vector $-\mathbf{a}$. $-\mathbf{a}$ is parallel to **a** but in the opposite direction.

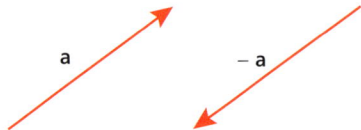

Terminology and notation

→ Worked example

Sketch the following vectors and find their magnitude:

a) $\begin{pmatrix} -1 \\ 3 \end{pmatrix}$ b) $2\mathbf{i} - 5\mathbf{j}$ c) $\begin{pmatrix} -3 \\ -4 \end{pmatrix}$

Using Pythagoras' theorem to find the magnitude of the vector

Solution

a) $\left\| \begin{pmatrix} -1 \\ 3 \end{pmatrix} \right\| = \sqrt{(-1)^2 + 3^2}$

$= \sqrt{10}$

$= 3.16$ (3 s.f.)

b) $|2\mathbf{i} - 5\mathbf{j}| = \sqrt{2^2 + (-5)^2}$

$= \sqrt{29}$

$= 5.39$ (3 s.f.)

c) $\left\| \begin{pmatrix} -3 \\ -4 \end{pmatrix} \right\| = \sqrt{((-3)^2 + (-4)^2)}$

$= 5$

→ Worked example

Write the vector **b**

a) as a column vector
b) using the unit vectors **i** and **j**.

'distance to the right' first and 'distance up' below

Solution

a) From the diagram, $\mathbf{b} = \begin{pmatrix} 2 \\ 1 \end{pmatrix}$.

b) $\mathbf{b} = 2\mathbf{i} + \mathbf{j}$

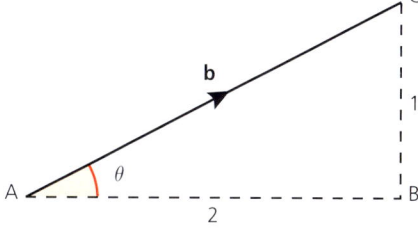

13 VECTORS IN TWO DIMENSIONS

Exercise 13.1

1 Express the following vectors in component form:

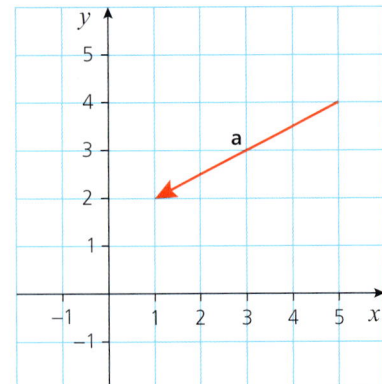
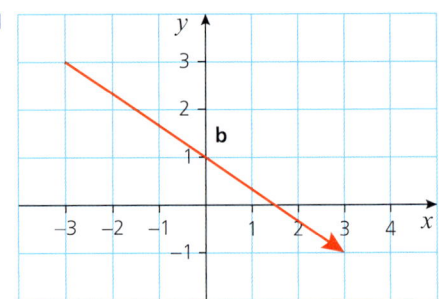
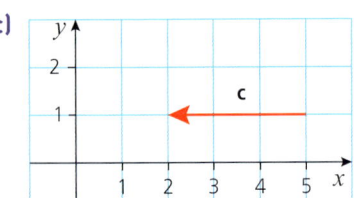
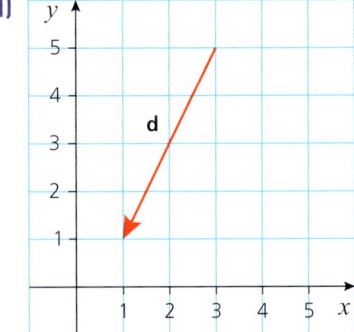

2 The coordinates of points P, Q, R and S are $(-1, -2)$, $(-2, 1)$, $(1, 2)$ and $(2, -1)$ respectively. The origin is the point O.
   a) Mark the points on a grid. Use equal scales on the two axes.
   b) Write as column vectors:
      i) $\overrightarrow{OR}$ ii) $\overrightarrow{RO}$
   c) Write as column vectors:
      i) $\overrightarrow{PR}$ ii) $\overrightarrow{QS}$
   d) Write down the lengths of the vectors:
      i) $\overrightarrow{PQ}$ ii) $\overrightarrow{QR}$ iii) $\overrightarrow{RS}$ iv) $\overrightarrow{SP}$
   e) Describe the quadrilateral $PQRS$.
3 Draw diagrams to illustrate each of the following vectors:
   a) $2\mathbf{i}$ b) $3\mathbf{j}$ c) $2\mathbf{i} + 3\mathbf{j}$ d) $2\mathbf{i} - 3\mathbf{j}$
4 For each of the following vectors
   i) draw a diagram ii) find its magnitude.
   a) $\begin{pmatrix} 0 \\ 4 \end{pmatrix}$ b) $\begin{pmatrix} -3 \\ 0 \end{pmatrix}$ c) $\begin{pmatrix} 5 \\ 7 \end{pmatrix}$ d) $\begin{pmatrix} 5 \\ -7 \end{pmatrix}$

Adding and subtracting vectors

5. $A(-1, 4)$, $B(2, 7)$ and $C(5, 0)$ form the vertices of a triangle.
   a) Draw the triangle on graph paper. Using your diagram, write the vectors representing the sides AB, BC and AC as column vectors.
   b) Which is the longest side of the triangle?
6. A, B, C and D have coordinates $(-3, -4)$, $(0, 2)$, $(5, 6)$ and $(2, 0)$ respectively.
   a) Draw the quadrilateral $ABCD$ on graph paper.
   b) Write down the position vectors of the points A, B, C and D.
   c) Write down the vectors $\overrightarrow{AB}$, $\overrightarrow{DC}$, $\overrightarrow{BC}$ and $\overrightarrow{AD}$.
   d) Describe shape $ABCD$.
7. $\overrightarrow{AB} = \mathbf{j}$, $\overrightarrow{BC}$ is the vector $2\mathbf{i} + 2\mathbf{j}$ and $\overrightarrow{CD} = \mathbf{i}$.
   a) Sketch the shape $ABCD$.
   b) Write $\overrightarrow{AD}$ as a column vector.
   c) Describe shape $ABCD$.

Multiplying a vector by a scalar

When a vector is multiplied by a scalar (i.e. a number) then its length is multiplied but its direction is unchanged.

For example, $3\mathbf{a} = \mathbf{a} + \mathbf{a} + \mathbf{a}$ gives a vector three times as long as $\mathbf{a}$ *in the same direction* and $2(2\mathbf{i} + 3\mathbf{j})$ gives a vector twice as long as $2\mathbf{i} + 3\mathbf{j}$ *in the same direction*.

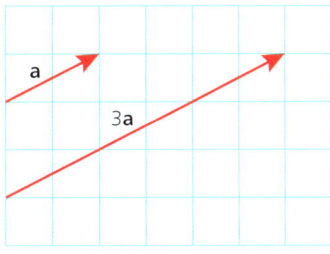
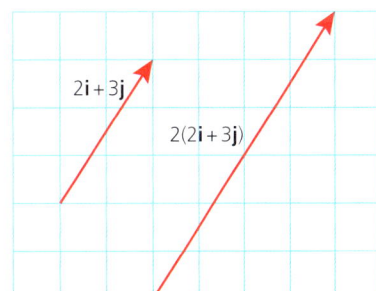

Adding and subtracting vectors

To add vectors written in component form, simply add the x-components together and the y-components together as shown in the example below.

→ Worked example

Add the vectors $2\mathbf{i} + 3\mathbf{j}$ and $\mathbf{i} - 2\mathbf{j}$

a) using algebra
b) by graphing them.

Collecting like terms →

Solution

a) $(2\mathbf{i} + 3\mathbf{j}) + (\mathbf{i} - 2\mathbf{j}) = (2 + 1)\mathbf{i} + (3 + (-2))\mathbf{j}$
$= 3\mathbf{i} + \mathbf{j}$ ← *This is called the resultant vector.*

13 VECTORS IN TWO DIMENSIONS

b)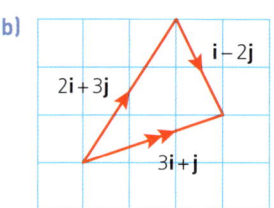

The resultant vector is shown by using two arrows.

To subtract one vector from another, add the equivalent negative vector. So, in the same way that $5 - 2 = 5 + (-2) = 3$,

$$\begin{pmatrix} 6 \\ 5 \end{pmatrix} - \begin{pmatrix} 4 \\ 2 \end{pmatrix} = \begin{pmatrix} 6 \\ 5 \end{pmatrix} + \begin{pmatrix} -4 \\ -2 \end{pmatrix}$$

$$= \begin{pmatrix} 6 + (-4) \\ 5 + (-2) \end{pmatrix}$$

$$= \begin{pmatrix} 2 \\ 3 \end{pmatrix}$$

Alternatively, you can simply subtract the second component from the first component in each case, for example, $\begin{pmatrix} 8 \\ -2 \end{pmatrix} - \begin{pmatrix} -1 \\ 4 \end{pmatrix} = \begin{pmatrix} 8 - (-1) \\ (-2) - 4 \end{pmatrix} = \begin{pmatrix} 9 \\ -6 \end{pmatrix}$

A very important result involves subtracting vectors.

Look at this diagram:

The position vector of A is $\overrightarrow{OA} = \mathbf{a}$.
The position vector of B is $\overrightarrow{OB} = \mathbf{b}$.

What can you say about the vector $\overrightarrow{AB}$ joining point A to point B?

Vector addition gives $\mathbf{a} + \overrightarrow{AB} = \mathbf{b}$

So, $\overrightarrow{AB} = \mathbf{b} - \mathbf{a}$

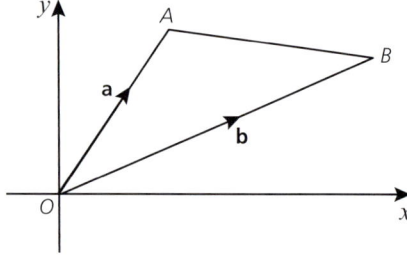

→ Worked example

The point P has position vector $-5\mathbf{i} + 3\mathbf{j}$.
The point Q has position vector $7\mathbf{i} - 8\mathbf{j}$.
Find the vector $\overrightarrow{PQ}$.

Solution

$\overrightarrow{PQ} = \mathbf{q} - \mathbf{p}$
$\overrightarrow{PQ} = 7\mathbf{i} - 8\mathbf{j} - (-5\mathbf{i} + 3\mathbf{j})$
$\overrightarrow{PQ} = 12\mathbf{i} - 11\mathbf{j}$

Zero and unit vectors

The **zero vector**, **0**, is the result of adding two vectors **a** and (−**a**). Since **a** + (−**a**) starts and finishes at the same place, **a** + (−**a**) = **0**.

It can be written as $0\mathbf{i} + 0\mathbf{j}$ or as $\begin{pmatrix} 0 \\ 0 \end{pmatrix}$.

To find the **unit vector** in the direction of any given vector, divide the given vector by its magnitude (length).

So, the unit vector in the direction of **a** is given by $\dfrac{\mathbf{a}}{|\mathbf{a}|}$.

Remember, in column form
$\mathbf{i}$ is written as $\begin{pmatrix} 1 \\ 0 \end{pmatrix}$ and
$\mathbf{j}$ is written as $\begin{pmatrix} 0 \\ 1 \end{pmatrix}$.

➡ Worked example

Find unit vectors parallel to:

a) $\begin{pmatrix} 3 \\ 4 \end{pmatrix}$

b) $2\mathbf{i} - 3\mathbf{j}$.

Solution

a) $\left| \begin{pmatrix} 3 \\ 4 \end{pmatrix} \right| = \sqrt{3^2 + 4^2}$

$= 5$

The required unit vector is $\dfrac{1}{5}\begin{pmatrix} 3 \\ 4 \end{pmatrix} = \begin{pmatrix} 0.6 \\ 0.8 \end{pmatrix}$

b) $|2\mathbf{i} - 3\mathbf{j}| = \sqrt{2^2 + (-3)^2}$

$= \sqrt{13}$

The required unit vector is $\dfrac{1}{\sqrt{13}}(2\mathbf{i} - 3\mathbf{j})$.

Remember to multiply each component of the vector by $\dfrac{1}{5}$.

This could also be written as $0.555\mathbf{i} - 1.11\mathbf{j}$ (3 s.f.) but it is often better to leave an answer in an exact form.

➡ Worked example

Find the unit vector in the direction of $2\mathbf{i} - 4\mathbf{j}$. Give your answer in simplest surd form.

Solution

$|2\mathbf{i} - 4\mathbf{j}| = \sqrt{2^2 + (-4)^2}$

$= \sqrt{20}$

$= 2\sqrt{5}$

The required unit vector is $\dfrac{1}{2\sqrt{5}}(2\mathbf{i} - 4\mathbf{j}) = \dfrac{1}{\sqrt{5}}\mathbf{i} - \dfrac{2}{\sqrt{5}}\mathbf{j}$.

Rationalising the denominator, this can also be written as $\dfrac{\sqrt{5}}{5}\mathbf{i} - \dfrac{2\sqrt{5}}{5}\mathbf{j}$.

13 VECTORS IN TWO DIMENSIONS

➡ Worked example

$\mathbf{a} = \begin{pmatrix} 3 \\ 4 \end{pmatrix}$ and $\mathbf{b} = \begin{pmatrix} 1 \\ 7 \end{pmatrix}$

Find: a) $2\mathbf{a} + 3\mathbf{b}$ b) $3\mathbf{a} - 2\mathbf{b}$

Notice that in $2\mathbf{a} + 3\mathbf{b}$, 2 and 3 are scalars multiplying the vectors $\mathbf{a}$ and $\mathbf{b}$.

Solution

a) $2\mathbf{a} + 3\mathbf{b} = 2\begin{pmatrix} 3 \\ 4 \end{pmatrix} + 3\begin{pmatrix} 1 \\ 7 \end{pmatrix}$

$= \begin{pmatrix} 6 \\ 8 \end{pmatrix} + \begin{pmatrix} 3 \\ 21 \end{pmatrix}$

$= \begin{pmatrix} 6+3 \\ 8+21 \end{pmatrix}$

$= \begin{pmatrix} 9 \\ 29 \end{pmatrix}$

b) $3\mathbf{a} - 2\mathbf{b} = 3\begin{pmatrix} 3 \\ 4 \end{pmatrix} - 2\begin{pmatrix} 1 \\ 7 \end{pmatrix}$

$= \begin{pmatrix} 9 \\ 12 \end{pmatrix} - \begin{pmatrix} 2 \\ 14 \end{pmatrix}$

$= \begin{pmatrix} 9-2 \\ 12-14 \end{pmatrix}$

$= \begin{pmatrix} 7 \\ -2 \end{pmatrix}$

➡ Worked example

$PQRS$ is a parallelogram, with $\overrightarrow{PQ} = 3\mathbf{i} + 4\mathbf{j}$ and $\overrightarrow{PS} = 5\mathbf{i}$.

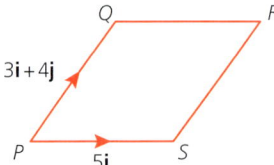

a) State the vectors that represent SR and QR.

b) Find the lengths of the sides of the parallelogram, and hence identify the type of parallelogram.

Solution

a) SR is parallel to PQ so is represented by the same vector, i.e. $3\mathbf{i} + 4\mathbf{j}$.

Similarly, QR is parallel to PS and so $\overrightarrow{QR} = 5\mathbf{i}$.

b) $PS = QR = |5\mathbf{i}|$

$= 5$ units

$PQ = SR = |3\mathbf{i} + 4\mathbf{j}|$

$= \sqrt{3^2 + 4^2}$

$= 5$ units

A parallelogram with all four sides equal is a rhombus.

Zero and unit vectors

There are many applications of vectors because many quantities have magnitude and direction. One of these is velocity and this is illustrated in the following example.

→ Worked example

Remember that bearings are measured clockwise from the north.

In this example, answers are given to 2 s.f. The unit vectors **i** and **j** are in the directions east and north.

The *Antares* is a sailing boat. It is travelling at a speed of 3 km h^{-1} on a bearing of 030°.

a) Find the components of this boat's velocity in the directions east and north.

The *Bellatrix* is another boat. It has velocity 2**i** − 1.5**j** in km h^{-1}.

b) Find the speed and direction of the *Bellatrix*.

Both boats start at the same place.

c) How far apart are they after 2 hours?

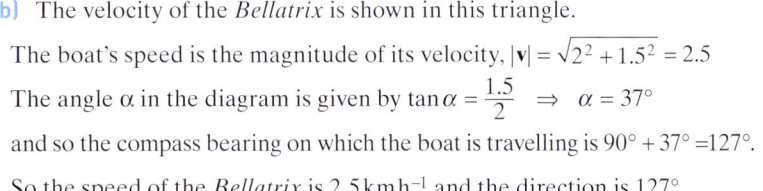

Solution

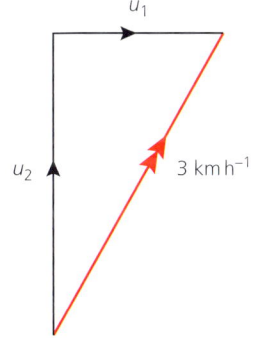

a) The components of the velocity of the *Antares* are shown on this right-angled triangle.

They are $u_1 = 3\sin 30° = 1.5$ in the direction east

and $u_2 = 3\cos 30° = 2.6$ in the direction north

So the velocity of the *Antares* is $1.5\mathbf{i} + 2.6\mathbf{j}$ km h^{-1}.

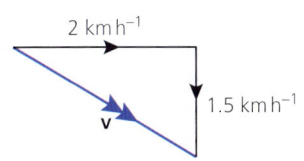

b) The velocity of the *Bellatrix* is shown in this triangle.

The boat's speed is the magnitude of its velocity, $|\mathbf{v}| = \sqrt{2^2 + 1.5^2} = 2.5$

The angle α in the diagram is given by $\tan \alpha = \frac{1.5}{2} \Rightarrow \alpha = 37°$

and so the compass bearing on which the boat is travelling is 90° + 37° = 127°.

So the speed of the *Bellatrix* is 2.5 km h^{-1} and the direction is 127°.

c) Assuming both boats started at the origin, their positions after 2 hours are

Antares $2 \times (1.5\mathbf{i} + 2.6\mathbf{j}) = 3\mathbf{i} + 5.2\mathbf{j}$

Bellatrix $2 \times (2\mathbf{i} - 1.5\mathbf{j}) = 4\mathbf{i} - 3\mathbf{j}$

So the displacement from the *Bellatrix* to the *Antares* is

$$(3\mathbf{i} + 5.2\mathbf{j}) - (4\mathbf{i} - 3\mathbf{j}) = -\mathbf{i} + 8.2\mathbf{j}$$

and the distance between the boats is $\sqrt{(-1)^2 + 7.2^2} = 8.3$ km (to 2 s.f.).

In this example, the velocity of the *Antares* was given in speed–direction form and was converted into components form. For the *Bellatrix* the reverse process was followed, converting from components into speed and direction. You need to be able to do both these conversions.

13 VECTORS IN TWO DIMENSIONS

Exercise 13.2

1. Simplify the following:
   a) $\begin{pmatrix} 2 \\ 3 \end{pmatrix} + \begin{pmatrix} 1 \\ 5 \end{pmatrix}$
   b) $\begin{pmatrix} 3 \\ -1 \end{pmatrix} + \begin{pmatrix} -1 \\ 3 \end{pmatrix}$
   c) $\begin{pmatrix} 5 \\ 0 \end{pmatrix} + \begin{pmatrix} 0 \\ 5 \end{pmatrix}$

2. Simplify the following:
   a) $(2\mathbf{i} + 3\mathbf{j}) - (3\mathbf{i} - 2\mathbf{j})$
   b) $3(2\mathbf{i} + 3\mathbf{j}) - 2(3\mathbf{i} - 2\mathbf{j})$

3. Given that $\mathbf{a} = 3\mathbf{i} + 4\mathbf{j}$, $\mathbf{b} = 2\mathbf{i} - 3\mathbf{j}$ and $\mathbf{c} = -\mathbf{i} + 5\mathbf{j}$, find the following vectors:
   a) $\mathbf{a} + \mathbf{b} + \mathbf{c}$
   b) $\mathbf{a} + \mathbf{b} - \mathbf{c}$
   c) $\mathbf{a} - \mathbf{b} + \mathbf{c}$
   d) $2\mathbf{a} + \mathbf{b} + 3\mathbf{c}$
   e) $\mathbf{a} - 2\mathbf{b} + 3\mathbf{c}$
   f) $2(\mathbf{a} + \mathbf{b}) - 3(\mathbf{b} - \mathbf{c})$
   g) $2(2\mathbf{a} + \mathbf{b} - \mathbf{c}) - 3(\mathbf{a} - 2\mathbf{b} + \mathbf{c})$

4. Sketch each of the following vectors and find their moduli:
   a) $3\mathbf{i} + 4\mathbf{j}$
   b) $3\mathbf{i} - 4\mathbf{j}$
   c) $7\mathbf{i}$
   d) $-7\mathbf{i}$
   e) $5\mathbf{i} + 3\mathbf{j}$
   f) $2\mathbf{i} - 7\mathbf{j}$
   g) $6\mathbf{i} - 6\mathbf{j}$
   h) $\mathbf{i} + \mathbf{j}$

5. Write the vectors joining each pair of points
   i) in the form $a\mathbf{i} + b\mathbf{j}$
   ii) as a column vector.
   a) (1, 4) to (3, 7)
   b) (1, 3) to (2, −4)
   c) (0, 0) to (3, 5)
   d) (−3, 7) to (7, −3)
   e) (−4, 2) to (0, 0)
   f) (−5, −2) to (−1, 0)

6.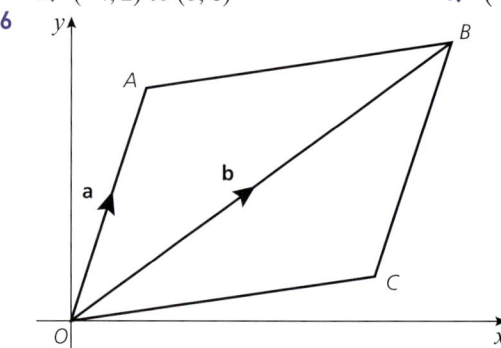

   The diagram shows a parallelogram $OABC$ with $\overrightarrow{OA} = \mathbf{a}$ and $\overrightarrow{OB} = \mathbf{b}$.
   Write the following vectors in terms of $\mathbf{a}$ and $\mathbf{b}$:
   a) $\overrightarrow{AB}$
   b) $\overrightarrow{BA}$
   c) $\overrightarrow{CB}$
   d) $\overrightarrow{BC}$
   e) $\overrightarrow{OB}$
   f) $\overrightarrow{BO}$
   g) $\overrightarrow{AC}$
   h) $\overrightarrow{CA}$

7. $ABCD$ is a rhombus where A is the point $(-1, -2)$.
   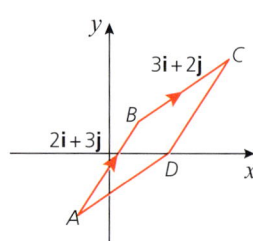
   a) Write the vectors $\overrightarrow{AD}$ and $\overrightarrow{DC}$ as column vectors.
   b) Write the diagonals $\overrightarrow{AC}$ and $\overrightarrow{BD}$ as column vectors.
   c) Use the properties of a parallelogram to find the coordinates of the point of intersection of the diagonals.
   d) Find the gradients of the diagonals and hence verify that the diagonals are perpendicular.

8. Find unit vectors parallel to each of the following:
   a) $3\mathbf{i} - 4\mathbf{j}$
   b) $5\mathbf{i} + 7\mathbf{j}$
   c) $\begin{pmatrix} 5 \\ 12 \end{pmatrix}$
   d) $\begin{pmatrix} 2 \\ -6 \end{pmatrix}$
   e) $5\mathbf{i}$
   f) $\begin{pmatrix} 1 \\ 1 \end{pmatrix}$

Zero and unit vectors

9 A is the point $(-3, -2)$, B is the point $(5, 4)$ and C is the point $(2, 8)$.
   a) Sketch the triangle ABC.
   b) Find the vectors representing the three sides AB, BC and CA in the form $x\mathbf{i} + y\mathbf{j}$.
   c) Find the lengths of each side of the triangle.
   d) What type of triangle is triangle ABC?

10 $ABCD$ is a kite and AC and BD meet at the origin O. A is the point $(-4, 0)$, B is $(0, 4)$ and D is $(0, -8)$.
   The diagonals of a kite are perpendicular and O is the midpoint of AC.
   a) Find each of the following in terms of $\mathbf{i}$ and $\mathbf{j}$:
      i) $\overrightarrow{OC}$ ii) $\overrightarrow{AB}$ iii) $\overrightarrow{BC}$
      iv) $\overrightarrow{AD}$ v) $\overrightarrow{CD}$ vi) $\overrightarrow{AC}$
   b) Find the lengths of the lines OC, AB, BC, AD, CD and AC.
   c) State two descriptions that are common to the triangles AOB, BOC and ABC.

11 $A (4, 4)$, $B (24, 19)$ and $C (48, 12)$ form the vertices of a triangle.
   a) Sketch the triangle.
   b) Write the vectors $\overrightarrow{AB}$, $\overrightarrow{BC}$ and $\overrightarrow{AC}$ as column vectors.
   c) Find the lengths of the sides of the triangle.
   d) What type of triangle is ABC?

12 Salman and Aloke are hiking on a flat level ground. Their starting point is taken as the origin and the unit vectors $\mathbf{i}$ and $\mathbf{j}$ are in the directions east and north. Salman walks with constant velocity $3\mathbf{i} + 6\mathbf{j}$ kilometres per hour. Aloke walks on a compass bearing of 300° at a steady speed of 6.5 kilometres per hour.
   i) Who is walking faster and by how much?
   ii) How far apart are they after 1½ hours?

13 Ama has her own small aeroplane. One afternoon, she flies for 1 hour with a velocity of $120\mathbf{i} + 160\mathbf{j}$ km h^{-1} where $\mathbf{i}$ and $\mathbf{j}$ are unit vectors in the directions east and north. Then she flies due north for 1 hour at the same speed. Finally, she returns to her starting point; flying in a straight line at the same speed.
   Find, to the nearest degree, the direction in which she travels on the final leg of her journey and, to the nearest minute, how long it takes her.

Past-paper questions

1 In this question $\mathbf{i}$ is a unit vector due East and $\mathbf{j}$ is a unit vector due North.

   At 12 00 hours, a ship leaves a port P and travels with a speed of 26 km h^{-1} in the direction $5\mathbf{i} + 12\mathbf{j}$.
   (i) Show that the velocity of the ship is $(10\mathbf{i} + 24\mathbf{j})$ km h^{-1}. [2]
   (ii) Write down the position vector of the ship, relative to P, at 16 00 hours. [1]

13 VECTORS IN TWO DIMENSIONS

(iii) Find the position vector of the ship, relative to P, t hours after 16 00 hours. [2]

At 16 00 hours, a speedboat leaves a lighthouse which has position vector $(120\mathbf{i} + 81\mathbf{j})$ km, relative to P, to intercept the ship. The speedboat has a velocity of $(-22\mathbf{i} + 30\mathbf{j})$ km h^{-1}.

(iv) Find the position vector, relative to P, of the speedboat t hours after 16 00 hours. [1]

(v) Find the time at which the speedboat intercepts the ship and the position vector, relative to P, of the point of interception. [4]

Cambridge O Level Additional Mathematics (4037)
Paper 12 Q10, June 2014
Cambridge IGCSE Additional Mathematics (0606)
Paper 12 Q10, June 2014

2 Relative to an origin O, the position vectors of the points A and B are $2\mathbf{i} - 3\mathbf{j}$ and $11\mathbf{i} + 42\mathbf{j}$ respectively.

(i) Write down an expression for $\overrightarrow{AB}$. [2]

The point C lies on AB such that $\overrightarrow{AC} = \frac{1}{3}\overrightarrow{AB}$.

(ii) Find the length of $\overrightarrow{OC}$. [4]

The point D lies on $\overrightarrow{OA}$ such that $\overrightarrow{DC}$ is parallel to $\overrightarrow{OB}$.

(iii) Find the position vector of D. [2]

Cambridge O Level Additional Mathematics (4037)
Paper 21 Q8, June 2012
Cambridge IGCSE Additional Mathematics (0606)
Paper 21 Q8, June 2012

3 Relative to an origin O, the position vectors of the points A and B are $\mathbf{i} - 4\mathbf{j}$ and $7\mathbf{i} + 20\mathbf{j}$ respectively. The point C lies on AB and is such that $\overrightarrow{AC} = \frac{2}{3}\overrightarrow{AB}$. Find the position vector of C and the magnitude of this vector. [5]

Cambridge O Level Additional Mathematics (4037)
Paper 21 Q3, June 2011
Cambridge IGCSE Additional Mathematics (0606)
Paper 21 Q3, June 2011

Now you should be able to:
★ understand and use vector notation
★ know and use position vectors and unit vectors
★ find the magnitude of a vector; add and subtract vectors and multiply vectors by scalars
★ compose and resolve velocities.

Key points

- A vector quantity has both a magnitude and a direction; a scalar quantity has magnitude only.
- Vectors are typeset in bold, for example **a**, or they may be written as lines with arrows along the top, for example $\overrightarrow{OA}$. When they are hand-written they are underlined, for example a.
- The length of a vector is also referred to as its magnitude or modulus. The length of the vector **a** is written as |**a**| or a and can be found using Pythagoras' theorem.
- A unit vector has length 1. Unit vectors in the directions x and y are denoted by **i** and **j** respectively.
- A vector can be written in component form, $x\mathbf{i} + y\mathbf{j}$ or $\begin{pmatrix} x \\ y \end{pmatrix}$, as in magnitude–direction form, (r, θ).
- The position vector $\overrightarrow{OA}$ of a point A is the vector joining the origin to A.
- The vector $\overrightarrow{AB}$ is given by **b** − **a** where **a** and **b** are the position vectors of A and B.

Review exercise 4

1. Arrangements containing 4 different letters from the word ALGORITHM are to be made. Find
   a the number of 4-letter arrangements if there are no restrictions [1]
   b the number of 4-letter arrangements which start with the letter G and end with the letter T. [1]
2. a Find how many different 4-digit numbers can be formed from the digits 1, 3, 4, 7 and 9 if each digit may be used just once. [1]
   b How many of these 4-digit numbers are odd? [1]
3. A team of 8 scientists are required for an expedition. They are to be selected from 12 ecologists and 6 meteorologists. Find the number of different teams that can be selected if
   a there are no restrictions [1]
   b the team contains all of the meteorologists [1]
   c the team contains at least 3 ecologists. [2]
4. A panel consisting of 7 people is to be assembled in order to carry out an investigation. The people are to be chosen from a group of 10 police officers and 15 civilians. Calculate the number of ways that the panel can be selected if
   a there are no restrictions [1]
   b there must be at least 3 civilians on the panel. [2]
   After the panel has been chosen, a chairperson and secretary must be selected from the 7 panel members.
   c Calculate the number of ways in which a chairperson and secretary can be selected. [1]
5. a Find the first four terms in the expansion of $(3+x)^8$ in ascending powers of x. [3]
   b Hence find the coefficient of x^3 in the expansion of $(2-x^2)(3+x)^8$. [3]
6. The tenth term of an arithmetic progression is -27 and the sum of the first five terms is 40.
   a Find the first term of the progression and the common difference. [4]
   b The nth term of the progression is -212. Find the value of n. [2]
7. a The sum of the first two terms of a geometric progression is 10 and the third term is 9.
      i Find the possible values of the common ratio and the first term. [5]
      ii Find the sum to infinity of the convergent progression. [1]
   b In an arithmetic progression, $u_1 = -10$ and $u_4 = 14$. Find $u_{100} + u_{101} + u_{102} + \ldots + u_{200}$, the sum of the 100th to the 200th terms of the progression. [4]

   Cambridge O Level Additional Mathematics (4037)
   Paper 22 Q13, February/March 2020
   Cambridge IGCSE Additional Mathematics (0606)
   Paper 22 Q13, February/March 2020

8. Relative to an origin O, the position vectors of the points A and B are $2\mathbf{i} - 5\mathbf{j}$ and $6\mathbf{i} + 9\mathbf{j}$ respectively. The point C lies on AB such that $\overrightarrow{AC} = \frac{4}{5}\overrightarrow{AB}$. Find the position vector of the point C and the magnitude of $\overrightarrow{OC}$. [5]

9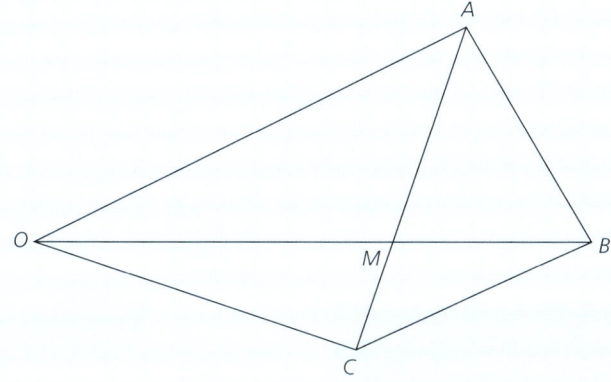

The diagram shows the quadrilateral $OABC$ such that $\overrightarrow{OA} = \mathbf{a}$, $\overrightarrow{OB} = \mathbf{b}$ and $\overrightarrow{OC} = \mathbf{c}$. It is given that $AM : MC = 2 : 1$ and $OM : MB = 3 : 2$.

i   Find $\overrightarrow{AC}$ in terms of **a** and **c**. [1]
ii Find $\overrightarrow{OM}$ in terms of **a** and **c**. [2]
iii Find $\overrightarrow{OM}$ in terms of **b**. [1]
iv Find $5\mathbf{a} + 10\mathbf{c}$ in terms of **b**. [2]
v   Find $\overrightarrow{AB}$ in terms of **a** and **c**, giving your answer in its simplest form. [2]

Cambridge O Level Additional Mathematics (4037)
Paper 12 Q6, February/March 2018
Cambridge IGCSE Additional Mathematics (0606)
Paper 12 Q6, February/March 2018

10 In this question, **i** is a unit vector due East and **j** is a unit vector due North. At 13 00 hours, a ship leaves port P and travels with a velocity of $(7\mathbf{i} - 24\mathbf{j})$ km h^{-1}.
 a State the speed of the ship. [1]
 b Find the bearing on which the ship is travelling. Give your answer to 3 significant figures. [2]
 c Find the position vector of the ship, relative to P, t hours after 17 00 hours. [2]
At 17 00 hours, a helicopter leaves its base which has position vector $(409\mathbf{i} + 141\mathbf{j})$, relative to P, in order to intercept the ship. The helicopter has velocity $(-120\mathbf{i} - 103\mathbf{j})$ km h^{-1}.
 d Find the time at which the helicopter intercepts the ship and its position vector, relative to P, at the point of interception. [5]

14 Differentiation

If I have seen further than others, it is by standing upon the shoulders of giants.

Isaac Newton (1642–1727)

Discussion point

Look at the planet Saturn in the image above. What connection did Newton make between an apple and the motion of the planets?

In Newton's early years, mathematics was not advanced enough to enable people to calculate the orbits of the planets round the sun. In order to address this, Newton invented calculus, the branch of mathematics that you will learn about in this chapter.

The gradient function

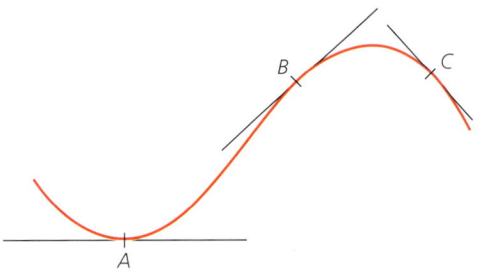

The curve in the diagram has a zero gradient at A, a positive gradient at B and a negative gradient at C.

The gradient function

Although you can calculate the gradient of a curve at a given point by drawing a tangent at that point and using two points on the tangent to calculate its gradient, this process is time-consuming and the results depend on the accuracy of your drawing and measuring. If you know the equation of the curve, you can use **differentiation** to calculate the gradient.

➡ Worked example

Work out the gradient of the curve $y = x^3$ at the general point (x, y).

Solution

Let P have the general value x as its x-coordinate, so P is the point (x, x^3).
Let the x-coordinate of Q be $(x + h)$ so Q is the point $((x + h), (x + h)^3)$.

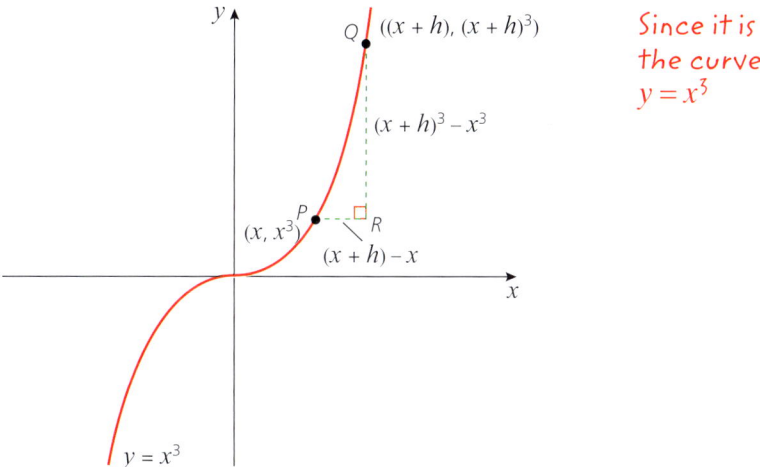

Since it is on the curve $y = x^3$

The gradient of the chord PQ is given by

$$\frac{QR}{PR} = \frac{(x+h)^3 - x^3}{(x+h) - x}$$

$$= \frac{x^3 + 3x^2h + 3xh^2 + h^3 - x^3}{h}$$

$$= \frac{3x^2h + 3xh^2 + h^3}{h}$$

$$= \frac{h(3x^2 + 3xh + h^2)}{h}$$

$$= 3x^2 + 3xh + h^2$$

As Q gets closer to P, h takes smaller and smaller values and the gradient approaches the value of $3x^2$, which is the gradient of the tangent at P. The gradient of the curve $y = x^3$ at the point (x, y) is equal to $3x^2$.

14 DIFFERENTIATION

*The gradient function is the gradient of the curve at the general point (x, y). The gradient function is also called the **derived function**.*

If the equation of the curve is written as $y = f(x)$, then the **gradient function** is written as $f'(x)$. Using this notation, the result in the previous example can be written as

$$f(x) = x^3 \quad \Rightarrow \quad f'(x) = 3x^2.$$

In the previous example, h was used to denote the difference between the x-coordinates of the points P and Q, where Q is close to P.

h is sometimes replaced by δx. The Greek letter δ (delta) is shorthand for 'a small change in' and so δx represents a small change in x, δy a small change in y and so on.

In the diagram the gradient of the chord PQ is $\frac{\delta y}{\delta x}$.

In the limit as δx tends towards 0, δx and δy both become infinitesimally small and the value obtained for $\frac{\delta y}{\delta x}$ approaches the gradient of the tangent at P.

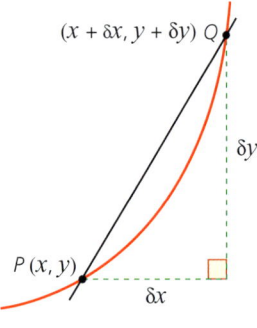

$$\lim \frac{\delta y}{\delta x} \text{ is written as } \frac{dy}{dx}.$$

Using this notation, you have a rule for differentiation.

$$y = x^n \quad \Rightarrow \quad \frac{dy}{dx} = nx^{n-1}$$

The gradient function, $\frac{dy}{dx}$, is sometimes called the **derivative** of y with respect to x. When you find it you have **differentiated** y with respect to x.

If the curve is written as $y = f(x)$, then the derivative is $f'(x)$.

If you are asked to differentiate a relationship in the form $y = f(x)$ in this book, this means differentiate with respect to x unless otherwise stated.

> **Note**
>
> There is nothing special about the letters x, y or f. If, for example, your curve represents time, t, on the horizontal axis and velocity, v, on the vertical axis, then the relationship could be referred to as $v = g(t)$. In this case v is a function of t and the gradient function is given by $\frac{dv}{dt} = g'(t)$.

The differentiation rule

Although it is possible to find the gradient from first principles which establishes a formal basis for differentiation, in practice you will use the differentiation rule introduced above;

$$y = x^n \quad \Rightarrow \quad \frac{dy}{dx} = nx^{n-1}.$$

You can also use this rule to differentiate (find the gradient of) equations that represent straight lines. For example, the gradient of the line $y = x$ is the same as $y = x^1$, so using the rule for differentiation, $\frac{dy}{dx} = 1 \times x^0 = 1$.

The gradient function

Lines of the form $y = c$ are parallel to the x-axis.

The gradient of the line $y = c$ where c is a constant is 0. For example, $y = 4$ is the same as $y = 4x^0$ so using the rule for differentiation, $\frac{dy}{dx} = 4 \times 0 \times x^{-1} = 0$. In general, differentiating any constant gives zero.

The rule can be extended further to include functions of the type $y = kx^n$ for any constant k, to give

This result is true for all powers of x, positive, negative and fractional.

$$y = kx^n \Rightarrow \frac{dy}{dx} = nkx^{n-1}.$$

You may find it helpful to remember the rule as

multiply by the power of x and reduce the power by 1.

➡ Worked example

For each function, find the gradient function.

a) $y = x^7$ b) $u = 4x^3$ c) $v = 5t^2$
d) $y = 4x^{-3}$ e) $P = 4\sqrt{t}$ f) $y = \frac{4x^3 - 5}{x^2}$

Solution

a) $y = x^7 \Rightarrow \frac{dy}{dx} = 7x^6$

b) $u = 4x^3 \Rightarrow \frac{du}{dx} = 4 \times 3x^2 = 12x^2$

c) $v = 5t^2$
$\Rightarrow \frac{dv}{dt} = 5 \times 2t = 10t$

d) $y = 4x^{-3}$
$\Rightarrow \frac{dy}{dx} = 4 \times (-3)x^{-3-1} = -12x^{-4}$

e) $P = 4t^{\frac{1}{2}}$

Using $\sqrt{t} = t^{\frac{1}{2}}$

$\Rightarrow \frac{dP}{dt} = 4 \times \frac{1}{2} t^{\frac{1}{2}-1}$
$= \frac{2}{t^{\frac{1}{2}}}$
$= \frac{2}{\sqrt{t}}$

f) $y = \frac{4x^3 - 5}{x^2} \Rightarrow y = \frac{4x^3}{x^2} - \frac{5}{x^2}$
$\Rightarrow y = 4x - 5x^{-2}$
$\Rightarrow \frac{dy}{dx} = 4 + 10x^{-3}$
$= 4 + \frac{10}{x^3}$

Sums and differences of functions

Many of the functions you will meet are sums or differences of simpler functions. For example, the function $(4x^3 + 3x)$ is the sum of the functions $4x^3$ and $3x$. To differentiate these functions, differentiate each part separately and then add the results together.

➡ Worked example

Differentiate $y = 4x^3 + 3x$.

Solution

$\frac{dy}{dx} = 12x^2 + 3$

14 DIFFERENTIATION

This example illustrates the general result that

$$y = f(x) + g(x) \implies \frac{dy}{dx} = f'(x) + g'(x).$$

Worked example

Given that $y = 2x^3 - 3x + 4$, find

a) $\frac{dy}{dx}$

b) the gradient of the curve at the point (2, 14).

Solution

a) $\frac{dy}{dx} = 6x^2 - 3$

b) At (2, 14), $x = 2$.

$\frac{dy}{dx} = 6 \times (2)^2 - 3 = 21$

Substituting $x = 2$ in the expression for $\frac{dy}{dx}$

Exercise 14.1

Differentiate the following functions using the rules

$$y = kx^n \implies \frac{dy}{dx} = nkx^{n-1}$$

and $\quad y = f(x) + g(x) \implies \frac{dy}{dx} = f'(x) + g'(x).$

1. a) $y = x^4$ b) $y = 2x^3$ c) $y = 5$ d) $y = 10x$
2. a) $y = x^{\frac{1}{2}}$ b) $y = 5\sqrt{x}$ c) $P = 7t^{\frac{3}{2}}$ d) $y = \frac{1}{5}x^{\frac{5}{2}}$
3. a) $y = 2x^5 + 4x^2$ b) $y = 3x^4 + 8x$ c) $y = x^3 + 4$ d) $y = x - 5x^3$
4. a) $f(x) = \frac{1}{x^2}$ b) $f(x) = \frac{6}{x^3}$
   c) $f(x) = 4\sqrt{x} - \frac{8}{\sqrt{x}}$ d) $f(x) = x^{\frac{1}{2}} - x^{-\frac{1}{2}}$
5. a) $y = x(x-1)$ b) $y = (x+1)(2x-3)$
   c) $y = \frac{x^3 + 5x}{x^2}$ d) $y = x\sqrt{x}$
6. Find the gradient of the curve $y = x^2 - 9$ at the points of intersection with the x- and y-axes.

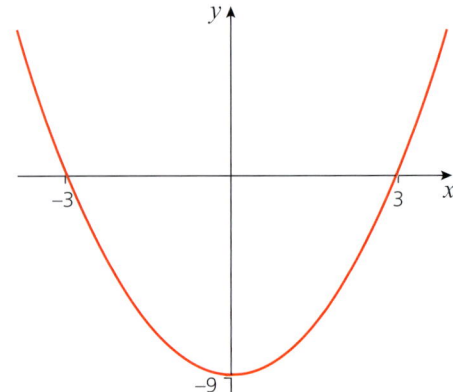

7 a) Copy the curve of $y = 4 - x^2$ and draw the graph of $y = x - 2$ on the same axes.
  b) Find the coordinates of the points where the two graphs intersect.
  c) Find the gradient of the curve at the points of intersection.

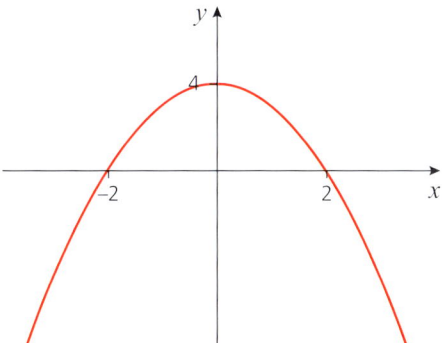

Stationary points

A **stationary point** is a point on a curve where the gradient is zero. This means that the tangents to the curve at these points are horizontal. The diagram shows a curve with four stationary points: A, B, C and D.

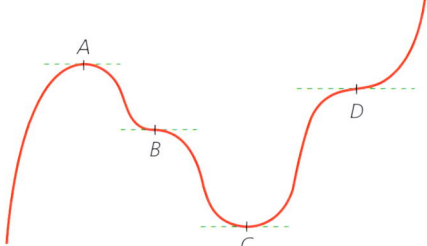

> **Note**
>
> Points where a curve 'twists' but doesn't have a zero gradient are also called points of inflexion. The tangent at a point of inflexion both touches and intersects the curve.
>
> This material goes beyond the syllabus.

The points A and C are **turning points** of the curve because as the curve passes through these points, it changes direction completely: at A the gradient changes from positive to negative and at C from negative to positive. A is called a **maximum** turning point, and C is a **minimum** turning point.

At B the curve does not turn: the gradient is negative both to the left and to the right of this point. B is a **stationary point of inflexion**.

Discussion point

What can you say about the gradient to the left and right of D?

14 DIFFERENTIATION

Maximum and minimum points

The graph shows the curve of $y = 4x - x^2$.

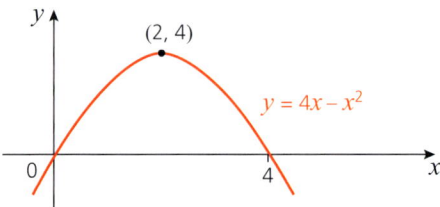

- The curve has a *maximum* point at (2, 4).
- The gradient $\frac{dy}{dx}$ at the maximum point is zero.
- The gradient is positive to the left of the maximum and negative to the right of it.

This is true for any maximum point.

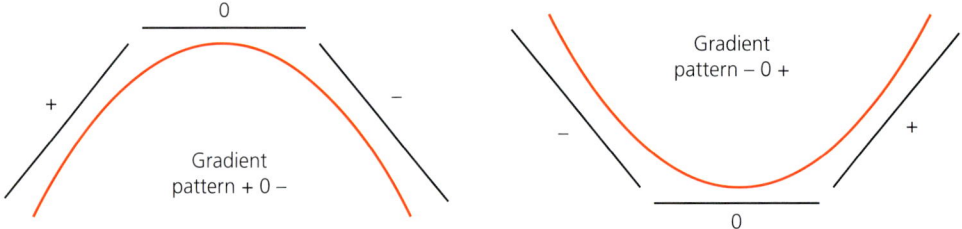

For any minimum turning point, the gradient

- is zero at that point
- goes from negative to zero to positive.

Once you can find the position of any stationary points, and what type of points they are, you can use this information to help you sketch graphs.

➡ Worked example

a) For the curve $y = x^3 - 12x + 3$
   i   find $\frac{dy}{dx}$ and the values of x for which $\frac{dy}{dx} = 0$
   ii classify the points on the curve with these values of x
   iii find the corresponding values of y
   iv sketch the curve.

b) Why can you be confident about continuing the sketch of the curve beyond the x-values of the turning points?

Stationary points

c) You did not need to find the coordinates of the points where the curve crosses the x-axis before sketching the graph. Why was this and under what circumstances would you find these points?

Solution

a) i $\frac{dy}{dx} = 3x^2 - 12$

When $\frac{dy}{dx} = 0$, $3x^2 - 12 = 0$

$\Rightarrow \quad 3(x^2 - 4) = 0$

$\Rightarrow \quad 3(x + 2)(x - 2) = 0$

$\Rightarrow \quad x = -2 \text{ or } x = 2$

Looking at the gradient pattern around $x = -2$ →

ii When $x = -3$, $\frac{dy}{dx} = 3(-3)^2 - 12 = 15$

When $x = -1$, $\frac{dy}{dx} = 3(-1)^2 - 12 = -9$

The gradient pattern is $+ \ 0 \ -$

$\Rightarrow$ maximum turning point at $x = -2$

Looking at the gradient pattern around $x = 2$ →

When $x = 1$, $\frac{dy}{dx} = 3(1)^2 - 12 = -9$

When $x = 3$, $\frac{dy}{dx} = 3(3)^2 - 12 = 15$

The gradient pattern is $- \ 0 \ +$

$\Rightarrow$ minimum turning point at $x = +2$

$(-2, 19)$ is a maximum and $(2, -13)$ a minimum.

iii When $x = -2$, $y = (-2)^3 - 12(-2) + 3 = 19$.
When $x = +2$, $y = (2)^3 - 12(2) + 3 = -13$.

The value of y when $x = 0$ tells you where the curve crosses the y-axis.

iv When $x = 0$, $y = (0)^3 - 12(0) + 3 = 3$.

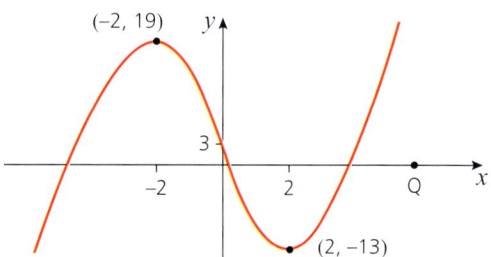

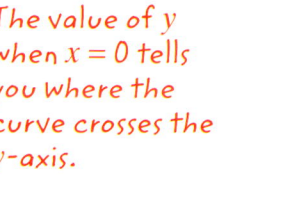

b) A cubic has at most 2 turning points and they have both been found. So the parts of the curve beyond them (to the left and to the right) just get steeper and steeper.

c) The sketch is showing the shape of the curve and this is not affected by where it crosses the axes. However, you can see from the equation that it crosses the y-axis at $(0, 3)$ and it is good practice to mark this in.

14 DIFFERENTIATION

> **Worked example**

Find all the turning points on the graph of $y = t^4 - 2t^3 + t^2 - 2$ and then sketch the curve.

Solution

$$\frac{dy}{dt} = 4t^3 - 6t^2 + 2t$$

$$\frac{dy}{dt} = 0 \Rightarrow 4t^3 - 6t^2 + 2t = 0$$
$$\Rightarrow 2t(2t^2 - 3t + 1) = 0$$
$$\Rightarrow 2t(2t - 1)(t - 1) = 0$$
$$\Rightarrow t = 0 \text{ or } t = 0.5 \text{ or } t = 1$$

Turning points occur when $\frac{dy}{dt} = 0$.

When $t = 0$, $y = (0)^4 - 2(0)^3 + (0)^2 - 2 = -2$.

When $t = 0.5$, $y = (0.5)^4 - 2(0.5)^3 + (0.5)^2 - 2 = -1.9375$.

When $t = 1$, $y = (1)^4 - 2(1)^3 + (1)^2 - 2 = -2$.

Plotting these points suggests that $(0.5, -1.9375)$ is a maximum turning point and $(0, -2)$ and $(1, -2)$ are minima, but you need more information to be sure. For example when $t = -1$, $y = +2$ and when $t = 2$, $y = +2$ so you know that the curve goes above the horizontal axis on both sides.

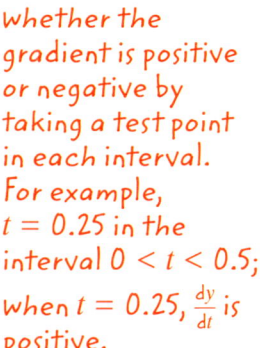

You can find whether the gradient is positive or negative by taking a test point in each interval. For example, $t = 0.25$ in the interval $0 < t < 0.5$; when $t = 0.25$, $\frac{dy}{dt}$ is positive.

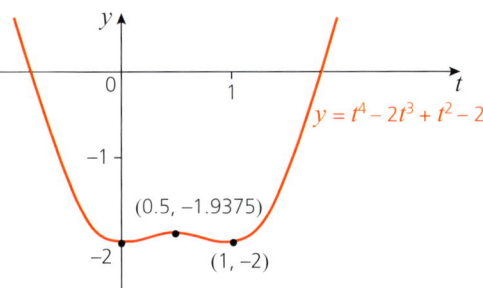

Exercise 14.2

You can use a graphic calculator to check your answers.

For each curve in questions 1–8:

i) find $\frac{dy}{dx}$ and the value(s) of x for which $\frac{dy}{dx} = 0$

ii) classify the point(s) on the curve with these x-values

iii) find the corresponding y-value(s)

iv) sketch the curve.

1. $y = 1 + x - 2x^2$
2. $y = 12x + 3x^2 - 2x^3$
3. $y = x^3 - 4x^2 + 9$
4. $y = x^2(x - 1)^2$
5. $y = x^4 - 8x^2 + 4$
6. $y = x^3 - 48x$
7. $y = x^3 + 6x^2 - 36x + 25$
8. $y = 2x^3 - 15x^2 + 24x + 8$

9 The graph of $y = px + qx^2$ passes through the point $(3, -15)$. Its gradient at that point is -14.
  a) Find the values of p and q.
  b) Calculate the maximum value of y and state the value of x at which it occurs.

10 a) Find the stationary points of the function $f(x) = x^2(3x^2 - 2x - 3)$ and distinguish between them.
  b) Sketch the curve $y = f(x)$.

Using second derivatives

In the same way as $\frac{dy}{dx}$ or $f'(x)$ is the gradient of the curve $y = f(x)$, $\frac{d}{dx}\left(\frac{dy}{dx}\right)$ or $f''(x)$ represents the gradient of the curve $y = f'(x)$.

This is also written as $\frac{d^2y}{dx^2}$ and is called the **second derivative**. You can find it by differentiating the function $\frac{dy}{dx}$.

Note that $\frac{d^2y}{dx^2}$ is not the same as $\left(\frac{dy}{dx}\right)^2$.

➔ Worked example

Find $\frac{d^2y}{dx^2}$ for $y = 4x^3 + 3x - 2$.

Solution

$\frac{dy}{dx} = 12x^2 + 3 \Rightarrow \frac{d^2y}{dx^2} = 24x$

In many cases, you can use the second derivative to determine if a stationary point is a maximum or minimum instead of looking at the value of $\frac{dy}{dx}$ on either side of the turning point.

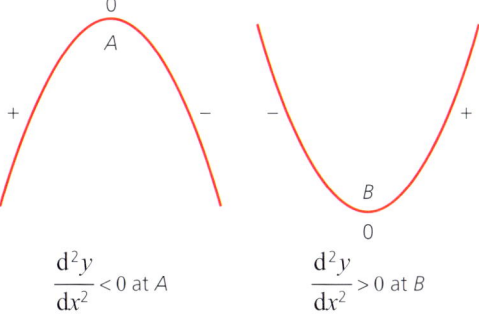

$\frac{d^2y}{dx^2} < 0$ at A $\frac{d^2y}{dx^2} > 0$ at B

At A, $\frac{dy}{dx} = 0$ and $\frac{d^2y}{dx^2} < 0$ showing that the gradient is zero and since $\frac{d^2y}{dx^2} < 0$, it is decreasing near that point, so must be going from positive to negative. This shows that A is a maximum turning point.

14 DIFFERENTIATION

At B, $\frac{dy}{dx} = 0$ and $\frac{d^2y}{dx^2} > 0$ showing that the gradient is zero and since $\frac{d^2y}{dx^2} > 0$, it is increasing near that point, so must be going from negative to positive. This shows that B is a minimum turning point.

Note that if $\frac{dy}{dx} = 0$ and $\frac{d^2y}{dx^2} = 0$ at the same point, you cannot make a decision about the type of turning point using this method.

➡ Worked example

For $y = 2x^3 - 3x^2 - 12x + 4$

a) Find $\frac{dy}{dx}$ and find the values of x when $\frac{dy}{dx} = 0$.

b) Find the value of $\frac{d^2y}{dx^2}$ at each stationary point and hence determine its nature.

c) Find the value of y at each of the stationary points.

d) Sketch the curve $y = 2x^3 - 3x^2 - 12x + 4$.

Solution

a) $\frac{dy}{dx} = 6x^2 - 6x - 12$
$= 6(x^2 - x - 2)$
$= 6(x+1)(x-2)$

So $\frac{dy}{dx} = 0$ when $x = -1$ and when $x = 2$.

b) $\frac{d^2y}{dx^2} = 12x - 6$

When $x = -1$, $\frac{d^2y}{dx^2} = -18 \Rightarrow$ a maximum

When $x = 2$, $\frac{d^2y}{dx^2} = 18 \Rightarrow$ a minimum

c) When $x = -1$, $y = 2(-1)^3 - 3(-1)^2 - 12(-1) + 4 = 11$

When $x = 2$, $y = 2(2)^3 - 3(2)^2 - 12(2) + 4 = -16$

d) The curve has a maximum turning point at $(-1, 11)$ and a minimum turning point at $(2, -16)$.

When $x = 0$, $y = 4$, so the curve crosses the y-axis at $(0, 4)$.

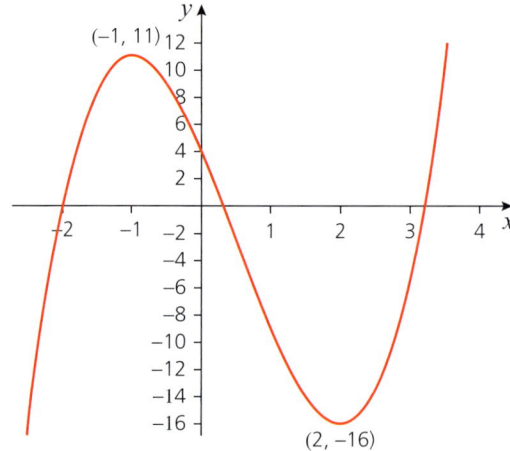

Using second derivatives

→ Worked example

Maria has made some sweets as a gift and makes a small box for them from a square sheet of card of side 24 cm. She cuts four identical squares of side x cm, one from each corner, and turns up the sides to make the box, as shown in the diagram.

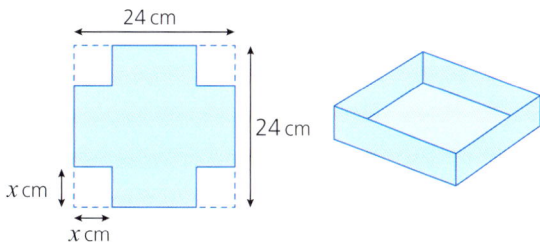

a) Write down an expression for the volume V of the box in terms of x.

b) Find $\frac{dV}{dx}$ and the values of x when $\frac{dV}{dx} = 0$.

c) Comment on this result.

d) Find $\frac{d^2V}{dx^2}$ and hence find the depth when the volume is a maximum.

Solution

a) The base of the box is a square of side $(24 - 2x)$ cm and the height is x cm, so $V = (24 - 2x)^2 \times x$

$= 4x(12 - x)^2 \text{ cm}^3$ ← *Taking a factor of 2 out of each bracket*

b) $V = 4x(144 - 24x + x^2)$

$= 576x - 96x^2 + 4x^3$

So $\frac{dV}{dx} = 576 - 192x + 12x^2$

$= 12(48 - 16x + x^2)$

$= 12(12 - x)(4 - x)$

So $\frac{dV}{dx} = 0$ when $x = 12$ and when $x = 4$.

c) When $x = 12$ there is no box, since the piece of cardboard was only a square of side 24 cm.

d) $\frac{d^2V}{dx^2} = -192 + 24x$ ← *Using $\frac{dV}{dx} = 576 - 192x + 12x^2$*

When $x = 4$, $\frac{d^2V}{dx^2} = -96$ which is negative.

Therefore the volume is a maximum when the depth $x = 4$ cm.

14 DIFFERENTIATION

Exercise 14.3

1. Find $\frac{dy}{dx}$ and $\frac{d^2y}{dx^2}$ for each of the following functions:
   a) $y = x^3 - 3x^2 + 2x - 6$
   b) $y = 3x^4 - 4x^3$
   c) $y = x^5 - 5x + 1$

2. For each of the following curves
   i) find any stationary points
   ii) use the second derivative test to determine their nature.
   a) $y = 2x^2 - 3x + 4$
   b) $y = x^3 - 2x^2 + x + 6$
   c) $y = 4x^4 - 2x^2 + 1$
   d) $y = x^5 - 5x$

3. For $y = 2x^3 - 3x^2 - 36x + 4$
   a) Find $\frac{dy}{dx}$ and the values of x when $\frac{dy}{dx} = 0$.
   b) Find the value of $\frac{d^2y}{dx^2}$ at each stationary point and hence determine its nature.
   c) Find the value of y at each stationary point.
   d) Sketch the curve.

4. A farmer has 160 m of fencing and wants to use it to form a rectangular enclosure next to a barn.

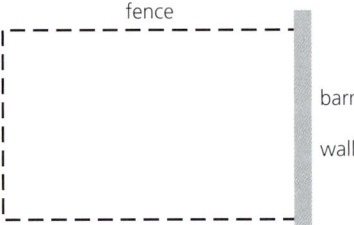

   Find the maximum area that can be enclosed and give its dimensions.

5. A cylinder has a height of h metres and a radius of r metres where $h + r = 3$.
   a) Find an expression for the volume of the cylinder in terms of r.
   b) Find the maximum volume.

6. A rectangle has sides of length x cm and y cm.
   a) If the perimeter is 24 cm, find the lengths of the sides when the area is a maximum, confirming that it is a maximum.
   b) If the area is 36 cm², find the lengths of the sides when the perimeter is a minimum, confirming that it is a minimum.

Equations of tangents and normals

Now that you know how to find the gradient of a curve at any point, you can use this to find the equation of the tangent at any given point on the curve.

Equations of tangents and normals

> ## Worked example

a) Find the equation of the tangent to the curve $y = 3x^2 - 5x - 2$ at the point $(1, -4)$.

b) Sketch the curve and show the tangent on your sketch.

Solution

a) $y = 3x^2 - 5x - 2 \Rightarrow \dfrac{dy}{dx} = 6x - 5$

At $(1, -4)$, $\dfrac{dy}{dx} = 6 \times 1 - 5$

Substituting $x = 1$ into this gradient function gives the gradient of the curve and therefore the tangent at this point.

and so $m = 1$

So the equation of the tangent is given by

$y - y_1 = m(x - x_1)$ ← $x_1 = 1, y_1 = -4$ and $m = 1$

$y - (-4) = 1(x - 1)$

$\Rightarrow \quad y = x - 5$ ← This is the equation of the tangent.

b) $y = 3x^2 - 5x - 2$ is a ∪-shaped quadratic curve that crosses the y-axis when $y = -2$ and x-axis when $3x^2 - 5x - 2 = 0$.

$3x^2 - 5x - 2 = 0 \Rightarrow (3x + 1)(x - 2) = 0$

$\Rightarrow x = -\dfrac{1}{3}$ or $x = 2$

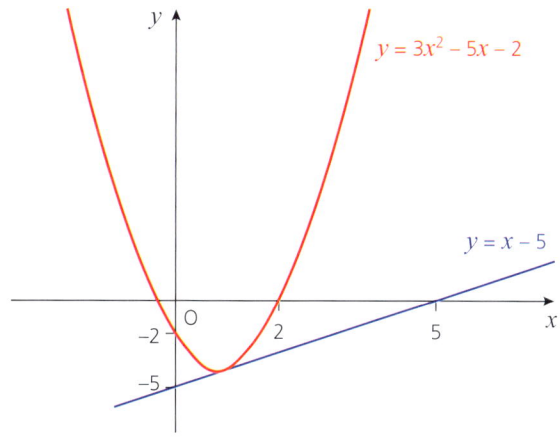

The **normal** to a curve at a given point is the straight line that is at right angles to the tangent at that point, as shown here.

Remember that for perpendicular lines $m_1 m_2 = -1$.

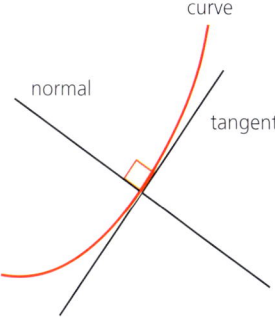

14 DIFFERENTIATION

➡ Worked example

Find the equation of the tangent and normal to the curve $y = 4x^2 - 2x^3$ at the point $(1, 2)$.
Draw a diagram showing the curve, the tangent and the normal.

Solution

It is slightly easier to use $y - y_1 = m(x - x_1)$ here than $y = mx + c$. If you substitute the gradient $m = 2$ and the point $(1, 2)$ into $y = mx + c$, you get $2 = 2 \times 1 + c$ and so $c = 0$. So the equation of the tangent is $y = 2x$.

$y = 4x^2 - 2x^3 \Rightarrow \dfrac{dy}{dx} = 8x - 6x^2$

At $(1, 2)$, the gradient is $\dfrac{dy}{dx} = 8 - 6 = 2$

The gradient of the tangent is $m_1 = 2$
So, using $\quad y - y_1 = m(x - x_1)$
the equation of the tangent is $y - 2 = 2(x - 1)$
$$y = 2x$$

The gradient of the normal is $m_2 = -\dfrac{1}{m_1} = -\dfrac{1}{2}$

So, using $\quad y - y_1 = m(x - x_1)$

the equation of the normal is $y - 2 = -\dfrac{1}{2}(x - 1)$

$$y = -\dfrac{x}{2} + \dfrac{5}{2}.$$

The curve, tangent and normal are shown on this graph.

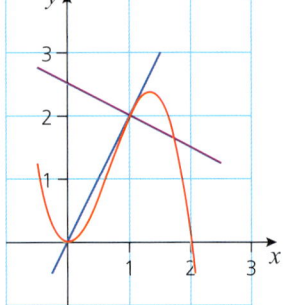

Note

The gradient at a particular point can be used to find the approximate change in y corresponding to a small change in x.

Discussion point

Consider the example above where $y = 4x^2 - 2x^3$. At the point $(1, 2)$, if the value of x increases by 0.001, what is the corresponding increase in y? What is the connection with the gradient at $(1, 2)$? What about at the points **i** $(2, 0)$ and **ii** $(0, 0)$?

➡ Worked example

a) Find the gradient of the curve $y = x^4 + 5x^3$ at $x = 1$.

b) Given that x increases from 1 to 1.0001, find the corresponding approximate change in y.

Solution

a) $y = x^4 + 5x^3$

$\Rightarrow \dfrac{dy}{dx} = 4x^3 + 15x^2$

So, when $x = 1$, $\dfrac{dy}{dx} = 4 \times 1^3 + 15 \times 1^2 = 19$

Equations of tangents and normals

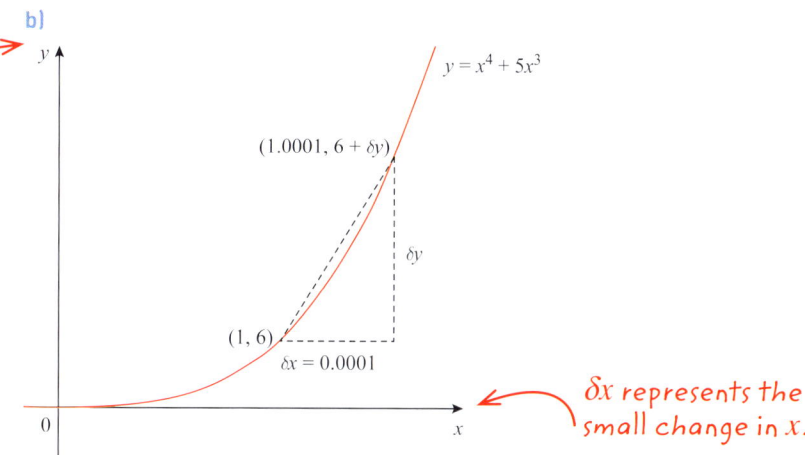

b)

δy represents the change in y corresponding to the small increase in x.

δx represents the small change in x.

$\dfrac{\delta y}{\delta x}$ is approximately equal to the gradient of the curve at $x = 1$.

$\Rightarrow \dfrac{\delta y}{\delta x} \approx 19$

$\Rightarrow \dfrac{\delta y}{0.0001} \approx 19$

$\Rightarrow \delta y \approx 0.0019$

So, y will increase by approximately 0.0019.

➡ Worked example

Find an approximation of $\sqrt{25.01}$ by considering the gradient of the graph $y = \sqrt{x}$ at the point $x = 25$.

Solution

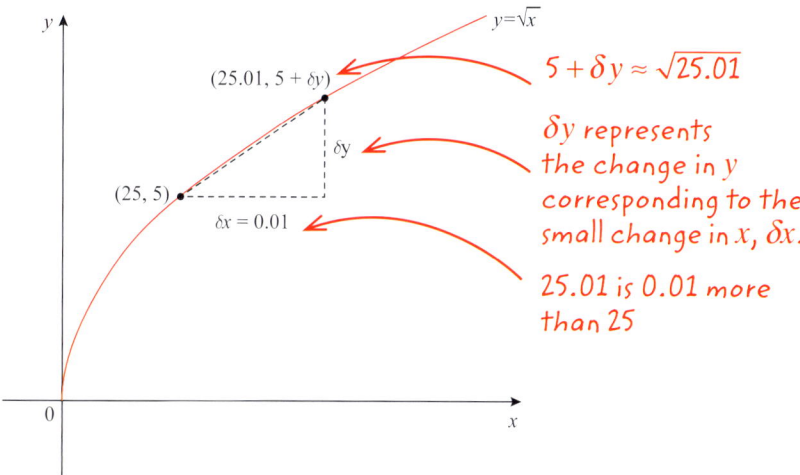

$5 + \delta y \approx \sqrt{25.01}$

δy represents the change in y corresponding to the small change in x, δx.

25.01 is 0.01 more than 25

14 DIFFERENTIATION

In order to be able to differentiate $\sqrt{x}$ you need to write it as a power of x, so $x^{\frac{1}{2}}$ ⟶ $\frac{\delta y}{\delta x}$ is approximately equal to the gradient of the curve at $x = 25$.

$$y = \sqrt{x} = x^{\frac{1}{2}}$$

$$\Rightarrow \frac{\delta y}{\delta x} = \frac{1}{2}x^{-\frac{1}{2}}$$

So, when $x = 25$, $\frac{\delta y}{\delta x} = \frac{1}{2} \times 25^{-\frac{1}{2}} = 0.1$

$$\Rightarrow \frac{\delta y}{\delta x} \approx 0.1$$

$$\Rightarrow \frac{\delta y}{0.01} \approx 0.1$$

$$\Rightarrow \delta y \approx 0.001$$

$5 + \delta y \approx \sqrt{25.01}$ ⟶ Hence, an approximation of $\sqrt{25.01}$ is 5.001.

Exercise 14.4

1. The sketch graph shows the curve of $y = 5x - x^2$. The marked point, P, has coordinates (3, 6). Find:

   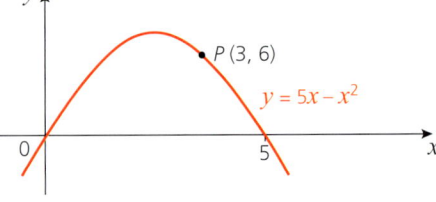

   a) the gradient function $\frac{dy}{dx}$
   b) the gradient of the curve at P
   c) the equation of the tangent at P
   d) the equation of the normal at P.

2. The sketch graph shows the curve of $y = 3x^2 - x^3$. The marked point, P, has coordinates (2, 4).

   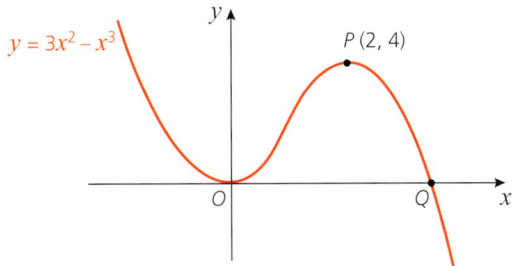

   a) Find:
      i) the gradient function $\frac{dy}{dx}$
      ii) the gradient of the curve at P
      iii) the equation of the tangent at P
      iv) the equation of the normal at P.
   b) The graph touches the x-axis at the origin O and crosses it at the point Q. Find:
      i) the coordinates of Q
      ii) the gradient of the curve at Q
      iii) the equation of the tangent at Q.
   c) Without further calculation, state the equation of the tangent to the curve at O.

3 The sketch graph shows the curve of $y = x^5 - x^3$.

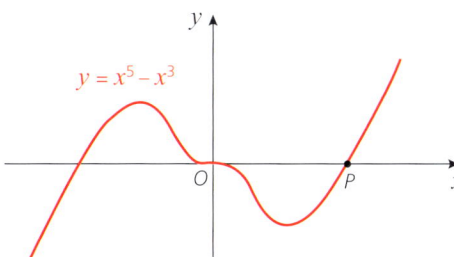

Find:
a) the coordinates of the point P where the curve crosses the positive x-axis
b) the equation of the tangent at P
c) the equation of the normal at P.

The tangent at P meets the y-axis at Q and the normal meets the y-axis at R.

d) Find the coordinates of Q and R and hence find the area of triangle PQR.

4 a) Given that $f(x) = x^3 - 3x^2 + 4x + 1$, find $f'(x)$.
   b) The point P is on the curve $y = f(x)$ and its x-coordinate is 2.
      i) Calculate the y-coordinate of P.
      ii) Find the equation of the tangent at P.
      iii) Find the equation of the normal at P.
   c) Find the values of x for which the curve has a gradient of 13.

5 The sketch graph shows the curve of $y = x^3 - 9x^2 + 23x - 15$.
The point P marked on the curve has its x-coordinate equal to 2.

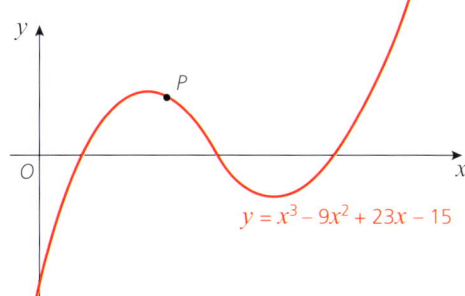

Find:
a) the gradient function $\frac{dy}{dx}$
b) the gradient of the curve at P
c) the equation of the tangent at P
d) the coordinates of another point on the curve, Q, at which the tangent is parallel to the tangent at P
e) the equation of the tangent at Q.

6 The point $(2, -8)$ is on the curve $y = x^3 - px + q$.
   a) Use this information to find a relationship between p and q.
   b) Find the gradient function $\frac{dy}{dx}$.

14 DIFFERENTIATION

Exercise 14.4 (cont)

The tangent to this curve at the point $(2, -8)$ is parallel to the x-axis.
  c) Use this information to find the value of p.
  d) Find the coordinates of the other point where the tangent is parallel to the x-axis.
  e) State the coordinates of the point P where the curve crosses the y-axis.
  f) Find the equation of the normal to the curve at the point P.

7 The sketch graph shows the curve of $y = x^2 - x - 1$.

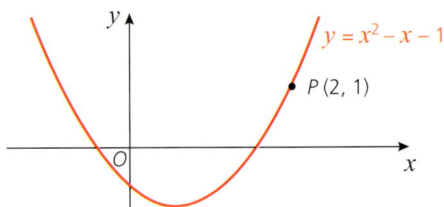

  a) Find the equation of the tangent at the point $P(2, 1)$.
  The normal at a point Q on the curve is parallel to the tangent at P.
  b) State the gradient of the tangent at Q.
  c) Find the coordinates of the point Q.

8 A curve has the equation $y = (x - 3)(7 - x)$.
  a) Find the gradient function $\frac{dy}{dx}$.
  b) Find the equation of the tangent at the point $(6, 3)$.
  c) Find the equation of the normal at the point $(6, 3)$.
  d) Which one of these lines passes through the origin?

9 A curve has the equation $y = 1.5x^3 - 3.5x^2 + 2x$.
  a) Show that the curve passes through the points $(0, 0)$ and $(1, 0)$.
  b) Find the equations of the tangents and normals at each of these points.
  c) Prove that the four lines in **b** form a rectangle.

10 a) Find the gradient of the curve $y = 3x^2 - 2x^3$ at $x = 3$.
   b) Given that x increases from 3 to 3.001, find the corresponding approximate change in y. State whether y increases or decreases.

11 Find an approximation of $\sqrt{16.01}$ by considering the gradient of the graph $y = \sqrt{x}$ at the point $x = 16$.

12 Estimate the value of $\sqrt[3]{27.1}$ by considering the gradient of the graph $y = \sqrt[3]{x}$ at an appropriate point.

Differentiating other functions of x

So far you have differentiated polynomials and other powers of x. Now this is extended to other expressions, starting with the three common trigonometrical functions. When doing this you will use the standard results that follow.

sin x, cos x and tan x

Deriving these results from first principles is beyond the scope of this book.

$y = \sin x \Rightarrow \dfrac{dy}{dx} = \cos x$

$y = \cos x \Rightarrow \dfrac{dy}{dx} = -\sin x$

$y = \tan x \Rightarrow \dfrac{dy}{dx} = \sec^2 x$ ← Recall $\sec x = \dfrac{1}{\cos x}$

Differentiating other functions of x

When differentiating any trigonometric function, the angle must be in **radians**.

➡ Worked example

Differentiate each of the following functions:

a) $y = \sin x - \cos x$

b) $y = 2\tan x + 3$

Solution

Using the results above →

a) $\dfrac{dy}{dx} = \cos x - (-\sin x)$

   $= \cos x + \sin x$

b) $y = 2\tan x + 3 \Rightarrow \dfrac{dy}{dx} = 2(\sec^2 x) + 0$ ← *Differentiating a constant always gives zero.*

   $= 2\sec^2 x$

➡ Worked example

a) Sketch the graph of $y = \sin\theta$ for $0 \leq \theta \leq 2\pi$.

b) i) Find the value of $\dfrac{dy}{d\theta}$ when $\theta = \dfrac{\pi}{2}$.

   ii) At which other point does $\dfrac{dy}{d\theta}$ have this value?

c) Use differentiation to find the value of $\dfrac{dy}{d\theta}$ when $\theta = \pi$.

Solution

a)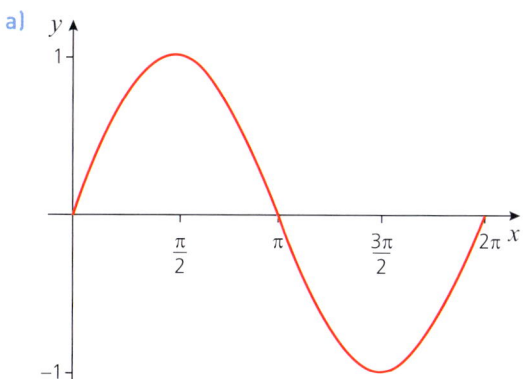

b) i) The tangent to the curve when $\theta = \dfrac{\pi}{2}$ is horizontal, so $\dfrac{dy}{d\theta} = 0$.

   ii) The gradient is also 0 when $\theta = \dfrac{3\pi}{2}$.

c) $y = \sin\theta \Rightarrow \dfrac{dy}{d\theta} = \cos\theta$

   When $\theta = \pi$, $\dfrac{dy}{d\theta} = \cos\pi = -1$.

14 DIFFERENTIATION

> **Worked example**

a) Find the turning point of the curve $y = \sin x - \cos x$ and determine its nature.

b) Sketch the curve for $0 \leq x \leq \pi$.

This means decide if it is a maximum or minimum point.

Solution

a) $y = \sin x - \cos x \Rightarrow \dfrac{dy}{dx} = \cos x + \sin x$

At the turning points $\cos x + \sin x = 0$
$\Rightarrow \qquad \sin x = -\cos x$ ← *Divide by $\cos x$*
$\Rightarrow \qquad \tan x = -1$
$\Rightarrow x = -\dfrac{\pi}{4}$ (not in the required range) or $x = \dfrac{3\pi}{4}$

When $x = \dfrac{3\pi}{4}$, $y = \sin\dfrac{3\pi}{4} - \cos\dfrac{3\pi}{4} = \sqrt{2}$

The turning point is at $\left(\dfrac{3\pi}{4}, \sqrt{2}\right)$.

When $x = \dfrac{\pi}{2}$ (to the left), $y = \sin\dfrac{\pi}{2} - \cos\dfrac{\pi}{2} = 1$.

$1 < \sqrt{2}$

When $x = \pi$ (to the right), $y = \sin\pi - \cos\pi = 1$.

So the point $\left(\dfrac{3\pi}{4}, \sqrt{2}\right)$ is a maximum turning point.

Check where the curve crosses the axes.

b) When $x = 0$, $y = \sin 0 - \cos 0 = -1$.

When $y = 0$, $0 = \sin x - \cos x$
$\Rightarrow \sin x = \cos x$ *Divide by $\cos x$*
$\Rightarrow \tan x = 1$
$\Rightarrow \qquad x = \dfrac{\pi}{4}$

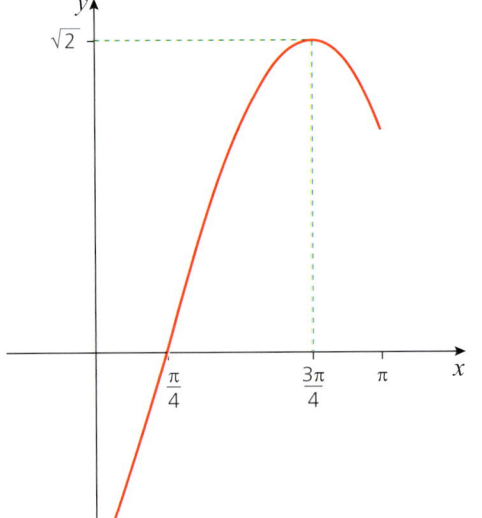

Differentiating other functions of x

➡ Worked example

For the curve $y = 2\cos\theta$ find:

a) the equation of the tangent at the point where $\theta = \frac{\pi}{3}$
b) the equation of the normal at the point where $\theta = \frac{\pi}{3}$.

Solution

a) $y = 2\cos\theta \implies \frac{dy}{d\theta} = -2\sin\theta$

When $\theta = \frac{\pi}{3}$, $y = 2\cos\frac{\pi}{3}$
$= 1$

and $\frac{dy}{d\theta} = -2\sin\frac{\pi}{3}$
$= -\sqrt{3}$

Using $y = mx + c$ ⟶ So the equation of the tangent is given by $y = -\theta\sqrt{3} + c$.

Substituting values for y and θ:

$1 = -\left(\frac{\pi}{3}\right)\sqrt{3} + c \implies c = 1 + \frac{\pi\sqrt{3}}{3}$

The equation of the tangent is therefore

$y = -\theta\sqrt{3} + 1 + \frac{\pi\sqrt{3}}{3}$.

b) The gradient of the normal $= -1 \div \frac{dy}{d\theta}$
$= -1 \div \left(-\sqrt{3}\right)$
$= \frac{1}{\sqrt{3}}$

Using $y = mx + c$ ⟶ The equation of the normal is given by $y = \frac{1}{\sqrt{3}}\theta + c$.

Substituting values for y and θ:

$1 = \frac{1}{\sqrt{3}}\left(\frac{\pi}{3}\right) + c \implies c = 1 - \frac{\pi}{3\sqrt{3}}$

The equation of the normal is therefore

$y = \frac{1}{\sqrt{3}}\theta + 1 - \frac{\pi}{3\sqrt{3}}$.

e^x and $\ln x$

Again, deriving these results from first principles is beyond the scope of this book.

You met exponential and logarithmic functions in Chapter 6. Here are the standard results for differentiating them.

$y = e^x \implies \frac{dy}{dx} = e^x$ ⟵ This is the only function where $y = \frac{dy}{dx}$.

$y = \ln x \implies \frac{dy}{dx} = \frac{1}{x}$

14 DIFFERENTIATION

> **Worked example**

Differentiate each of the following functions:

a) $y = 5\ln x$

b) $y = \ln(5x)$

c) $y = 2e^x + \ln(2x)$

Solution

a) $y = 5\ln x \Rightarrow \dfrac{dy}{dx} = 5\left(\dfrac{1}{x}\right)$

$\qquad\qquad\qquad\qquad = \dfrac{5}{x}$

b) $y = \ln(5x) \Rightarrow y = \ln 5 + \ln x$ ln 5 is a number so differentiating it gives zero.

$\qquad\qquad\Rightarrow \dfrac{dy}{dx} = \dfrac{1}{x}$

c) $y = 2e^x + \ln(2x) \Rightarrow \dfrac{dy}{dx} = 2e^x + \dfrac{1}{x}$

Part **b** shows an important result. Since $\ln(ax) = \ln a + \ln x$ for all values where $a > 0$,

$$y = \ln(ax) \Rightarrow \dfrac{dy}{dx} = \dfrac{1}{x}.$$

> **Worked example**

a) Find the turning point of the curve $y = 2x - \ln x$ and determine its nature.

b) Sketch the curve for $0 < x \leq 3$.

Solution

a) $y = 2x - \ln x \Rightarrow \dfrac{dy}{dx} = 2 - \dfrac{1}{x}$

$\qquad \dfrac{dy}{dx} = 0 \Rightarrow 2 = \dfrac{1}{x}$

$\qquad\qquad\qquad \Rightarrow x = 0.5$

When $x = 0.5$, $2x - \ln x = 1.7$ (1 d.p.).

When $x = 0.3$ (to the left), $2x - \ln x = 1.8$ (1 d.p.).

When $x = 1.0$ (to the right), $2x - \ln x = 2$ (1 d.p.).

Therefore the point $(0.5, 1.7)$ is a minimum turning point.

Differentiating other functions of x

b)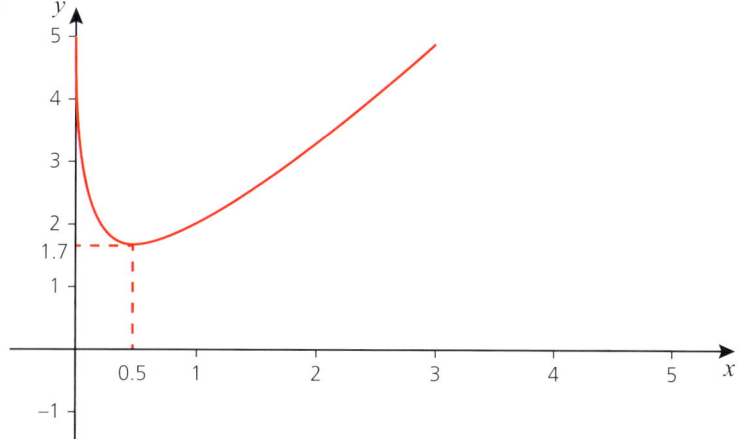

Note

In this graph the *y*-axis is an asymptote. The curve gets nearer and nearer to it but never quite reaches it.

Notice that $\ln x$ is not defined for $x \leqslant 0$, and as $x \to 0$, $\ln x \to -\infty$ so $2x - \ln x \to +\infty$.

→ Worked example

For the curve $y = 2e^x + 5$ find the equation of:

a) the tangent at the point where $x = -1$

b) the normal at the point where $x = -1$.

Solution

a) $y = 2e^x + 5 \quad \Rightarrow \quad \dfrac{dy}{dx} = 2e^x$

When $x = -1$, $y = 2e^{-1} + 5$

$$= \frac{2}{e} + 5$$

$$\frac{dy}{dx} = 2e^{-1}$$

Using $y = mx + c$ → So the equation of the tangent is given by $y = 2e^{-1}x + c$.

Substituting values for *y* and *x*:

$$\frac{2}{e} + 5 = 2e^{-1}(-1) + c$$

$$= -\frac{2}{e} + c$$

$$\Rightarrow c = \frac{4}{e} + 5$$

The equation of the tangent is therefore

$y = \dfrac{2}{e}x + \dfrac{4}{e} + 5$.

b) The gradient of the normal $= -1 \div \dfrac{dy}{dx}$

$$= -1 \div \left(\frac{2}{e}\right)$$

$$= -\frac{e}{2}$$

Using $y = mx + c$ → The equation of the normal is given by $y = -\dfrac{e}{2}x + c$.

14 DIFFERENTIATION

Substituting values for y and x:

$$\frac{2}{e} + 5 = -\frac{e}{2}(-1) + c$$

$$\Rightarrow c = \frac{2}{e} + 5 - \frac{e}{2}$$

The equation of the normal is therefore

$$y = -\frac{e}{2}x + \frac{2}{e} + 5 - \frac{e}{2}.$$

Differentiating products and quotients of functions

Deriving these results from first principles is beyond the scope of this book.

Sometimes you meet functions like $y = x^2 e^x$, which are the product of two functions, x^2 and e^x. To differentiate such functions you use the **product rule**.

When $u(x)$ and $v(x)$ are two functions of x

» $\quad y = uv \quad \Rightarrow \quad \dfrac{dy}{dx} = u\dfrac{dv}{dx} + v\dfrac{du}{dx}$

A shorthand form of $y = u(x) \times v(x)$

➡ Worked example

Differentiate $y = (x^2 + 1)(2x - 3)$

a) by expanding the brackets

b) by using the product rule.

Solution

a) $y = (x^2 + 1)(2x - 3)$
$ = 2x^3 - 3x^2 + 2x - 3$
$\Rightarrow \dfrac{dy}{dx} = 6x^2 - 6x + 2$

b) Let $u = (x^2 + 1)$ and $v = (2x - 3)$

$\dfrac{du}{dx} = 2x$ and $\dfrac{dv}{dx} = 2$

Product rule: $\dfrac{dy}{dx} = u\dfrac{dv}{dx} + v\dfrac{du}{dx}$

So $\dfrac{dy}{dx} = (x^2 + 1)(2) + (2x - 3)(2x)$

$\phantom{So \dfrac{dy}{dx}} = 2x^2 + 2 + 4x^2 - 6x$

$\phantom{So \dfrac{dy}{dx}} = 6x^2 - 6x + 2$

Differentiating products and quotients of functions

In this example you had a choice of methods; both gave you the same answer. In the next example there is no choice; you must use the product rule.

➡ Worked example

Differentiate each of the following functions:

a) $y = x^2 e^x$ b) $y = x^3 \sin x$ c) $y = (2x^3 - 4)(e^x - 1)$

Solution

a) Let $u = x^2$ and $v = e^x$

$\dfrac{du}{dx} = 2x$ and $\dfrac{dv}{dx} = e^x$

Product rule: $\dfrac{dy}{dx} = u\dfrac{dv}{dx} + v\dfrac{du}{dx}$

So $\dfrac{dy}{dx} = x^2 e^x + e^x(2x)$

$= xe^x(x+2)$

b) Let $u = x^3$ and $v = \sin x$

$\dfrac{du}{dx} = 3x^2$ and $\dfrac{dv}{dx} = \cos x$

Product rule: $\dfrac{dy}{dx} = u\dfrac{dv}{dx} + v\dfrac{du}{dx}$

So $\dfrac{dy}{dx} = x^3 \cos x + (\sin x)(3x^2)$

$= x^2(x\cos x + 3\sin x)$

c) Let $u = (2x^3 - 4)$ and $v = (e^x - 1)$

$\dfrac{du}{dx} = 6x^2$ and $\dfrac{dv}{dx} = e^x$

Product rule: $\dfrac{dy}{dx} = u\dfrac{dv}{dx} + v\dfrac{du}{dx}$

So $\dfrac{dy}{dx} = (2x^3 - 4)(e^x) + (e^x - 1)(6x^2)$

$= 2x^3 e^x - 4e^x + 6x^2 e^x - 6x^2$

Sometimes you meet functions like $y = \dfrac{e^x}{x^2 + 1}$ where one function, in this case e^x, is divided by another, $x^2 + 1$. To differentiate such functions you use the **quotient rule**.

For $y = \dfrac{u}{v}$

$\Rightarrow \dfrac{dy}{dx} = \dfrac{v\dfrac{du}{dx} - u\dfrac{dv}{dx}}{v^2}$

➡ Worked example

Differentiate $y = \dfrac{x^3 + 3}{x^2}$

a) by simplifying first

b) by using the quotient rule.

14 DIFFERENTIATION

Solution

a) $y = \dfrac{x^3 + 3}{x^2}$

$= (x^3 + 3)x^{-2}$

$= x + 3x^{-2}$

So $\dfrac{dy}{dx} = 1 - 6x^{-3}$

b) Let $u = x^3 + 3$ and $v = x^2$

$\dfrac{du}{dx} = 3x^2$ and $\dfrac{dv}{dx} = 2x$

Quotient rule: $\dfrac{dy}{dx} = \dfrac{v\dfrac{du}{dx} - u\dfrac{dv}{dx}}{v^2}$

$\dfrac{dy}{dx} = \dfrac{x^2(3x^2) - (x^3 + 3)2x}{(x^2)^2}$

$= \dfrac{3x^4 - 2x^4 - 6x}{x^4}$

$= \dfrac{x^4 - 6x}{x^4}$

$= 1 - 6x^{-3}$

This quotient rule is longer in this case, but is useful when it is not possible to simplify first.

➡ Worked example

Differentiate each of the following functions:

a) $y = \dfrac{2x^3 + 3}{x^2 - 1}$

b) $y = \dfrac{e^x}{x^2}$

Solution

a) $y = \dfrac{2x^3 + 3}{x^2 - 1}$

Let $u = 2x^3 + 3$ and $v = x^2 - 1$

$\dfrac{du}{dx} = 6x^2$ and $\dfrac{dv}{dx} = 2x$

Quotient rule: $\dfrac{dy}{dx} = \dfrac{v\dfrac{du}{dx} - u\dfrac{dv}{dx}}{v^2}$

$\dfrac{dy}{dx} = \dfrac{(x^2 - 1)6x^2 - (2x^3 + 3)2x}{(x^2 - 1)^2}$

$= \dfrac{6x^4 - 6x^2 - 4x^4 - 6x}{(x^2 - 1)^2}$

$= \dfrac{2x^4 - 6x^2 - 6x}{(x^2 - 1)^2}$

$= \dfrac{2x(x^3 - 3x - 3)}{(x^2 - 1)^2}$

b) $y = \dfrac{e^x}{x^2}$

Let $u = e^x$ and $v = x^2$

$\dfrac{du}{dx} = e^x$ and $\dfrac{dv}{dx} = 2x$

Quotient rule: $\dfrac{dy}{dx} = \dfrac{v\dfrac{du}{dx} - u\dfrac{dv}{dx}}{v^2}$

$\dfrac{dy}{dx} = \dfrac{x^2(e^x) - e^x(2x)}{(x^2)^2}$

$= \dfrac{x^2 e^x - 2xe^x}{(x^2)^2}$

$= \dfrac{xe^x(x - 2)}{x^4}$

$= \dfrac{e^x(x - 2)}{x^3}$

Differentiating composite functions

Sometimes you will need to differentiate an expression that is a function of a function.

For example, look at $y = \sqrt{x^2 + 1}$; the function 'Take the positive square root of', denoted by $\sqrt{}$, is applied to the function $(x^2 + 1)$. In such cases you use the **chain rule**.

You know how to differentiate $y = \sqrt{x}$ and you know how to differentiate $y = x^2 + 1$ but so far you have not met the case where two functions like this are combined into one.

The first step is to make a substitution.

Let $u = x^2 + 1$.

So now you have to differentiate $y = \sqrt{u}$ or $y = u^{\frac{1}{2}}$.

You know the right-hand side becomes $\frac{1}{2}u^{\frac{1}{2}-1}$ or $\frac{1}{2}u^{-\frac{1}{2}}$.

What about the left-hand side?

The differentiation is with respect to u rather than x and so you get $\frac{dy}{du}$ rather than $\frac{dy}{dx}$ that you actually want.

To go from $\frac{dy}{du}$ to $\frac{dy}{dx}$, you use the **chain rule**,

$$\frac{dy}{dx} = \frac{dy}{du} \times \frac{du}{dx}.$$

You made the substitution $u = x^2 + 1$ and differentiating this gives

$$\frac{du}{dx} = 2x.$$

So now you can substitute for both $\frac{dy}{du}$ and $\frac{du}{dx}$ in the chain rule and get

$$\frac{dy}{dx} = \frac{1}{2}u^{-\frac{1}{2}} \times 2x = xu^{-\frac{1}{2}}$$

This isn't quite the final answer because the right-hand side includes the letter u whereas it should be given entirely in terms of x.

Substituting $u = x^2 + 1$, gives $\frac{dy}{dx} = x(x^2 + 1)^{-\frac{1}{2}}$ and this is now the answer.

However it can be written more neatly as

$$\frac{dy}{dx} = \frac{x}{\sqrt{x^2 + 1}}.$$

Notice how the awkward function, $\sqrt{x^2 + 1}$, has reappeared in the final answer.

This is an important method and with experience you will find short cuts that will mean you don't have to write everything out in full as it has been here.

14 DIFFERENTIATION

> ### Worked example
>
> Given that $y = (2x-3)^4$, find $\dfrac{dy}{dx}$.
>
> ### Solution
>
> Let $u = (2x-3)$ so $y = u^4$
>
> $\dfrac{dy}{du} = 4u^3$
>
> $\phantom{\dfrac{dy}{du}} = 4(2x-3)^3$
>
> $\dfrac{du}{dx} = 2$
>
> Using $\dfrac{dy}{dx} = \dfrac{dy}{du} \times \dfrac{du}{dx}$ $\Rightarrow$ $\dfrac{dy}{dx} = 4(2x-3)^3 \times 2$
>
> $\phantom{\dfrac{dy}{dx}} = 8(2x-3)^3$

You can use the chain rule in conjunction with the product rule or the quotient rule as shown in the following example.

> ### Worked example
>
> Find $\dfrac{dy}{dx}$ when $y = (2x+1)(x+2)^{10}$.
>
> ### Solution
>
> Let $u = (2x+1)$ and $v = (x+2)^{10}$
>
> Using the chain rule to find $\dfrac{dv}{dx}$ → Then $\dfrac{du}{dx} = 2$ and $\dfrac{dv}{dx} = 10(x+2)^9 \times 1$
>
> Using the product rule $\dfrac{dy}{dx} = u\dfrac{dv}{dx} + v\dfrac{du}{dx}$
>
> $\dfrac{dy}{dx} = (2x+1) \times 10(x+2)^9 + (x+2)^{10} \times 2$
>
> $\phantom{\dfrac{dy}{dx}} = 10(2x+1)(x+2)^9 + 2(x+2)^{10}$
>
> $\phantom{\dfrac{dy}{dx}} = 2(x+2)^9 [5(2x+1) + (x+2)]$ ← Taking $2(x+2)^9$ out as a common factor
>
> $\phantom{\dfrac{dy}{dx}} = 2(x+2)^9 (11x+7)$

Note
The chain rule can also be used when solving problems that involve **connected rates of change**.

> ### Worked example
>
> The radius r of a sphere increases at a rate of 0.1 metres per second.
>
> a) Find $\dfrac{dV}{dt}$ where V is the volume of the sphere in cubic metres and t is time in seconds.
>
> b) Find the rate at which the volume is increasing when $r = 2$ metres.
>
> ### Solution
>
> a) The volume V of a sphere is given by $V = \dfrac{4}{3}\pi r^3$.
>
> Differentiating $V = \dfrac{4}{3}\pi r^3$ with respect to r → $\Rightarrow \dfrac{dV}{dr} = 4\pi r^2$
>
> From the question you have $\dfrac{dr}{dt} = 0.1$

Differentiating composite functions

By the chain rule

$$\frac{dV}{dt} = \frac{dV}{dr} \times \frac{dr}{dt}$$

$$\Rightarrow \frac{dV}{dt} = 4\pi r^2 \times 0.1$$

$$\frac{dV}{dt} = 0.4\pi r^2$$

This is the rate at which the volume is increasing with respect to time.

b) When $r = 2$, $\frac{dV}{dt} = 0.4 \times \pi \times 2^2 = 5.03$ cubic metres per second (2 d.p.).

Exercise 14.5

1 Differentiate each of the following functions:
   a) $y = 3\sin x - 2\tan x$ b) $y = 5\sin\theta - 6$ c) $y = 2\cos\theta - 2\sin\theta$
   d) $y = 4\ln x$ e) $y = \ln 4x$ f) $y = 3e^x$
   g) $y = 2e^x - \ln 2x$

2 Use the product rule to differentiate each of the following functions:
   a) $y = x\sin x$ b) $y = x\cos x$ c) $y = x\tan x$
   d) $y = e^x \sin x$ e) $y = e^x \cos x$ f) $y = e^x \tan x$

3 Use the quotient rule to differentiate each of the following functions:
   a) $y = \frac{\sin x}{x}$ b) $y = \frac{x}{\sin x}$ c) $y = \frac{\cos x}{x^2}$
   d) $y = \frac{x^2}{\cos x}$ e) $y = \frac{x}{\tan x}$ f) $y = \frac{\tan x}{x}$

4 Use the chain rule to differentiate each of the following functions:
   a) $y = (x + 3)^4$ b) $y = (2x + 3)^4$ c) $y = (x^2 + 3)^4$
   d) $y = \sqrt{x + 3}$ e) $y = \sqrt{2x + 3}$ f) $y = \sqrt{x^2 + 3}$

5 Use an appropriate method to differentiate each of the following functions:
   a) $y = \frac{\sin x}{1 + \cos x}$ b) $y = \frac{1 + \cos x}{\sin x}$
   c) $y = \sin x(1 + \cos x)$ d) $y = \cos x(1 + \sin x)$
   e) $y = \sin x(1 + \cos x)^2$ f) $y = \cos x(1 + \sin x)^2$

6 Use an appropriate method to differentiate each of the following functions:
   a) $y = e^x \ln x$ b) $y = \frac{e^x}{\ln x}$ c) $y = \frac{\ln x}{e^x}$

7 Use an appropriate method to differentiate each of the following functions:
   a) $y = e^{-x} \sin x$ b) $y = \frac{e^{-x}}{\sin x}$ c) $y = \frac{\sin x}{e^{-x}}$

8 A curve has the equation $y = \sin x - \cos x$ where x is measured in radians.
   a) Show that the curve passes through the points $(0, -1)$ and $(\pi, 1)$.
   b) Find the equations of the tangents and normals at each of these points.

9 A curve has the equation $y = 2\tan x - 1$ where x is measured in radians.
   a) Show that the curve passes through the points $(0, -1)$ and $\left(\frac{\pi}{4}, 1\right)$.
   b) Find the equations of the tangents and normals at each of these points.

10 A curve has the equation $y = 2\ln x - 1$.
   a) Show that the curve passes through the point $(e, 1)$.
   b) Find the equations of the tangent and normal at this point.

14 DIFFERENTIATION

Exercise 14.5 (cont)

11 A curve has the equation $y = e^x - \ln x$.
   a) Sketch the curves $y = e^x$ and $y = \ln x$ on the same axes and explain why this implies that $e^x - \ln x$ is always positive.
   b) Show that the curve $y = e^x - \ln x$ passes through the point (1, e).
   c) Find the equations of the tangent and normal at this point.

12 The surface area S of a sphere of radius r is given by the formula $S = 4\pi r^2$.
   a) Find the rate at which the surface area is increasing when the radius of the sphere is increasing at a rate of 0.25 cm per second.
   b) Find the rate at which the surface area is increasing when $r = 3.5$ cm, giving your answer to 2 significant figures.

Past-paper questions

1 The diagram shows a cuboid with a rectangular base of sides x cm and $2x$ cm. The height of the cuboid is y cm and its volume is 72 cm³.

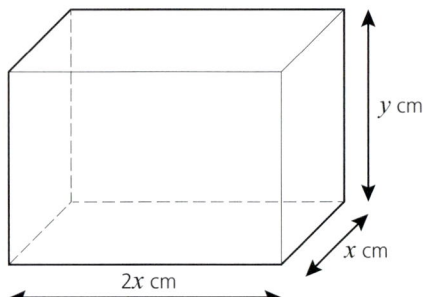

   (i) Show that the surface area A cm² of the cuboid is given by
   $A = 4x^2 + \dfrac{216}{x}$. [3]
   (ii) Given that x can vary, find the dimensions of the cuboid when A is a minimum. [4]
   (iii) Given that x increases from 2 to $2 + p$, where p is small, find, in terms of p, the corresponding approximate change in A, stating whether this change is an increase or a decrease. [3]

Cambridge O Level Additional Mathematics (4037)
Paper 11 Q12–OR, June 2011
Cambridge IGCSE Additional Mathematics (0606)
Paper 11 Q12–OR, June 2011

2 Find $\dfrac{dy}{dx}$ when
   (i) $y = \cos 2x \sin\left(\dfrac{x}{3}\right)$, [4]
   (ii) $y = \dfrac{\tan x}{1 + \ln x}$ [4]

Cambridge O Level Additional Mathematics (4037)
Paper 21 Q10, June 2014
Cambridge IGCSE Additional Mathematics (0606)
Paper 21 Q10, June 2014

Differentiating composite functions

3 (i) Differentiate $4x^3 \ln(2x+1)$ with respect to x. [3]

(ii) Given that $y = \dfrac{2x}{\sqrt{x+2}}$, show that $\dfrac{dy}{dx} = \dfrac{x+4}{(\sqrt{x+2})^3}$. [4]

Cambridge O Level Additional Mathematics (4037)
Paper 12 Q9 a & b (i), November 2013
Cambridge IGCSE Additional Mathematics (0606)
Paper 12 Q9 a & b (i), November 2013

Now you should be able to:
★ understand the idea of a derived function
★ use the notations $f'(x)$, $f''(x)$, $\dfrac{dy}{dx}$, $\dfrac{d^2y}{dx^2} \left[= \dfrac{d}{dx}\left(\dfrac{dy}{dx}\right)\right]$, dx, $dx \to 0$, $\dfrac{\delta y}{\delta x}$
★ know and use the derivatives of the standard functions x^n (for any rational n), $\sin x$, $\cos x$, $\tan x$, e^x, $\ln x$.
★ differentiate products and quotients of functions
★ use differentiation to find gradients, tangents and normals
★ use differentiation to find stationary points
★ apply differentiation to connected rates of change, small increments and approximations
★ apply differentiation to practical problems involving maxima and minima
★ use the first and second derivative tests to discriminate between maxima and minima.

Key points

✓ $y = kx^n \Rightarrow \dfrac{dy}{dx} = nkx^{n-1}$ and $y = c \Rightarrow \dfrac{dy}{dx} = 0$,
where n is a positive integer and k and c are constants.

✓ $y = f(x) + g(x) \Rightarrow \dfrac{dy}{dx} = f'(x) + g'(x)$

✓ $\dfrac{dy}{dx} = 0$ at a stationary point. The nature of the stationary point can be determined by looking at the sign of the gradient immediately either side of it or by considering the sign of $\dfrac{d^2y}{dx^2}$.

• If $\dfrac{d^2y}{dx^2} < 0$, the point is a maximum.

• If $\dfrac{d^2y}{dx^2} > 0$, the point is a minimum.

• If $\dfrac{d^2y}{dx^2} = 0$, the point could be a maximum, a minimum or a point of inflexion. Check the values of $\dfrac{dy}{dx}$ on either side of the point to determine its nature.

14 DIFFERENTIATION

✓ For the tangent and normal at (x_1, y_1)
  - the gradient of the tangent, $m_1 = \dfrac{dy}{dx}$
  - the gradient of the normal, $m_2 = -\dfrac{1}{m_1}$
  - the equation of the tangent is $y - y_1 = m_1(x - x_1)$
  - the equation of the normal is $y - y_1 = m_2(x - x_1)$.

✓ Derivatives of other functions:

Function	Derivative $\dfrac{dy}{dx}$
$\sin x$	$\cos x$
$\cos x$	$-\sin x$
$\tan x$	$\sec^2 x$
e^x	e^x
$\ln x$	$\dfrac{1}{x}$

✓ The product rule $\dfrac{dy}{dx} = u\dfrac{dv}{dx} + v\dfrac{du}{dx}$.

✓ The quotient rule $\dfrac{dy}{dx} = \dfrac{v\dfrac{du}{dx} - u\dfrac{dv}{dx}}{v^2}$.

✓ The chain rule $\dfrac{dy}{dx} = \dfrac{dy}{du} \times \dfrac{du}{dx}$.

15 Integration

Growth is painful. Change is painful. But nothing is as painful as staying stuck where you do not belong.

N.R. Narayana Murthy (born 1946)

Discussion point

Mita is a long-distance runner. She carries a speed meter, which tells her what her speed is at various times during a race.

Time (hours)	0	$\frac{1}{2}$	1	$1\frac{1}{2}$	2	$2\frac{1}{2}$	3
Speed (metres per second)	4.4	4.4	4.4	4.6	5.0	5.2	0

What race do you think she was running?

How would you estimate her time?

15 INTEGRATION

Integration is the process of getting from a differential equation to the general solution.

Integration involves using the rate of change of a quantity to find its total value at the end of an interval, for example using the speed of a runner to find the distance travelled at any time. The process is the reverse of differentiation.

Look at the **differential equation** $\frac{dy}{dx} = 3x^2$.

Since $\frac{dy}{dx} = 3x^2$ for x^3, $x^3 + 7$ and $x^3 - 3$, these expressions are all solutions of this equation.

The **general solution** of this differential equation is given as $y = x^3 + c$, where c is an **arbitrary constant** that can take any value (positive, negative or zero).

A solution containing an arbitrary constant gives a family of curves, as shown below. Each curve corresponds to a particular value of c.

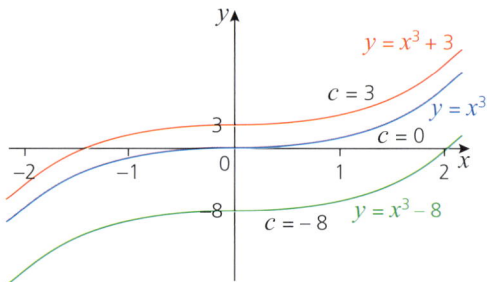

Suppose that you are also given that the solution curve passes through the point (1, 4). Substituting these coordinates in $y = x^3 + c$ gives

$$4 = 1^3 + c \implies c = 3$$

So the equation of the curve is $y = x^3 + 3$. ← *This is called the particular solution.*

This example shows that if you know a point on a curve in the family, you can find the value of c and therefore the particular solution of a differential equation.

The rule for differentiation is usually given as

$$y = x^n \implies \frac{dy}{dx} = nx^{n-1}.$$

It can also be given as

$$y = x^{n+1} \implies \frac{dy}{dx} = (n+1)x^n$$

which is the same as

$$y = \frac{1}{n+1}x^{n+1} \implies \frac{dy}{dx} = x^n.$$

Reversing this gives you the rule for integrating x^n. This is usually written using the integral symbol, $\int$.

Note that if you are asked to integrate an expression $f(x)$, this will mean integrate with respect to x unless otherwise stated.

$$\int x^n \, dx = \frac{x^{n+1}}{n+1} + c \quad \text{for } n \neq -1$$

Integration

The integral when $n = -1$ is a special case. If you try to apply the general rule, $n + 1$ is zero on the bottom line and so the expression you get is undefined. Instead you use the result that:

$$\int x^{-1} \, dx = \int \frac{1}{x} \, dx = \ln|x| + c$$

You will use this result later in the chapter.

Notice the use of dx on the left-hand side. This tells you that you are integrating with respect to x. So in this case you would read the left-hand side as 'The integral of x^n with respect to x'.

You may find it helpful to remember the rule as
» **add 1 to the power**
» **divide by the new power**
» **add a constant.**

> *Remember to include the arbitrary constant, c, until you have enough information to find a value for it.*

➜ Worked example

Integrate each of the following:

a) x^6 b) $5x^4$ c) 7 d) $4\sqrt{x}$

Solution

a) $\frac{x^7}{7} + c$ b) $5 \times \frac{x^5}{5} + c = x^5 + c$

c) 7 can be thought of as $7x^0$ so applying the rule gives $7x + c$

d) $4\sqrt{x} = 4x^{\frac{1}{2}}$ so applying the rule gives $\frac{4x^{\frac{3}{2}}}{\frac{3}{2}} + c = \frac{8}{3}x^{\frac{3}{2}} + c$

➜ Worked example

Given that $\frac{dy}{dx} = 6x^2 + 2x - 5$

a) Find the general solution of this differential equation.
b) Find the equation of the curve with this gradient function that passes through the point $(1, 7)$.

Solution

a) $y = 6 \times \frac{x^3}{3} + 2 \times \frac{x^2}{2} - 5x + c$ *(By integration)*

$= 2x^3 + x^2 - 5x + c$

b) Since $(1, 7)$ is a point on the graph

$7 = 2(1)^3 + 1^2 - 5 + c$

$\Rightarrow c = 9$

$\Rightarrow y = 2x^3 + x^2 - 5x + 9$

15 INTEGRATION

→ Worked example

Find $f(x)$ given that $f'(x) = 2x + 4$ and $f(2) = -4$.

Solution

$f'(x) = 2x + 4$

By integration → $f(x) = \frac{2x^2}{2} + 4x + c$

$\qquad = x^2 + 4x + c$

$f(2) = -4 \Rightarrow -4 = (2)^2 + 4(2) + c$

$\Rightarrow c = -16$

$\Rightarrow f(x) = x^2 + 4x - 16$

→ Worked example

A curve passes through $(3, 5)$. The gradient of the curve is given by $\frac{dy}{dx} = x^2 - 4$.

a) Find y in terms of x.

b) Find the coordinates of any stationary points of the graph of y.

c) Sketch the curve.

Solution

a) $\frac{dy}{dx} = x^2 - 4 \Rightarrow y = \frac{x^3}{3} - 4x + c$

When $x = 3$, $\quad 5 = 9 - 12 + c$

$\Rightarrow c = 8$

So the equation of the curve is $y = \frac{x^3}{3} - 4x + 8$.

b) $\frac{dy}{dx} = 0$ at all stationary points.

Substituting these values into the equation to find y

$\Rightarrow x^2 - 4 = 0$

$\Rightarrow (x+2)(x-2) = 0$

$\Rightarrow x = -2$ or $x = 2$

The stationary points are $(-2, 13\frac{1}{3})$ and $(2, 2\frac{2}{3})$.

c) It crosses the y-axis at $(0, 8)$.

The curve is a cubic with a positive x^3 term with two turning points, so it has this shape:

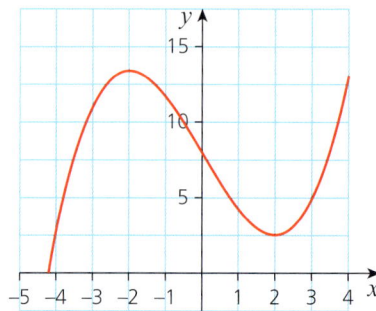

Integration

> **Worked example**

Find $\int (x^3 - 2x^2)\,dx$.

Solution

$$\int (x^3 - 2x^2)\,dx = \frac{x^4}{4} - \frac{2x^3}{3} + c$$

> **Worked example**

Find $\int (2x+1)(x-4)\,dx$

You need to multiply out the brackets before you can integrate.

Solution

$$\int (2x+1)(x-4)\,dx = \int (2x^2 - 7x - 4)\,dx$$

$$= \frac{2x^3}{3} - \frac{7x^2}{2} - 4x + c$$

Exercise 15.1

1. Find y in each of the following cases:
   a) $\frac{dy}{dx} = 4x + 2$
   b) $\frac{dy}{dx} = 6x^2 - 5x - 1$
   c) $\frac{dy}{dx} = 3 - 5x^3$
   d) $\frac{dy}{dx} = (x-2)(3x+2)$

2. Find $f(x)$ given that
   a) $f'(x) = 5x + 3$
   b) $f'(x) = x^4 + 2x^3 - x + 8$
   c) $f'(x) = (x-4)(x^2+2)$
   d) $f'(x) = (x-7)^2$

3. Find the following integrals:
   a) $\int 5\,dx$
   b) $\int 5x^3\,dx$
   c) $\int (2x - 3)\,dx$
   d) $\int (3x^3 - 4x + 3)\,dx$

4. Find the following integrals:
   a) $\int (3 - x)^2\,dx$
   b) $\int (2x + 1)(x - 3)\,dx$
   c) $\int (x + 1)^2\,dx$
   d) $\int (2x - 1)^2\,dx$

5. Find the equation of the curve $y = f(x)$ that passes through the specified point for each of the following gradient functions:
   a) $\frac{dy}{dx} = 2x - 3$; $(2, 4)$
   b) $\frac{dy}{dx} = 4 + 3x^3$ $(4, -2)$
   c) $\frac{dy}{dx} = 5x - 6$; $(-2, 4)$
   d) $f'(x) = x^2 + 1$; $(-3, -3)$
   e) $f'(x) = (x+1)(x-2)$; $(6, -2)$
   f) $f'(x) = (2x+1)^2$; $(1, -1)$

15 INTEGRATION

Exercise 15.1 (cont)

6 Find the equation of the curve $y = f(x)$ that passes through the specified point for each of the following gradient functions:
  a) $\frac{dy}{dx} = 2\sqrt{x} - 1$;   (1, 1)
  b) $f'(x) = x - \sqrt{x}$;   (4, 2)

7 You are given that $\frac{dy}{dx} = 2x + 3$.
  a) Find $\int (2x + 3)\, dx$.
  b) Find the general solution of the differential equation.
  c) Find the equation of the curve with gradient function $\frac{dy}{dx}$ and that passes through (2, −1).
  d) Hence show that (−1, −13) lies on the curve.

8 The curve C passes through the point (3, 21) and its gradient at any point is given by $\frac{dy}{dx} = 3x^2 - 4x + 1$.
  a) Find the equation of the curve C.
  b) Show that the point (−2, −9) lies on the curve.

9 a) Find $\int (4x - 1)\, dx$.
  b) Find the general solution of the differential equation $\frac{dy}{dx} = 4x - 1$.
  c) Find the particular solution that passes through the point (−1, 4).
  d) Does this curve pass above, below or through the point (2, 4)?

10 The curve $y = f(x)$ passes through the point (2, −4) and $f'(x) = 2 - 3x^2$. Find the value of $f(-1)$.

11 A curve, C, has stationary points at the points where $x = 0$ and where $x = 2$.
  a) Explain why $\frac{dy}{dx} = x^2 - 2x$ is a possible expression for the gradient of C. Give a different possible expression for $\frac{dy}{dx}$.
  b) The curve passes through the point (3, 2).
     Given that $\frac{dy}{dx}$ is $x^2 - 2x$, find the equation of C.

Definite integrals

So far, all the integrals you have met have been **indefinite integrals** such as $\int 3x^2\, dx$; the resulting expressions for y have all finished with '$+ c$'. You may or may not have been given additional information to enable you to find a value for c.

By contrast, a **definite integral** has two limits.

$$\int_1^3 3x^2\, dx$$

← This is the **upper limit**.
← This is the **lower limit**.

To find the value of a definite integral, you integrate it and substitute in the values of the limits. Then you subtract the value of the integral at the lower limit from the value of the integral at the upper limit.

Integration

➔ Worked example

Find $\int_1^3 3x^2 \, dx$.

Solution

Subtracting the value at $x = 1$ from the value at $x = 3$

$\int 3x^2 \, dx = x^3 + c$

$(3^3 + c) - (1^3 + c) = 26$ so $\int_1^3 3x^2 \, dx = 26$

Notice how the c is eliminated when you simplify this expression.

When evaluating definite integrals, it is common practice to omit the c and write

$$\int_1^3 3x^2 \, dx = [x^3]_1^3 = [3^3] - [1^3] = 26.$$

The definite integral is defined as

$$\int_a^b f'(x) \, dx = [f(x)]_a^b = f(b) - f(a).$$

➔ Worked example

'Evaluate' means 'find the numerical value of'.

Evaluate $\int_1^4 (x^2 + 3) \, dx$.

Solution

$\int_1^4 (x^2 + 3) \, dx = \left[\dfrac{x^3}{3} + 3x \right]_1^4$

$= \left(\dfrac{4^3}{3} + 3 \times 4 \right) - \left(\dfrac{1^3}{3} + 3 \times 1 \right)$

$= 30$

➔ Worked example

Evaluate $\int_{-1}^3 (x+1)(x-3) \, dx$.

Solution

Notice how you need to expand $(x+1)(x+3)$ before integrating it.

$\int_{-1}^3 (x+1)(x-3) \, dx = \int_{-1}^3 (x^2 - 2x - 3) \, dx$

$= \left[\dfrac{x^3}{3} - x^2 - 3x \right]_{-1}^3$

$= \left(\dfrac{3^3}{3} - 3^2 - 3 \times 3 \right) - \left(\dfrac{(-1)^3}{3} - (-1)^2 - 3 \times (-1) \right)$

$= -10 \dfrac{2}{3}$

15 INTEGRATION

Exercise 15.2 Evaluate the following definite integrals:

1. $\int_1^2 3x^2 \, dx$

2. $\int_1^4 4x^3 \, dx$

3. $\int_{-1}^1 6x^2 \, dx$

4. $\int_1^5 4 \, dx$

5. $\int_2^4 (x^2 + 1) \, dx$

6. $\int_{-2}^3 (2x + 5) \, dx$

7. $\int_2^5 (4x^3 - 2x + 1) \, dx$

8. $\int_5^6 (x^2 - 5) \, dx$

9. $\int_1^3 (x^2 - 3x + 1) \, dx$

10. $\int_{-1}^2 (x^2 + 3) \, dx$

11. $\int_{-4}^{-1} (16 - x^2) \, dx$

12. $\int_1^3 (x+1)(3-x) \, dx$

13. $\int_2^4 (3x(x+2)) \, dx$

14. $\int_{-1}^1 (x+1)(x-1) \, dx$

15. $\int_{-1}^2 (x + 4x^2) \, dx$

16. $\int_{-1}^1 x(x-1)(x+1) \, dx$

17. $\int_{-1}^3 (x^3 + 2) \, dx$

18. $\int_{-3}^1 (9 - x^2) \, dx$

Finding the area between a graph and the x-axis

The area under the curve $y = f(x)$ between $x = a$ and $x = b$, the shaded region in the graph below, is given by a definite integral: $\int_a^b f(x) \, dx$.

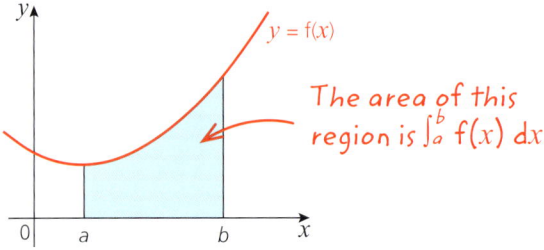

The area of this region is $\int_a^b f(x) \, dx$.

Finding the area between a graph and the x-axis

→ Worked example

Find the area of the shaded region under the curve $y = 4 - x^2$.

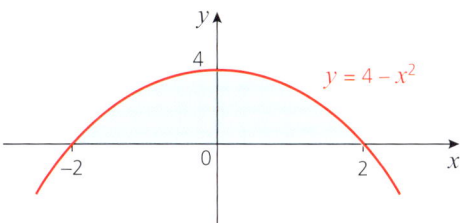

Solution

Area $= \int_{-2}^{2} (4 - x^2) \, dx = \left[4x - \dfrac{x^3}{3} \right]_{-2}^{2}$

$= \left[4 \times 2 - \dfrac{2^3}{3} \right] - \left[4 \times (-2) - \dfrac{(-2)^3}{3} \right]$

$= 10\dfrac{2}{3}$ units^2

Exercise 15.3

Find the area of each of the shaded regions:

1
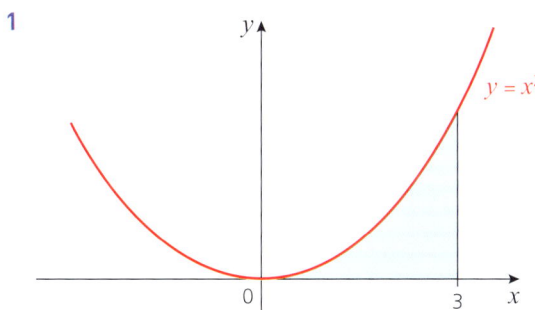

2
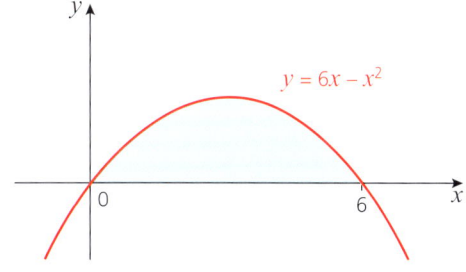

3
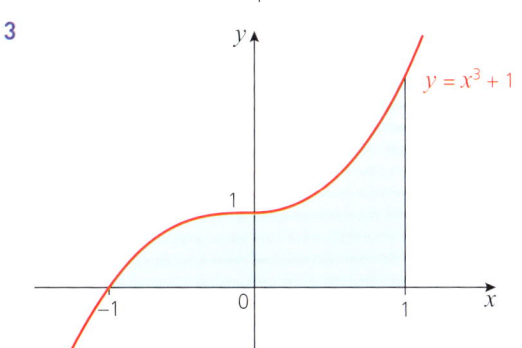

4
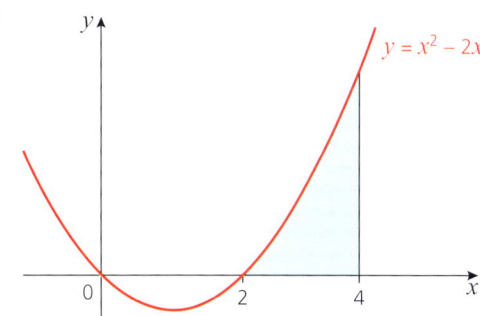

15 INTEGRATION

Exercise 15.3 (cont)

5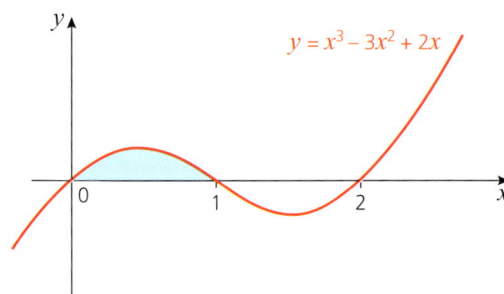

6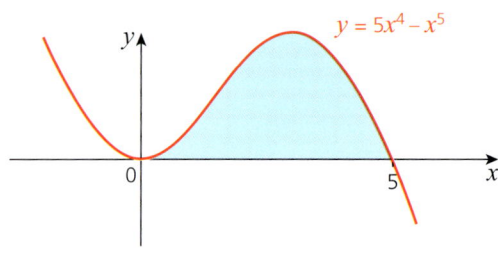

7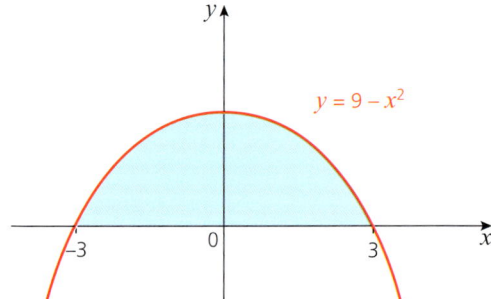

8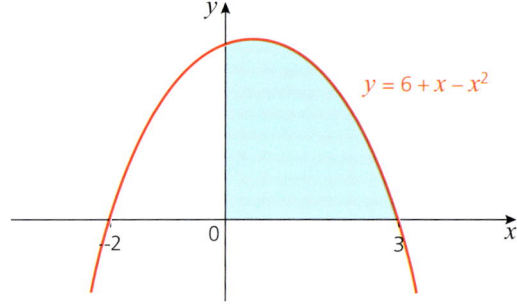

9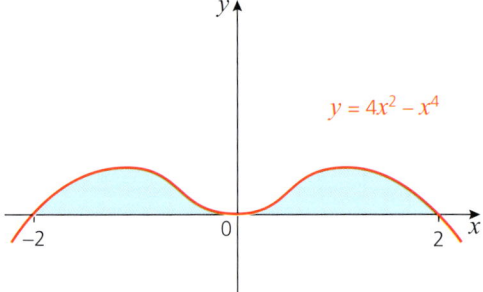

10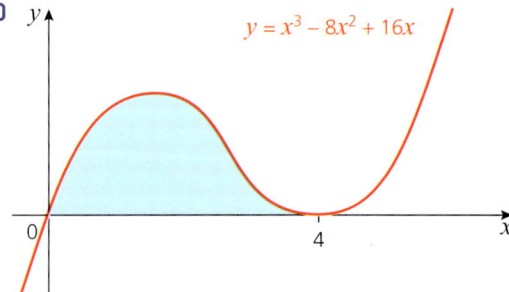

Finding the area between a graph and the x-axis

So far all the areas you have found have been above the x-axis. The next example involves a region that is below the x-axis.

→ Worked example

The diagram shows the line $y = x$ and two regions marked A and B.

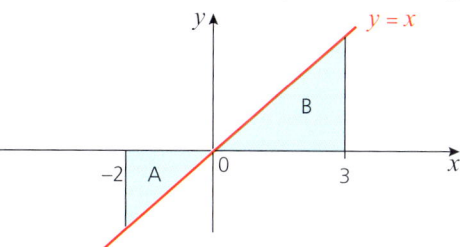

a) Calculate the areas of A and B using the formula for the area of a triangle.

b) Evaluate $\int_{-2}^{0} x\,dx$ and $\int_{0}^{3} x\,dx$. What do you notice?

c) Evaluate $\int_{-2}^{3} x\,dx$. What do you notice?

Solution

a) Area of A = $\frac{1}{2} \times 2 \times 2 = 2$ square units.

Area of B = $\frac{1}{2} \times 3 \times 3 = 4.5$ square units.

b) $\int_{-2}^{0} x\,dx = \left(\frac{x^2}{2}\right)_{-2}^{0}$

$= 0 - (2)$

$= -2$

$\int_{0}^{3} x\,dx = \left(\frac{x^2}{2}\right)_{0}^{3}$

$= 4.5 - 0$

$= 4.5$

The areas have the same numerical values as the integral but when the area is below the x-axis, the integral is negative.

c) $\int_{-2}^{3} x\,dx = \left(\frac{x^2}{2}\right)_{-2}^{3} = 4.5 - (2)$

$= 4.5 - (2)$

$= 2.5$

The areas above and below the x-axis have cancelled each other out.

This example shows you how using integration gives a negative answer for the area of a region below the x-axis. In some contexts this will make sense and in others it won't, so you always have to be careful.

15 INTEGRATION

→ Worked example

The curve $y = x(x-2)(x+2)$ is drawn on the axes.

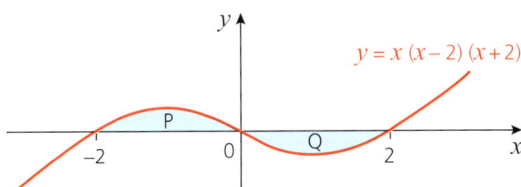

a) Use integration to find the areas of each of the shaded regions P and Q.

b) Evaluate $\int_{-2}^{2} x(x-2)(x+2)\,dx$.

c) What do you notice?

Solution

a) Area of P:
$$\int_{-2}^{0} x(x-2)(x+2)\,dx = \int_{-2}^{0} (x^3 - 4x)\,dx$$
$$= \left[\frac{x^4}{4} - 2x^2\right]_{-2}^{0}$$
$$= 0 - \left[\frac{(-2)^4}{4} - 2\times(-2)^2\right]$$
$$= 4$$

So P has an area of 4 units².

Area of Q:
$$\int_{0}^{2} x(x-2)(x+2)\,dx = \int_{0}^{2} (x^3 - 4x)\,dx$$
$$= \left[\frac{x^4}{4} - 2x^2\right]_{0}^{2}$$
$$= \left[\frac{2^4}{4} - 2\times 2^2\right] - 0$$
$$= -4$$

The areas of P and Q are the same since the curve has rotational symmetry about the origin.

So Q also has an area of 4 units².

b)
$$\int_{-2}^{2} x(x-2)(x+2)\,dx = \int_{-2}^{2} (x^3 - 4x)\,dx$$
$$= \left[\frac{x^4}{4} - 2x^2\right]_{-2}^{2}$$
$$= \left[\frac{2^4}{4} - 2\times 2^2\right] - \left[\frac{(-2)^4}{4} - 2\times(-2)^2\right]$$
$$= 0$$

Always draw a sketch graph when you are going to calculate areas. This will avoid any cancelling out of areas above and below the x-axis.

c) The areas of P and Q have 'cancelled out'.

Finding the area between a graph and the x-axis

You can also use integration to find the area of a region enclosed between a line and a curve or the area enclosed between two curves.

➡ Worked example

The diagram shows a sketch of the curve $y = x^2 - 2x + 1$ and the line $y = x + 1$.

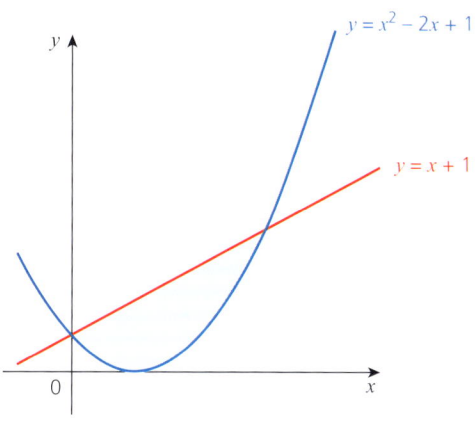

a) Find the coordinates of the points of intersection of the line and the curve.

b) Find the area of the shaded region.

Solution

a) $y = x^2 - 2x + 1$

$y = x + 1$

$\Rightarrow x^2 - 2x + 1 = x + 1$

$\Rightarrow x^2 - 3x = 0$

$\Rightarrow x(x - 3) = 0$

So, $x = 0$ or $x = 3$

Substituting the x-values into $y = x + 1$ ⟶ $x = 0 \Rightarrow y = 1$ and $x = 3 \Rightarrow y = 4$

Hence the curves intersect at (0, 1) and (3, 4).

b)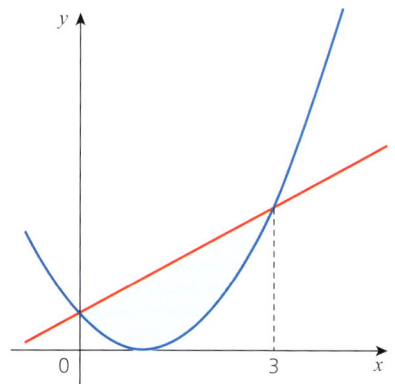

15 INTEGRATION

To find the area of the shaded region, you should subtract the area under the curve from the area under the line, between $x = 0$ and $x = 3$.

So, the area of the shaded region is given by

Area under the line $\longrightarrow \int_0^3 (x+1)\,dx - \int_0^3 (x^2 - 2x + 1)\,dx \longleftarrow$ *Area under the curve*

The limits of the integrals are the x-values of the points of intersection.

$$= \int_0^3 (3x - x^2)\,dx \longleftarrow \text{The integrals can be combined.}$$

$$= \left[\frac{3}{2}x^2 - \frac{1}{3}x^3\right]_0^3$$

$$= \frac{3}{2} \times 3^2 - \frac{1}{3} \times 3^3$$

$$= \frac{9}{2} \text{ units}^2$$

→ Worked example

The diagram shows a sketch of the curves $y = x^2$ and $y = 4x - x^2$.

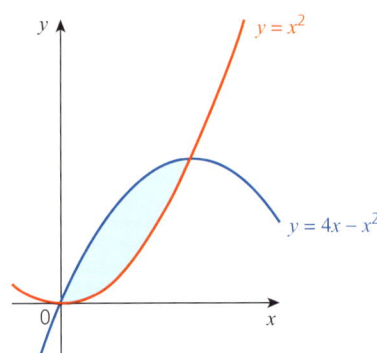

a) Find the coordinates of the points of intersection of the two curves.
b) Find the area of the shaded region between the two curves.

Solution
a) $y = x^2$
$y = 4x - x^2$
$\Rightarrow x^2 = 4x - x^2$
$\Rightarrow 2x^2 - 4x = 0$
$\Rightarrow 2x(x - 2) = 0$

So, $x = 0$ or $x = 2$

Substituting the x-values into $y = x^2$ $\longrightarrow$ $x = 0 \Rightarrow y = 0$ and $x = 2 \Rightarrow y = 4$

Hence the curves intersect at $(0, 0)$ and $(2, 4)$.

Finding the area between a graph and the x-axis

b)

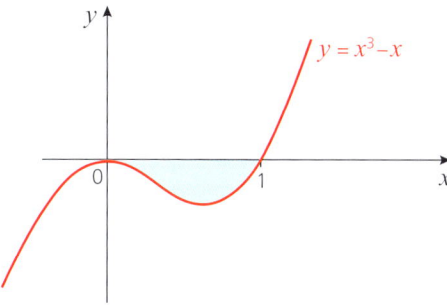

To find the area of the shaded region, you should subtract the area under the bottom curve from the area under the top curve, between $x = 0$ and $x = 2$.

So, the area of the shaded region is given by

Top curve − bottom curve →
$$\int_0^2 (4x - x^2)\,dx - \int_0^2 x^2\,dx$$

Combining the integrals →
$$= \int_0^2 (4x - 2x^2)\,dx$$

$$= \left[2x^2 - \frac{2}{3}x^3\right]_0^2$$

$$= 2 \times 2^2 - \frac{2}{3} \times 2^3$$

$$= \frac{8}{3} \text{ units}^2$$

Exercise 15.4

1. The sketch shows the curve $y = x^3 - x$. Calculate the area of the shaded region.

15 INTEGRATION

Exercise 15.4 (cont)

2 The sketch shows the curve $y = x^3 - 4x^2 + 3x$.
   a) Calculate the area of each shaded region.
   b) State the area enclosed between the curve and the x-axis.

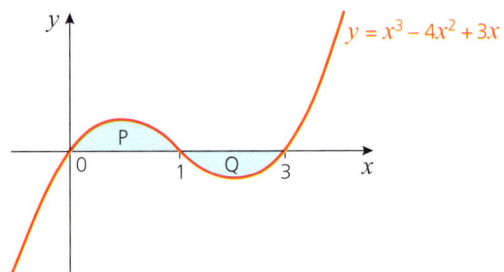

3 The sketch shows the curve $y = x^4 - 2x$.
   a) Find the coordinates of the point A.
   b) Calculate the area of the shaded region.

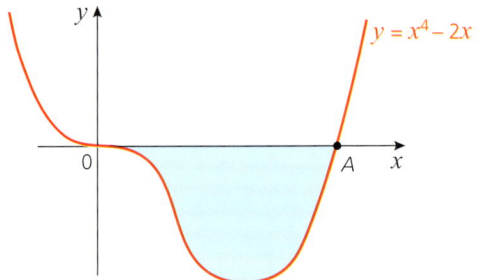

4 The sketch shows the curve $y = x^3 + x^2 - 6x$.
   Work out the area between the curve and the x-axis.

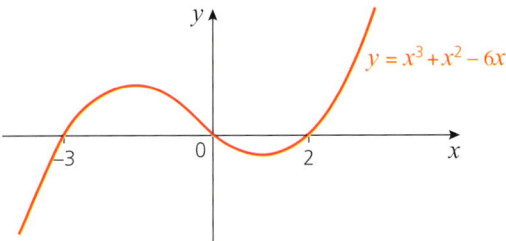

5 a) Sketch the curve $y = x^2$ for $-3 < x < 3$.
   b) Shade the area bounded by the curve, the lines $x = -1$ and $x = 2$ and the x-axis.
   c) Find, by integration, the area of the region you have shaded.

6 a) Sketch the curve $y = x^2 - 2x$ for $-1 < x < 3$.
   b) For what values of x does the curve lie below the x-axis?
   c) Find the area between the curve and the x-axis.

7 a) Sketch the curve $y = x^3$ for $-3 < x < 3$.
   b) Shade the area between the curve, the x-axis and the line $x = 2$.
   c) Find, by integration, the area of the region you have shaded.
   d) Without any further calculation, state, with reasons, the value of
   $$\int_{-2}^{2} x^3 \, dx.$$

8 a) Shade, on a suitable sketch, the region with an area given by
$$\int_{-1}^{2} (x^2 + 1)\,dx.$$
  b) Evaluate this integral.

9 a) Evaluate $\int_{1}^{4} (2x + 1)\,dx$.
  b) Interpret this integral on a sketch graph.

10 The diagram shows a sketch of the curve $y = x^2$ and the line $y = x + 2$.

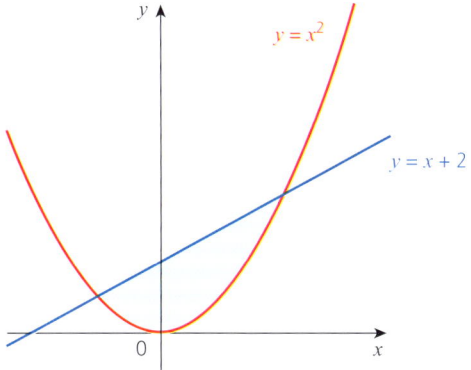

  a) Find the coordinates of the points of intersection of the line and the curve.
  b) Find, by integration, the area of the shaded region.

11 a) On the same axes, sketch the line $y = 2x + 8$ and the curve $y = 16 - x^2$ for $-4 \leq x \leq 4$.
  b) Find the area of the region enclosed by the line $y = 2x + 8$ and the curve $y = 16 - x^2$.

12 The diagram shows a sketch of the curves $y = (x - 3)^2$ and $y = 9 - x^2$.

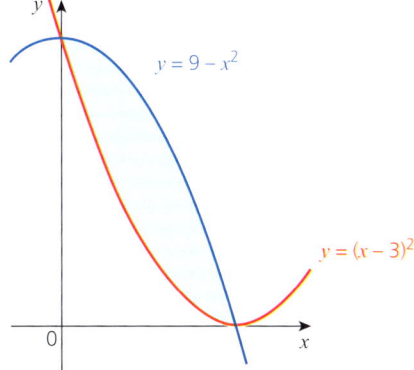

  a) Find the coordinates of the points of intersection of the two curves.
  b) Find the area of the shaded region.

15 INTEGRATION

Integrating other functions of x

As with differentiation, there are a number of special cases when integrating. The proofs are beyond the scope of this book, but you will need to know and be able to use these results.

Integrating $\frac{1}{x}$ (i.e. x^{-1}) does not follow the general rule since that would give $\frac{x^0}{0}$ which is undefined.

Differentiation $\Rightarrow$	Basic integral $\Rightarrow$	Generalised integral
$y = x^n \Rightarrow \frac{dy}{dx} = nx^{n-1}$	$\int x^n dx = \frac{x^{n+1}}{n+1} + c$, for $n \neq -1$	$\int (ax+b)^n dx = \frac{1}{a}\frac{(ax+b)^{n+1}}{n+1} + c$, for $n \neq -1$
$y = \sin x \Rightarrow \frac{dy}{dx} = \cos x$	$\int \cos x\, dx = \sin x + c$	$\int \cos(ax+b) dx = \frac{1}{a}\sin(ax+b) + c$
$y = \cos x \Rightarrow \frac{dy}{dx} = -\sin x$	$\int \sin x\, dx = -\cos x + c$	$\int \sin(ax+b) dx = -\frac{1}{a}\cos(ax+b) + c$
$y = \tan x \Rightarrow \frac{dy}{dx} = \sec^2 x$	$\int \sec^2 x\, dx = \tan x + c$	$\int \sec^2(ax+b) dx = \frac{1}{a}\tan(ax+b) + c$
$y = e^x \Rightarrow \frac{dy}{dx} = e^x$	$\int e^x dx = e^x + c$	$\int e^{(ax+b)} dx = \frac{1}{a}e^{(ax+b)} + c$
$y = \ln x \Rightarrow \frac{dy}{dx} = \frac{1}{x}$	$\int \frac{1}{x} dx = \ln|x| + c$	$\int \frac{1}{ax+b} dx = \frac{1}{a}\ln|ax+b| + c$

➡ Worked example

Find the following indefinite integrals:

a) $\int \frac{1}{2x-3} dx$ b) $\int (2x-3)^4 dx$ c) $\int (2x-3)^{\frac{1}{2}} dx$

d) $\int e^{2x-3} dx$ e) $\int \sin(2x-3) dx$ f) $\int \cos(2x-3) dx$

Solution

a) Using $\int \frac{1}{ax+b} dx = \frac{1}{a}\ln|ax+b| + c$

   gives $\int \frac{1}{2x-3} dx = \frac{1}{2}\ln|2x-3| + c$

b) Using $\int (ax+b)^n dx = \frac{1}{a}\frac{(ax+b)^{n+1}}{n+1} + c$

   gives $\int (2x-3)^4 dx = \frac{1}{2}\frac{(2x-3)^5}{5} + c = \frac{(2x-3)^5}{10} + c$

c) Using $\int (ax+b)^n dx = \frac{1}{a}\frac{(ax+b)^{n+1}}{n+1} + c$

   gives $\int (2x-3)^{\frac{1}{2}} dx = \frac{1}{2}\frac{(2x-3)^{\frac{3}{2}}}{\frac{3}{2}} + c = \frac{1}{3}(2x-3)^{\frac{3}{2}} + c$

Integrating other functions of x

d) Using $\int e^{ax+b} dx = \frac{1}{a} e^{ax+b} + c$

gives $\int e^{2x-3} dx = \frac{1}{2} e^{2x-3} + c$

e) Using $\int \sin(ax+b) dx = -\frac{1}{a} \cos(ax+b) + c$

gives $\int \sin(2x-3) dx = -\frac{1}{2} \cos(2x-3) + c$

f) Using $\int \cos(ax+b) dx = \frac{1}{a} \sin(ax+b) + c$

gives $\int \cos(2x-3) dx = \frac{1}{2} \sin(2x-3) + c$

→ Worked example

When integrating trigonometric functions, the angles must be in radians

Evaluate the following definite integrals:

a) $\int_2^3 \frac{1}{2x+1} dx$ b) $\int_2^3 (2x+1)^4 dx$ c) $\int_2^3 e^{2x+1} dx$

d) $\int_0^1 (2x+1)^{\frac{1}{2}} dx$ e) $\int_0^1 (2x+1)^{-2} dx$ f) $\int_0^{\frac{\pi}{3}} \sin\left(2x + \frac{\pi}{6}\right) dx$

g) $\int_0^{\frac{\pi}{3}} \cos\left(2x + \frac{\pi}{6}\right) dx$ h) $\int_0^{\frac{\pi}{12}} \sec^2\left(2x + \frac{\pi}{6}\right) dx$

Solution

Using $\int \frac{1}{ax+b} dx$
$= \frac{1}{a} \ln|ax+b| + c$

a) $\int_2^3 \frac{1}{2x+1} dx = \left[\frac{1}{2} \ln|2x+1|\right]_2^3$

$= \frac{1}{2} \ln 7 - \frac{1}{2} \ln 5$

$= \frac{1}{2}(\ln 7 - \ln 5)$

$= \frac{1}{2} \ln \frac{7}{5}$

Using $\int (ax+b)^n dx$
$= \frac{1}{a} \frac{(ax+b)^{n+1}}{n+1} + c$

b) $\int_2^3 (2x+1)^4 dx = \left[\frac{1}{2} \frac{(2x+1)^5}{5}\right]_2^3$

$= \frac{1}{2}\left[\frac{7^5}{5} - \frac{5^5}{5}\right]$

$= 1368.2$

Using $\int e^{ax+b} dx$
$= \frac{1}{a} e^{ax+b} + c$

c) $\int_2^3 e^{2x+1} dx = \left[\frac{1}{2} e^{2x+1}\right]_2^3$

$= \frac{1}{2}(e^7 - e^5)$

$= 474.1$

15 INTEGRATION

d) $\int_0^1 (2x+1)^{\frac{1}{2}} dx = \left[\frac{1}{2} \frac{(2x+1)^{\frac{3}{2}}}{\frac{3}{2}} \right]_0^1$

$= \frac{1}{3}\left[3^{\frac{3}{2}} - 1^{\frac{3}{2}} \right]$

$= 1.40 \text{ (3 s.f.)}$

e) $\int_0^1 (2x+1)^{-2} dx = \left[\frac{1}{2} \frac{(2x+1)^{-1}}{-1} \right]_0^1$

$= -\frac{1}{2}[3^{-1} - 1^{-1}]$

$= \frac{1}{3}$

Using $\int \sin(ax+b) dx$
$= -\frac{1}{a}\cos(ax+b) + c$

f) $\int_0^{\frac{\pi}{3}} \sin\left(2x + \frac{\pi}{6}\right) dx = \left[-\frac{1}{2}\cos\left(2x + \frac{\pi}{6}\right) \right]_0^{\frac{\pi}{3}}$

$= -\frac{1}{2}\cos\frac{5\pi}{6} + \frac{1}{2}\cos\frac{\pi}{6}$

$= \frac{\sqrt{3}}{2}$

Using $\int \cos(ax+b) dx$
$= \frac{1}{a}\sin(ax+b) + c$

g) $\int_0^{\frac{\pi}{3}} \cos\left(2x + \frac{\pi}{6}\right) dx = \left[\frac{1}{2}\sin\left(2x + \frac{\pi}{6}\right) \right]_0^{\frac{\pi}{3}}$

$= \frac{1}{2}\sin\frac{5\pi}{6} - \frac{1}{2}\sin\frac{\pi}{6}$

$= 0$

Using $\int \sec^2(ax+b) dx$
$= \frac{1}{a}\tan(ax+b) + c$

h) $\int_0^{\frac{\pi}{12}} \sec^2\left(2x + \frac{\pi}{6}\right) dx = \left[\frac{1}{2}\tan\left(2x + \frac{\pi}{6}\right) \right]_0^{\frac{\pi}{12}}$

$= \frac{1}{2}\tan\frac{\pi}{3} - \frac{1}{2}\tan\frac{\pi}{6}$

$= \frac{\sqrt{3}}{3}$

Exercise 15.5

1 Find the following indefinite integrals:

a) $\int \frac{1}{3x+1} dx$

b) $\int (3x+1)^4 dx$

c) $\int e^{3x+1} dx$

d) $\int \sin(3x+1) dx$

e) $\int \cos(3x+1) dx$

f) $\int \frac{3}{x-3} dx$

g) $\int (2x-1)^3 dx$

h) $\int 4e^{2x-3} dx$

i) $\int 3\sin(3x) dx$

j) $\int 4\cos\left(\frac{x}{2}\right) dx$

k) $\int (x-2)^{\frac{3}{2}} dx$

l) $\int (2x-1)^{\frac{3}{2}} dx$

Integrating other functions of x

2 Evaluate the following definite integrals:

a) $\int_2^4 \dfrac{1}{3x+1}\,dx$ b) $\int_2^4 (3x+1)^4\,dx$ c) $\int_2^4 e^{3x+1}\,dx$

d) $\int_0^{\frac{\pi}{3}} \sin\left(3x+\dfrac{\pi}{3}\right)dx$ e) $\int_0^{\frac{\pi}{3}} \cos\left(3x+\dfrac{\pi}{3}\right)dx$ f) $\int_4^8 \dfrac{4}{x-2}\,dx$

g) $\int_{-1}^3 (2x+3)^4\,dx$ h) $\int_0^2 10e^{-2x}\,dx$ i) $\int_0^{\frac{\pi}{2}} \sin\left(2x-\dfrac{\pi}{4}\right)dx$

j) $\int_0^{\frac{\pi}{2}} \cos\left(2x-\dfrac{\pi}{4}\right)dx$ k) $\int_0^{\frac{\pi}{36}} \sec^2\left(3x+\dfrac{\pi}{4}\right)dx$

Past-paper questions

1 (a) A curve is such that $\dfrac{dy}{dx} = ae^{1-x} - 3x^2$, where a is a constant. At the point $(1, 4)$, the gradient of the curve is 2.

   (i) Find the value of a. [1]

   (ii) Find the equation of the curve. [5]

 (b) (i) Find $\int (7x+8)^{\frac{1}{3}}\,dx$. [2]

   (ii) Hence evaluate $\int_0^8 (7x+8)^{\frac{1}{3}}\,dx$. [2]

Cambridge O Level Additional Mathematics (4037)
Paper 11 Q10, June 2011
Cambridge IGCSE Additional Mathematics (0606)
Paper 11 Q10, June 2011

2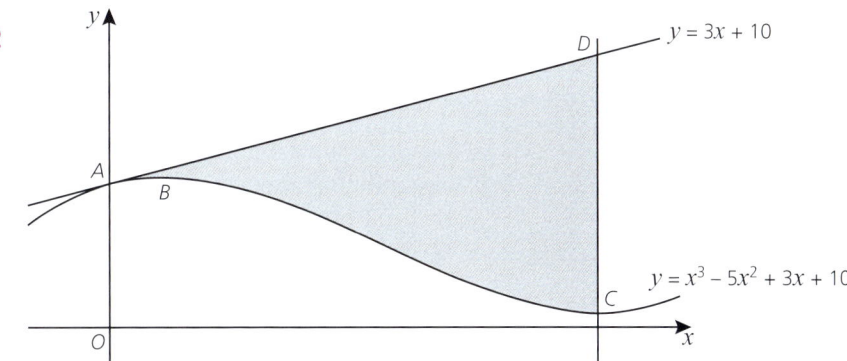

The diagram shows parts of the line $y = 3x + 10$ and the curve $y = x^3 - 5x^2 + 3x + 10$. The line and the curve both pass through the point A on the y-axis. The curve has a maximum at the point B and a minimum at the point C. The line through C, parallel to the y-axis, intersects the line $y = 3x + 10$ at the point D.

15 INTEGRATION

(i) Show that the line AD is a tangent to the curve at A. [2]
(ii) Find the x-coordinate of B and of C. [3]
(iii) Find the area of the shaded region $ABCD$, showing all your working. [5]

Cambridge O Level Additional Mathematics (4037)
Paper 11 Q9, June 2015
Cambridge IGCSE Additional Mathematics (0606)
Paper 11 Q9, June 2015

3 The diagram below shows part of the curve $y = 3x - 14 + \dfrac{32}{x^2}$ cutting the x-axis at the points P and Q.

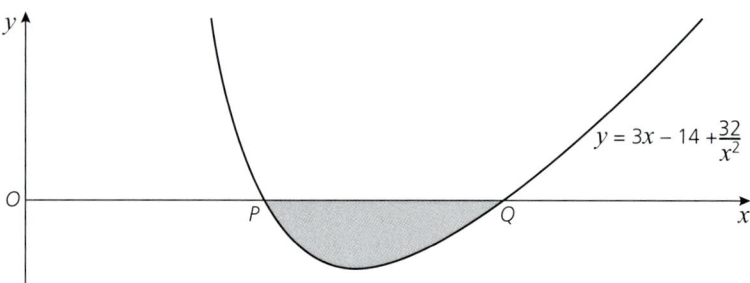

(i) State the x-coordinates of P and Q. [1]

(ii) Find $\int \left(3x - 14 + \dfrac{32}{x^2}\right) dx$ and hence determine the area of the shaded region. [4]

Cambridge O Level Additional Mathematics (4037)
Paper 22 Q12 (iii) & (iv), November 2014
Cambridge IGCSE Additional Mathematics (0606)
Paper 22 Q12 (iii) & (iv), November 2014

Now you should be able to:
★ understand integration as the reverse process of differentiation
★ integrate sums of terms in powers of x, including $\dfrac{1}{x}$ and $\dfrac{1}{(ax+b)}$
★ integrate functions of the form:
  ● $(ax+b)^n$ for any rational n
  ● $\sin(ax+b)$
  ● $\cos(ax+b)$
  ● $\sec^2(ax+b)$
  ● e^{ax+b}
★ evaluate definite integrals and apply integration to the evaluation of plane areas.

Integrating other functions of x

Key points

✔ $\dfrac{dy}{dx} = x^n \Rightarrow y = \dfrac{x^{n+1}}{n+1} + c$ for $n \neq -1$

This is an indefinite integral.

✔ $\int_a^b x^n \, dx = \left[\dfrac{x^{n+1}}{n+1} \right]_a^b = \dfrac{b^{n+1} - a^{n+1}}{n+1}$ for $n \neq -1$

This is a definite integral.

✔ The area of a region between a curve $y = f(x)$ and the x-axis is given by $\int_a^b y \, dx$.

Area of A $= \int_a^b y \, dx = \int_a^b f(x) \, dx$

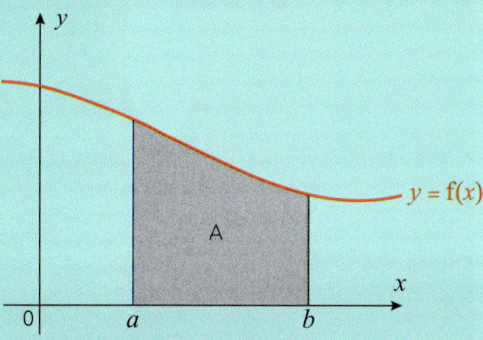

✔ Areas below the x-axis give rise to negative values for the integral.
✔ Integrals of other functions where c is a constant:

Function $y = f(x)$	Integral $\int y \, dx$
$\dfrac{1}{x}$	$\ln|x| + c$
$\dfrac{1}{ax+b}$	$\dfrac{1}{a} \ln|ax+b| + c$
$(ax+b)^n$	$\dfrac{1}{a} \dfrac{(ax+b)^{n+1}}{n+1} + c$
e^{ax+b}	$\dfrac{1}{a} e^{ax+b} + c$
$\sin(ax+b)$	$-\dfrac{1}{a} \cos(ax+b) + c$
$\cos(ax+b)$	$\dfrac{1}{a} \sin(ax+b) + c$
$\sec^2(ax+b)$	$\dfrac{1}{a} \tan(ax+b) + c$

311

16 Kinematics

Do not look at stars as bright spots only. Try to take in the vastness of the universe.

Maria Mitchell, First Professor of Astronomy,
Vasser College (1818–1889)

Discussion point

A spacecraft leaves the Earth on a journey to Jupiter. Its initial direction is directly towards Jupiter. Will it travel in a straight line?

Displacement and velocity

You know that $\frac{dy}{dx}$ represents the rate of change of y with respect to x. It gives the gradient of the x–y graph, where x is plotted on the horizontal axis and y on the vertical axis.

The following graph represents the distance, s metres, travelled by a cyclist along a country road in time, t seconds. Time is measured along the horizontal axis and distance from the starting point is measured on the vertical axis. When he reaches E the cyclist takes a short break and then returns home along the same road.

Displacement and velocity

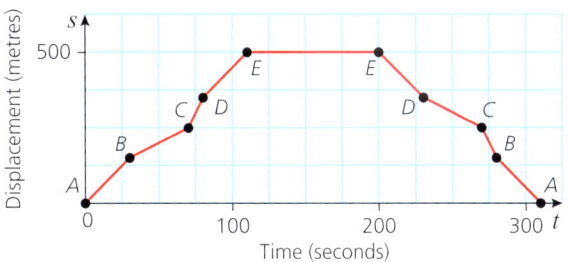

Speed is given by the gradient of the distance–time graph. In this graph the axes are labelled s and t, rather than y and x, so the gradient (representing the speed) is given by $\frac{ds}{dt}$.

A graph showing **displacement** (the distance from the starting point) looks quite different from one showing the total distance travelled.

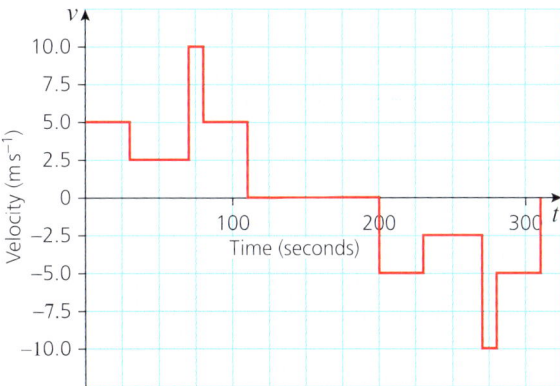

Velocity is given by the gradient of a displacement–time graph.

Acceleration is $\left(\frac{\text{change in velocity}}{\text{time taken}}\right)$ and this is given by the gradient of a velocity–time graph.

Motion in a straight line

*You will be treating each object as a **particle**, i.e. something with a mass but no dimension.*

In the work that follows you will use displacement, which measures position, rather than distance travelled.

Before doing anything else you need to make two important decisions:

1. where you will take your origin
2. which direction you will take as positive.

Some options are:

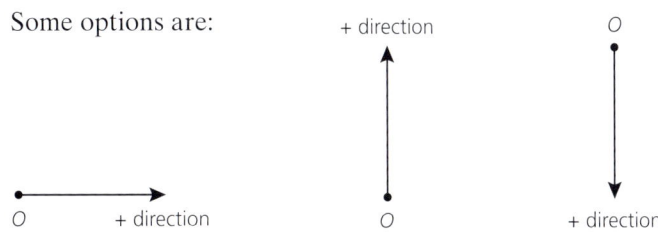

16 KINEMATICS

Think about the motion of a tennis ball that is thrown up vertically and allowed to fall to the ground, as in the diagram below. Assume that the ball leaves your hand at a height of 1 m above the ground and rises a further 2 m to the highest point. At this point the ball is *instantaneously at rest.*

This means it is about to change direction through 180°.

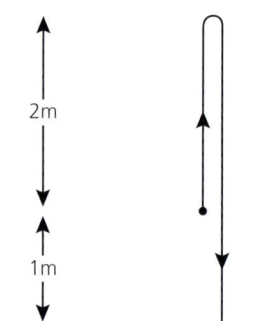

The displacement–time graph of the ball's flight is shown below. For this graph, displacement is measured from ground level with upwards as the positive direction.

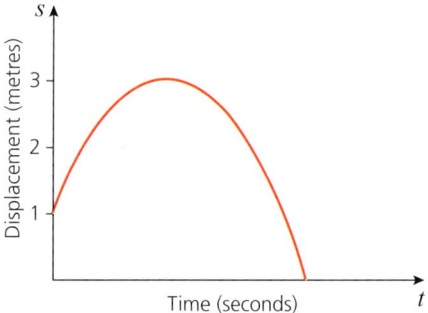

> **Note**
>
> Be careful not to confuse the terms velocity and speed. Speed has magnitude (size) but no direction. Velocity has direction and magnitude. For example, taking upwards as the positive direction,
> - a speed of 3 m s^{-1} upwards is a velocity +3 m s^{-1}
> - a speed of 3 m s^{-1} downwards is a velocity of −3 m s^{-1}.
>
>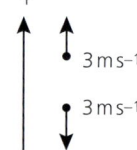

Displacement and velocity

The table gives the terms that you will be using, together with their definitions, units and the letters that are commonly used to represent those quantities.

Quantity	Definition	S.I. unit	Unit symbol	Notation
Time	Measured from a fixed origin	second	s	t
Distance	Distance travelled in a given time	metre	m	x, y, s
Speed	Rate of change of distance	metre per second	$m\,s^{-1}$	$v = \dfrac{dx}{dt}$ etc.
Displacement	Distance from a fixed origin	metre	m	x, y, s, h
Velocity	Rate of change of displacement	metre per second	$m\,s^{-1}$	$v = \dfrac{ds}{dt}$
Acceleration	Rate of change of velocity	metre per second per second	$m\,s^{-2}$	$a = \dfrac{dv}{dt} = \dfrac{d^2s}{dt^2}$ etc.

Like velocity, acceleration can also be either positive or negative. A negative acceleration is an object either moving in the positive direction and slowing down, or moving in the negative direction and speeding up.

Motion with variable acceleration: the general case

If the motion involves variable acceleration, you must use calculus. You should know and be able to use these relationships.

$$v = \frac{ds}{dt} \qquad a = \frac{dv}{dt} = \frac{d^2s}{dt^2}$$

These relationships are used in the next two examples.

➜ Worked example

a) The displacement in metres, s, of a sports car from its initial position during the first 4 seconds is given by
   $s = 12t^2 - t^3$.
   Find:
   i   an expression for the velocity in terms of t
   ii the initial velocity
   iii the velocity after 4 seconds
   iv an expression for the acceleration in terms of t
   v   the acceleration after 4 seconds.

b) The national speed limit in Great Britain is 70 mph.
   At the end of 4 seconds, would the driver of this sports car be breaking the British national speed limit?

315

Solution

a) i $v = \dfrac{ds}{dt}$
   $= 24t - 3t^2$

   ii When $t = 0$, $v = 0$
   The initial velocity is $0\,\text{m s}^{-1}$.

   iii When $t = 4$, $v = 24 \times 4 - 3 \times 4^2 = 48$
   The velocity after 4 seconds is $48\,\text{m s}^{-1}$.

   iv $a = \dfrac{dv}{dt} = 24 - 6t$

   v When $t = 4$, $a = 24 - 6 \times 4 = 0$
   The acceleration after 4 seconds is $0\,\text{m s}^{-2}$.

b) $48\,\text{m s}^{-1} = \dfrac{48 \times 60 \times 60}{1000} = 172.8\,\text{km h}^{-1}$

   $172.8\,\text{km h}^{-1} \approx \dfrac{5}{8} \times 172.8 = 108\,\text{mph}$

   The driver would be breaking the British speed limit.

→ Worked example

A particle travels in a straight line such that t seconds after passing through a fixed point O, its displacement s metres is given by $s = 5 + 2t^3 - 3t^2$.

a) Find:
   i expressions for the velocity and acceleration in terms of t
   ii the times when it is at rest.

b) Draw for $0 \leqslant t \leqslant 2$:
   i the displacement–time graph
   ii the distance–time graph
   iii the velocity–time graph
   iv the speed–time graph
   v the acceleration–time graph.

c) Find:
   i how far it is from O when it is at rest
   ii the initial acceleration of the particle.

Solution

a) i $v = \dfrac{ds}{dt} = 6t^2 - 6t$

   Notice that the acceleration varies with time.

   $a = \dfrac{dv}{dt} = 12t - 6$

   ii The particle is at rest when $v = 0$.
   $\Rightarrow \quad 6t^2 - 6t = 0$
   $\Rightarrow \quad 6t(t - 1) = 0$
   $\Rightarrow \quad t = 0 \text{ or } t = 1$

   So the particle is at rest initially and after 1 second.

b) i

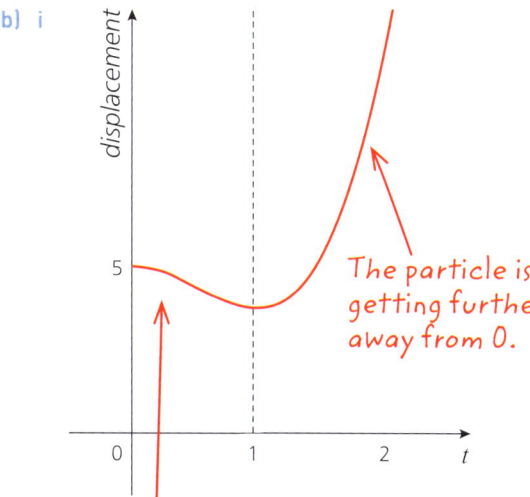

The particle is getting close to 0.

The particle is getting further away from 0.

ii

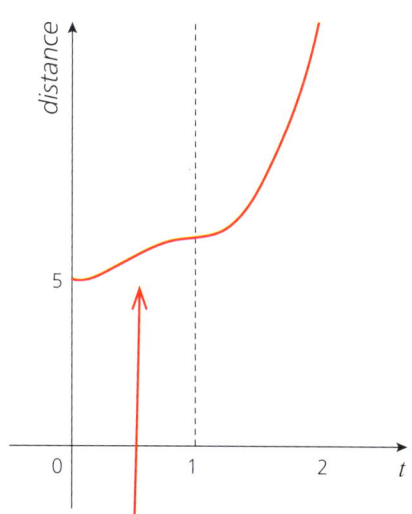

The graph increases here as distance does not depend on the direction that the particle travels.

iii

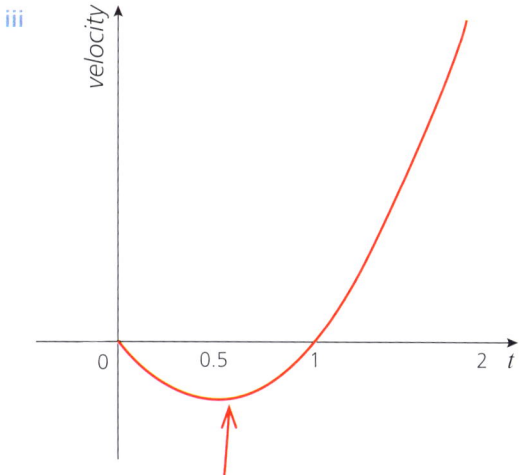

The negative values of the velocity show that the particle is moving towards 0.

iv

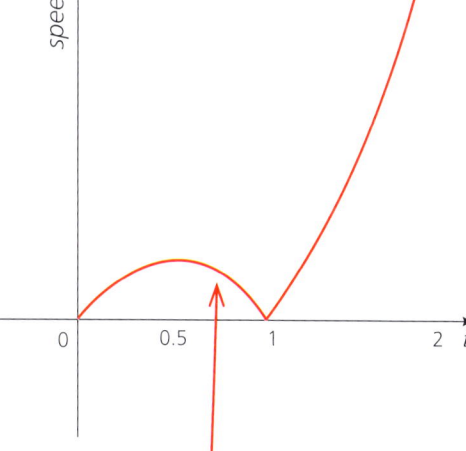

Speed cannot be negative. Notice how the corresponding section of the velocity–time graph has been reflected in the x-axis.

16 KINEMATICS

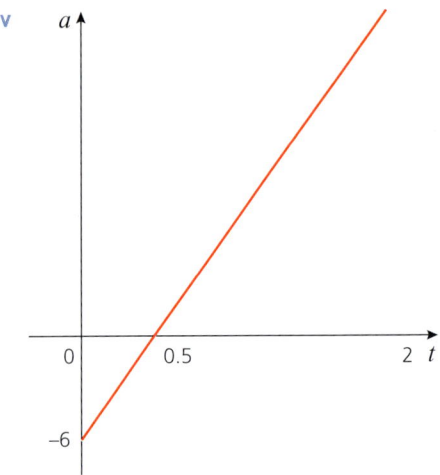

c) i When $t = 0$, $s = 5$.
When $t = 1$, $s = 5 + 2 - 3 = 4$.
The particle is at rest initially when it is 5 m from O
and after 1 second when it is instantaneously at rest 4 m from O.

ii When $t = 0$, $a = -6$. The initial acceleration is $-6\,\mathrm{m\,s^{-2}}$.

Discussion point

How would you interpret the negative acceleration in the above example?

Exercise 16.1

1 A particle moves in a straight line such that at time t seconds after passing through a fixed point O, its displacement s metres is given by $s = 4t^3 - 6t^2 + 2$.
   a) Write expressions for the velocity and acceleration of the particle in terms of t.
   b) Given the displacement–time graph, draw the corresponding distance–time graph for $0 \leq t \leq 2$.

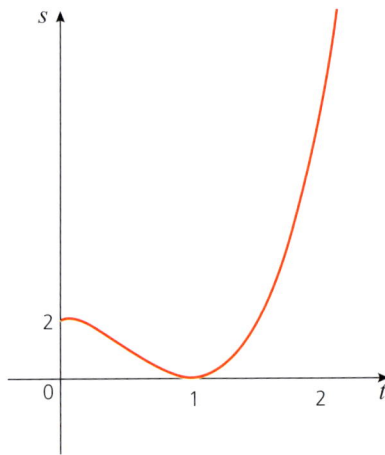

c) Draw for $0 \leq t \leq 2$:
  i) the velocity–time graph
  ii) the speed–time graph
  iii) the acceleration–time graph.

2. In each of the following cases $t \geq 0$. The quantities are given in SI units, so distances are in metres and times in seconds:
   i) find expressions for the velocity and acceleration at time t
   ii) use these expressions to find the initial position, velocity and acceleration
   iii) find the time and position when the velocity is zero.
   a) $s = 5t^2 - t + 3$
   b) $s = 3t - t^3$
   c) $s = t^4 - 4t - 6$
   d) $s = 4t^3 - 3t + 5$
   e) $s = 5 - 2t^2 + t$

3. A particle is projected in a straight line from a point O. After t seconds its displacement, s metres, from O is given by $s = 3t^2 - t^3$.
   a) Write expressions for the velocity and acceleration at time t.
   b) Find the times when the body is instantaneously at rest.
   c) What distance is travelled between these times?
   d) Find the velocity when $t = 4$ and interpret your result.
   e) Find the initial acceleration.

4. A ball is thrown upwards and its height, h metres, above ground after t seconds is given by $h = 1 + 4t - 5t^2$.
   a) From what height was the ball projected?
   b) Write an expression for the velocity of the ball at time t.
   c) When is the ball instantaneously at rest?
   d) What is the greatest height reached by the ball?
   e) After what length of time does the ball hit the ground?
   f) Sketch the graph of h against t.
   g) At what speed is the ball travelling when it hits the ground?

5. In the early stages of its motion the height of a rocket, h metres, is given by $h = \frac{1}{6}t^4$, where t seconds is the time after launch.
   a) Find expressions for the velocity and acceleration of the rocket at time t.
   b) After how long is the acceleration of the rocket $72\,\text{m s}^{-2}$?
   c) Find the height and velocity of the rocket at this time.

6. The velocity of a moving object at time t seconds is given by $v\,\text{m s}^{-1}$, where $v = 15t - 2t^2 - 25$.
   a) Find the times when the object is instantaneously at rest.
   b) Find the acceleration at these times.
   c) Find the velocity when the acceleration is zero.
   d) Sketch the graph of v against t.

16 KINEMATICS

Finding displacement from velocity and velocity from acceleration

In the previous section, you used the result $v = \frac{ds}{dt}$; in other words when s was given as an expression in t, you differentiated to find v. Therefore when v is given as an expression in t, integrating v gives an expression for s,

$$s = \int v \, dt.$$

Similarly, you can reverse the result $a = \frac{dv}{dt}$ to give

$$v = \int a \, dt.$$

➡ Worked example

A particle P moves in a straight line so that at time t seconds its acceleration is $(6t + 2) \, \text{m s}^{-2}$.

The acceleration is not constant.

P passes through a point O at time $t = 0$ with a velocity of $3 \, \text{m s}^{-1}$.
Find:

a) the velocity of P in terms of t

b) the distance of P from O when $t = 2$.

Solution

a) $v = \int a \, dt$

$= \int (6t + 2) \, dt$

$= 3t^2 + 2t + c$

When $t = 0$, $v = 3$

$\Rightarrow \quad c = 3.$

c represents the initial velocity.

Therefore $v = 3t^2 + 2t + 3$.

b) $s = \int v \, dt$

$= \int (3t^2 + 2t + 3) \, dt$

$= t^3 + t^2 + 3t + k$

When $t = 0$, $s = 0$

k is the value of the displacement when $t = 0$.

$\Rightarrow \quad k = 0$

$\Rightarrow \quad s = t^3 + t^2 + 3t$

When $t = 2$, $s = 8 + 4 + 6 = 18$.

When $t = 2$ the particle is $18 \, \text{m}$ from O.

Finding displacement from velocity and velocity from acceleration

→ Worked example

The acceleration of a particle, $a\,\text{m s}^{-2}$, at time t seconds is given by $a = 6 - 2t$.

When $t = 0$, the particle is at rest at a point 4 m from the origin O.

This tells you that the acceleration varies with time.

a) Find expressions for the velocity and displacement in terms of t.

b) Find when the particle is next at rest, and its displacement from O at that time.

Solution

a) $v = \int a\,\mathrm{d}t$

$ = \int (6 - 2t)\,\mathrm{d}t$

$ = 6t - t^2 + c$

When $t = 0$, $v = 0$ (given) $\Rightarrow c = 0$

Therefore $v = 6t - t^2$.

$s = \int v\,\mathrm{d}t$

$ = \int (6t - t^2)\,\mathrm{d}t$

$ = 3t^2 - \dfrac{t^3}{3} + k$

When $t = 0$, $s = 4$ (given) $\Rightarrow k = 4$

Therefore $s = 3t^2 - \dfrac{t^3}{3} + 4$.

b) The particle is at rest when $v = 0 \Rightarrow 6t - t^2 = 0$
$\Rightarrow t(6 - t) = 0$
$\Rightarrow t = 0 \text{ or } t = 6$

The particle is next at rest after 6 seconds.

When $t = 6$, $s = 3 \times 6^2 - \dfrac{6^3}{3} + 4 = 40$

The particle is 40 m from O after 6 seconds.

→ Worked example

A particle is projected along a straight line.

Its velocity, $v\,\text{m s}^{-1}$, after t seconds is given by $v = 2t + 3$.

a) Sketch the graph of v against t.

b) Find the distance the particle moves in the third second.

Solution

a) $v = 2t + 3$ is a straight line with gradient 2 that passes through (0, 3).

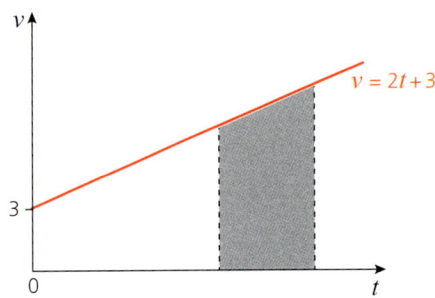

b) **Method 1**

The graph shows that the velocity is always positive, so the velocity and speed are the same. The distance travelled is equal to the area under the graph.

The third second starts when $t = 2$ and finishes when $t = 3$.

Using the formula for the area of a trapezium,

$$\text{distance} = \tfrac{1}{2}(7+9) \times 1 = 8\,\text{m}.$$

Method 2

The area under a graph can also be found using integration.

$$\begin{aligned}
\text{Distance} &= \int_a^b v \; dt \\
&= \int_2^3 (2t + 3) \; dt \\
&= \left[t^2 + 3t \right]_2^3 \\
&= [9 + 9] - [4 + 6] \\
&= 8\,\text{m}
\end{aligned}$$

Discussion point

- Which method did you prefer to use in the previous example?
- Which method would you need to use if v was given by $v = 3t^2 + 2$?
  - Is acceleration constant in this case? How can you tell?
- Could you have used the constant acceleration (*suvat*) equations?
- Can you use calculus when acceleration is constant?

Exercise 16.2

1. Find expressions for the velocity, v, and displacement, s, at time t in each of the following cases:
   a) $a = 2 - 6t$; when $t = 0$, $v = 1$ and $s = 0$
   b) $a = 4t$; when $t = 0$, $v = 4$ and $s = 3$
   c) $a = 12t^2 - 4$; when $t = 0$, $v = 2$ and $s = 1$
   d) $a = 2$; when $t = 0$, $v = 2$ and $s = 4$
   e) $a = 4 + t$; when $t = 0$, $v = 1$ and $s = 3$

Finding displacement from velocity and velocity from acceleration

2 A particle *P* sets off from the origin, *O*, with a velocity of $9\,\text{m s}^{-1}$ and moves along the *x*-axis.
   At time *t* seconds, its acceleration is given by $a = (6t - 12)\,\text{m s}^{-2}$.
   a) Find expressions for the velocity and displacement at time *t*.
   b) Find the time when the particle returns to its starting point.

3 A particle *P* starts from rest at a fixed origin *O* when $t = 0$.
   The acceleration $a\,\text{m s}^{-2}$ at time *t* seconds is given by $a = 6t - 6$.
   a) Find the velocity of the particle after 1 second.
   b) Find the time after leaving the origin when the particle is next instantaneously at rest, and the distance travelled to this point.

4 The speed, $v\,\text{m s}^{-1}$, of a car during braking is given by $v = 30 - 5t$, where *t* seconds is the time since the brakes were applied.
   a) Sketch a graph of *v* against *t*.
   b) How long does the car take to stop?
   c) How far does it travel while braking?

5 A particle *P* moves in a straight line, starting from rest at the point *O*.
   t seconds after leaving *O*, the acceleration, $a\,\text{m s}^{-2}$, of *P* is given by $a = 4 + 12t$.
   a) Find an expression for the velocity of the particle at time *t*.
   b) Calculate the distance travelled by *P* in the third second.

6 The velocity, $v\,\text{m s}^{-1}$, of a particle *P* at time *t* seconds is given by $v = t^3 - 4t^2 + 4t + 2$.
   P moves in a straight line.
   a) Find an expression for the acceleration, $a\,\text{m s}^{-2}$, in terms of *t*.
   b) Find the times at which the acceleration is zero, and say what is happening between these times.
   c) Find the distance travelled in the first three seconds.

Past-paper questions

1 A particle *P* moves in a straight line such that, *t* s after leaving a point *O*, its velocity $v\,\text{m s}^{-1}$ is given by $v = 36t - 3t^2$ for $t \geqslant 0$.
   (i) Find the value of *t* when the velocity of *P* stops increasing. [2]
   (ii) Find the value of *t* when *P* comes to instantaneous rest. [2]
   (iii) Find the distance of *P* from *O* when *P* is at instantaneous rest. [3]
   (iv) Find the speed of *P* when *P* is again at *O*. [4]

   Cambridge O Level Additional Mathematics (4037)
   Paper 12 Q12, June 2013
   Cambridge IGCSE Additional Mathematics (0606)
   Paper 12 Q12, June 2013

2 A particle travels in a straight line so that, *t* s after passing through a fixed point *O*, its velocity, $v\,\text{m s}^{-1}$, is given by $v = 3 + 6\sin 2t$.
   (i) Find the velocity of the particle when $t = \dfrac{\pi}{4}$. [1]
   (ii) Find the acceleration of the particle when $t = 2$. [3]
   The particle first comes to instantaneous rest at the point *P*.
   (iii) Find the distance *OP*. [5]

   Cambridge O Level Additional Mathematics (4037)
   Paper 23 Q9, November 2013
   Cambridge IGCSE Additional Mathematics (0606)
   Paper 23 Q9, November 2013

16 KINEMATICS

3 A particle travels in a straight line so that, t s after passing through a fixed point O, its displacement s m from O is given by $s = \ln(t^2 + 1)$.
 (i) Find the value of t when $s = 5$. [2]
 (ii) Find the distance travelled by the particle during the third second. [2]
 (iii) Show that, when $t = 2$, the velocity of the particle is $0.8\,\text{m}\,\text{s}^{-1}$. [2]
 (iv) Find the acceleration of the particle when $t = 2$. [3]

Cambridge O Level Additional Mathematics (4037)
Paper 13 Q10, November 2010
Cambridge IGCSE Additional Mathematics (0606)
Paper 13 Q10, November 2010

Now you should be able to:
★ apply differentiation and integration to kinematics problems that involve displacement, velocity and acceleration of a particle moving in a straight line with variable or constant acceleration
★ make use of the above relationships to draw and use the following graphs:
 • displacement–time
 • distance–time
 • velocity–time
 • speed–time
 • acceleration–time.

Key points

Quantity	Definition	SI unit	Unit symbol	Notation
Time	Measured from a fixed origin	second	s	t
Distance	Distance travelled in a given time	metre	m	x, y, s
Speed	Rate of change of distance	metre per second	$\text{m}\,\text{s}^{-1}$	$v = \dfrac{dx}{dt}$ etc.
Displacement	Distance from a fixed origin	metre	m	x, y, s, h
Velocity	Rate of change of displacement	metre per second	$\text{m}\,\text{s}^{-1}$	$v = \dfrac{ds}{dt}$
Acceleration	Rate of change of velocity	metre per second per second	$\text{m}\,\text{s}^{-2}$	$a = \dfrac{dv}{dt} = \dfrac{d^2s}{dt^2}$ etc.

✔ For a **displacement–time** graph, the gradient is the velocity.
✔ For a **velocity–time** graph the gradient is the acceleration and the area under the graph is the displacement.
✔ For a **distance–time** graph the gradient is the speed.
✔ For general motion:
 • $s = \int v\,dt$ (Displacement is the area under a velocity–time graph.)
 • $v = \int a\,dt$

Review exercise 5

 1 Find $\dfrac{dy}{dx}$ when

  a) $y = \ln(2x+1)\sin 4x$ [4]

  b) $y = \dfrac{e^x}{x^3}$ [4]

2 Given that $y = \dfrac{6x}{\sqrt{x-4}}$:

  a) Find $\dfrac{dy}{dx}$. [4]

  b) Find the equation of the tangent to the curve when $x = 13$. Give your answer in the form $ax + by + c = 0$. [4]

3

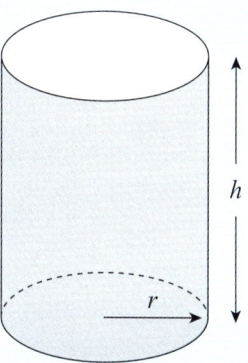

A container is a circular cylinder, open at one end, with a base radius of r cm and a height of h cm. The volume of the container is $1000\,\text{cm}^3$. Given that r and h can vary and that the total outer surface area of the container has a minimum value, find this value. [8]

Cambridge O Level Additional Mathematics (4037)
Paper 22 Q11, February/March 2020
Cambridge IGCSE Additional Mathematics (0606)
Paper 22 Q11, February/March 2020

4 **a)** Find $\displaystyle\int (2x+3)^2 \, dx$. [2]

  b) Hence evaluate $\displaystyle\int_{-1}^{2} (2x+3)^2 \, dx$. [2]

 5 A curve is such that when $x = 0$, both $y = 6$ and $\dfrac{dy}{dx} = -2$.

Given that $\dfrac{d^2y}{dx^2} = 8\sin 2x + 1$, find the equation of the curve. [7]

REVIEW EXERCISE 5

6 **i** Show that $5 + 4\tan^2\left(\frac{x}{3}\right) = 4\sec^2\left(\frac{x}{3}\right) + 1$. [1]

ii Given that $\frac{d}{dx}\left(\tan\left(\frac{x}{3}\right)\right) = \frac{1}{3}\sec^2\left(\frac{x}{3}\right)$, find $\int \sec^2\left(\frac{x}{3}\right) dx$. [1]

iii

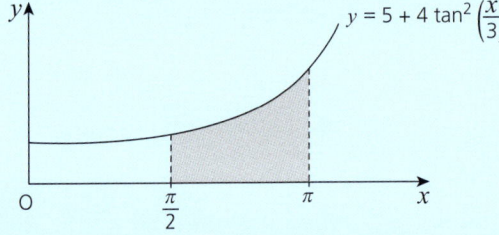

The diagram shows part of the curve $y = 5 + 4\tan^2\left(\frac{x}{3}\right)$. Using the results from parts **i** and **ii**, find the exact area of the shaded region enclosed by the curve, the x-axis and the lines $x = \frac{\pi}{2}$ and $x = \pi$. [5]

Cambridge O Level Additional Mathematics (4037)
Paper 11 Q9, May/June 2017
Cambridge IGCSE Additional Mathematics (0606)
Paper 11 Q9, May/June 2017

7 a)

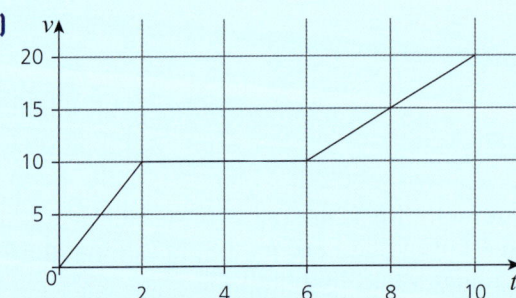

The diagram shows the velocity–time graph of a particle P moving in a straight line with velocity $v\,\text{m s}^{-1}$ at time t seconds after leaving a fixed point.
  i Write down the value of the acceleration of P when $t = 5$. [1]
  ii Find the distance travelled by the particle P between $t = 0$ and $t = 10$. [2]
b) A particle Q moves such that its velocity, $v\,\text{m s}^{-1}$, t seconds after leaving a fixed point, is given by $v = 3\sin 2t - 1$.
  i Find the speed of Q when $t = \frac{7\pi}{12}$. [2]
  ii Find the least value of t for which the acceleration of Q is zero. [3]

Cambridge O Level Additional Mathematics (4037)
Paper 12 Q8, February/March 2019
Cambridge IGCSE Additional Mathematics (0606)
Paper 12 Q8, February/March 2019

Mathematical notation

Miscellaneous symbols

$=$	is equal to	$\leqslant$	is less than or equal to
$\neq$	is not equal to	$>$	is greater than
$\equiv$	is identical to or is congruent to	$\geqslant$	is greater than or equal to
$\approx$	is approximately equal to	∞	infinity
$\sim$	is distributed as	$\Rightarrow$	implies
$\cong$	is isomorphic to	$\Leftarrow$	is implied by
$\propto$	is proportional to	$\Leftrightarrow$	implies and is implied by (is equivalent to)
$<$	is less than		

Operations

$a + b$	a plus b
$a - b$	a minus b
$a \times b, ab$	a multiplied by b
$a \div b, \frac{a}{b}$	a divided by b
$\sum_{i=1}^{n} a_i$	$a_1 + a_2 + \ldots + a_n$
$\sqrt{a}$	the non-negative square root of a, for $a \in \mathbb{R}, a \geqslant 0$
$\sqrt[n]{a}$	the (real) nth root of a, for $a \in \mathbb{R}$, where $\sqrt[n]{a} \geqslant 0$ for $a \geqslant 0$
$\|a\|$	the modulus of a
$n!$	n factorial
$\binom{n}{r}$, ^nC_r	the binomial coefficient $\frac{n!}{r!(n-r)!}$ for $n, r \in \mathbb{Z}$ and $0 \leqslant r \leqslant n$

Functions

$f(x)$	the value of the function f at x
$f : A \to B$	f is a function under which each element of set A has an image in set B
$f : x \mapsto y$	the function f maps the element x to the element y
f^{-1}	the inverse function of the one-one function f
gf	the composite function of f and g, which is defined by $gf(x) = g(f(x))$

MATHEMATICAL NOTATION

$\lim_{x \to a} f(x)$	the limit of $f(x)$ as x tends to a
$\Delta x, \delta x$	an increment of x
$\dfrac{dy}{dx}$	the derivative of y with respect to x
$\dfrac{d^n y}{dx^n}$	the nth derivative of y with respect to x
$f'(x), f''(x), \ldots, f^n(x)$	the first, second, ..., nth derivatives of $f(x)$ with respect to x
$\int y\, dx$	the indefinite integral of y with respect to x
$\int_a^b y\, dx$	the definite integral of y with respect to x between the limits $x = a$ and $x = b$
$\dot{x}, \ddot{x}, \ldots$	the first, second, ... derivatives of x with respect to t

Exponential and logarithmic functions

e	base of natural logarithms
$e^x, \exp(x)$	exponential function of x
$\log_a x$	logarithm to the base a of x
$\ln x$	natural logarithm of x
$\lg x, \log_{10} x$	logarithm of x to base 10

Circular functions

$\left.\begin{array}{l}\sin, \cos, \tan \\ \operatorname{cosec}, \sec, \cot\end{array}\right\}$	the circular functions
$\left.\begin{array}{l}\sin^{-1}, \cos^{-1}, \tan^{-1} \\ \operatorname{cosec}^{-1}, \sec^{-1}, \cot^{-1}\end{array}\right\}$	the inverse circular functions

Vectors

a	the vector **a**
$\overrightarrow{AB}$	the vector represented in magnitude and direction by the directed line segment AB
â	a unit vector in the direction of **a**
i, j	unit vectors in the directions of the Cartesian coordinate axes
$\begin{pmatrix} x \\ y \end{pmatrix}$	the vector $x\mathbf{i} + y\mathbf{j}$
$|\mathbf{a}|, a$	the magnitude of **a**

Answers

The questions, with the exception of those from past question papers, and all example answers that appear in this book were written by the authors. Cambridge Assessment International Education bears no responsibility for the example answers to questions taken from its past question papers which are contained in this publication.

Review chapter

1 C 2 B 3 A 4 D
5 A 6 C 7 D 8 D
9 A 10 C 11 B 12 A
13 B 14 D 15 B 16 D
17 C 18 D 19 A 20 D

Chapter 1 Functions

Discussion point Page 3
The amount of petrol, in litres, is measured and the cost is calculated from this.

Discussion point Page 3
The numbers not appearing as final digits are 2, 3, 7, 8. (The Outputs are the final digits of the square numbers.)

Exercise 1.1 Page 12

1 a 13
  b −2
  c 4
  d 5.5 or $\frac{11}{2}$
2 a 36
  b 16
  c 4
  d $\frac{25}{4}$
3 a 13
  b 28
  c 1
  d $\frac{4}{3}$
4 a 4
  b −2
  c 2
  d $\frac{13}{6}$

5 a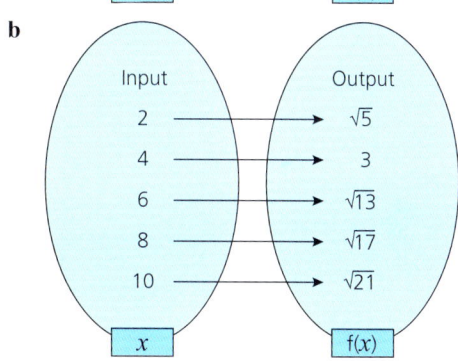
  b
  c Any numbers less than $(-\frac{1}{2})$ must be excluded as an input.
6 a {1, 4, 7, 10, 13}
  b $\{-3, -\frac{5}{2}, -2, -\frac{3}{2}, -1\}$
  c $\{y : y \geq 0\}$
  d $\{y : y \geq 6\}$
7 a 0
  b $x < 1$
  c 1.5
  d $\{x : x < -\sqrt{2} \text{ or } x > \sqrt{2}\}$
8 a $f^{-1}(x) = \frac{x+2}{7}$
  b $g^{-1}(x) = \frac{2x-4}{3}$
  c $h^{-1}(x) = \sqrt{x} + 1$ for $x \geq 0$
  d $f^{-1}(x) = \sqrt{x-4}$ for $x \geq 4$

9 a $f^{-1}(x) = \dfrac{x+4}{3}$

  b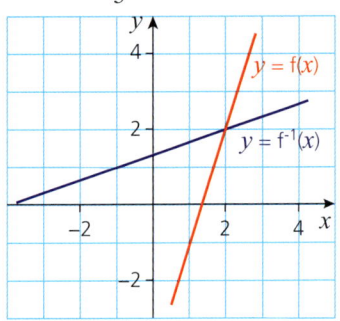

10 a See the curve $y = f(x)$ on the graph in part c below.

  b $f^{-1}(-5) = 3$
  $f^{-1}(0) = 2$
  $f^{-1}(3) = 1$
  $f^{-1}(4) = 0$

  c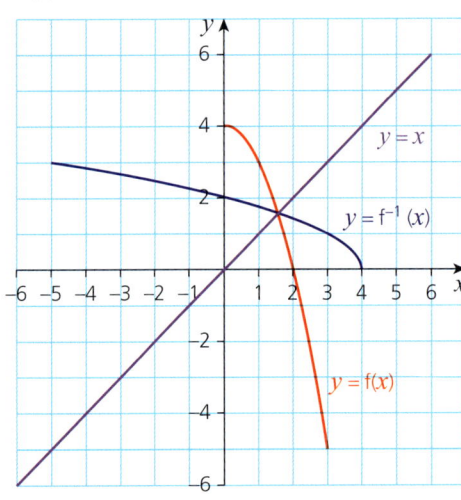

  d Domain is $-5 \le x \le 4$, range is $0 \le f^{-1}(x) \le 3$

Exercise 1.2 Page 17

1 a 14
  b $3x^2 + 2$
  c $4x^2$
  d $12(x^2) + 2$

2 a $\sqrt{17}$
  b -1
  c $\sqrt{9 - 2x}$
  d $4 - \sqrt{2x + 1}$

3 a $x + 8$
  b $8x^4$
  c $\dfrac{2x + 1}{2x + 3}$
  d $\dfrac{1}{(4x^2 + 32x + 65)}$

4 a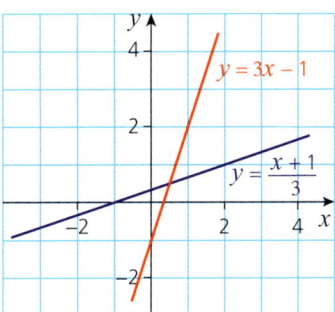

  b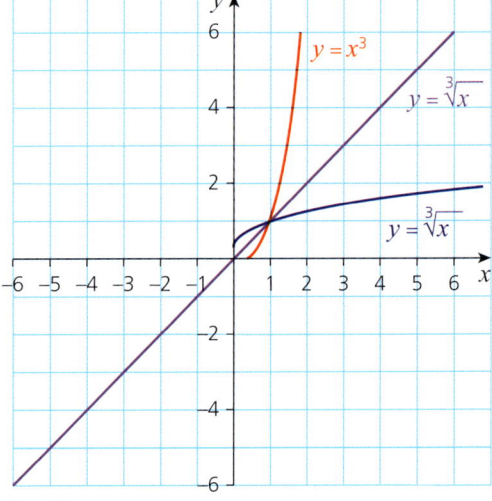

5 a -1 or 7
  b -4 or 3
  c -1 or $\dfrac{7}{3}$
  d -4 or 0

6 a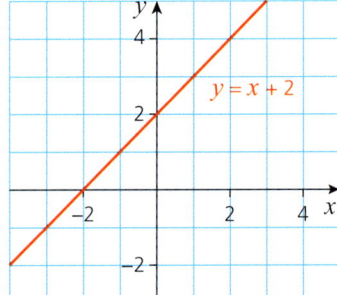

  b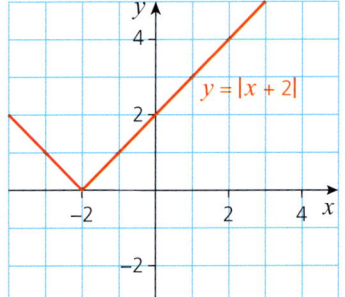

330

c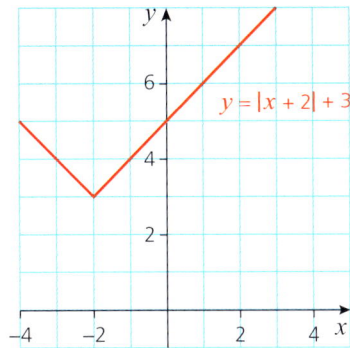

d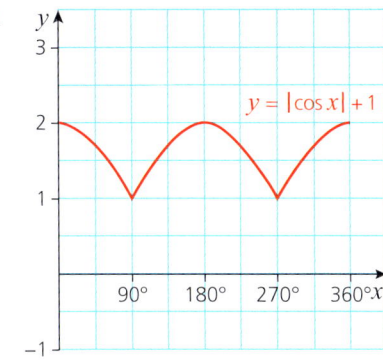

7 a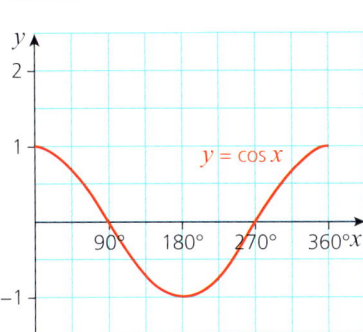

b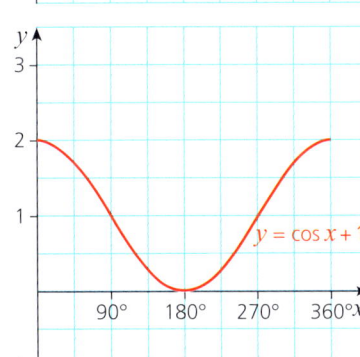

c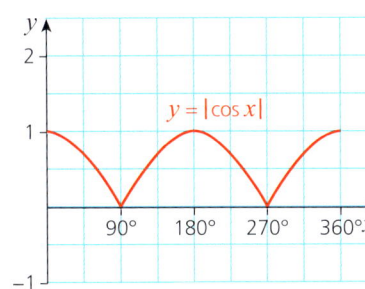

8 The equation for graph 2 is $y = |2x - 1|$
The equation for graph 3 is $y = |2x - 1| + 3$

9 a $y = x^2 + 1$
b $y = \sqrt{x - 1}$

10 a i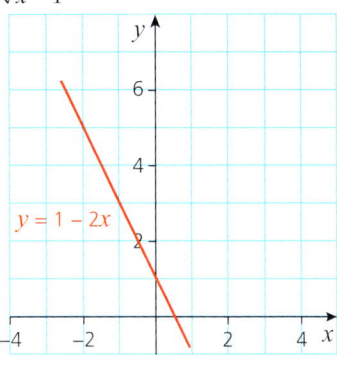

ii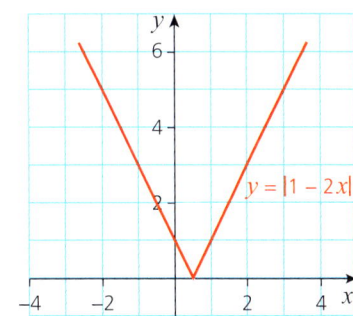

iii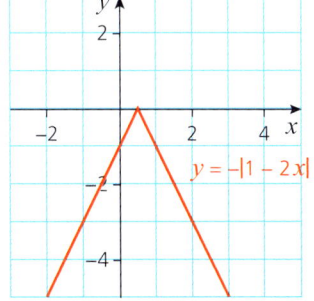

iv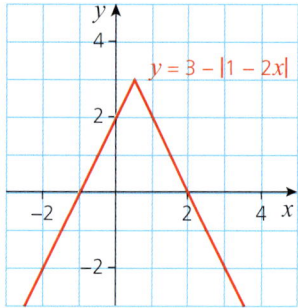

b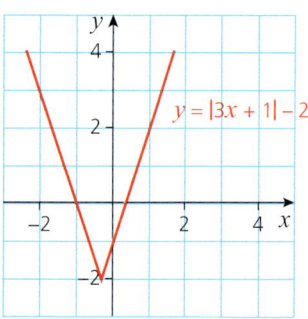

11 a i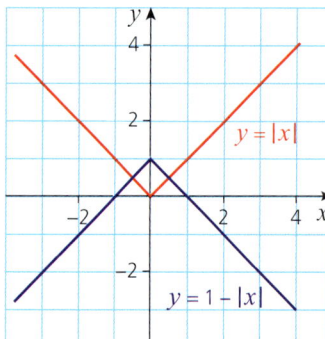

ii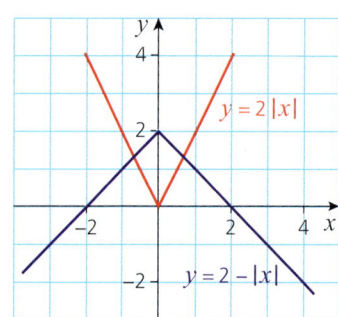

iii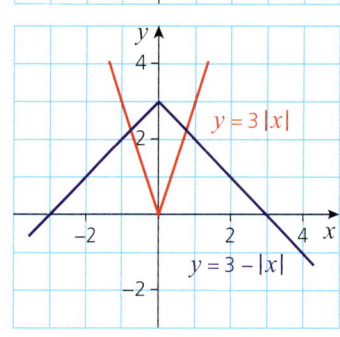

b i $\left(\pm\frac{1}{2}, \frac{1}{2}\right)$

ii $\left(\pm\frac{2}{3}, \frac{4}{3}\right)$

iii $\left(\pm\frac{3}{4}, \frac{9}{4}\right)$

Past-paper questions Page 18

1 i $\frac{6}{4}$

ii $\frac{4x}{3x+1}$

iii $g^{-1}(x) = x^2 - 1$
(Domain) $x > 0$
(Range) $g^{-1}(x) > -1$

iv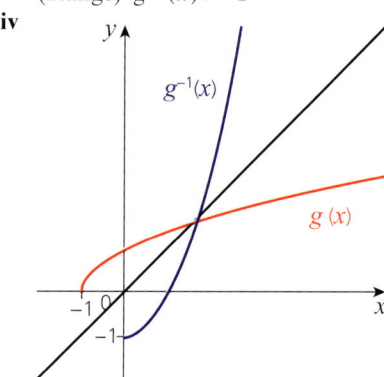

2 i and ii

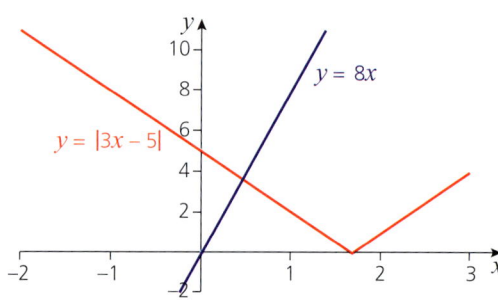

iii $8x = \pm(3x - 5)$

leading to $x = \frac{5}{11}$ or 0.455

Chapter 2 Quadratic functions

Discussion point Page 20

3.22 and 0.78

Exercise 2.1 Page 28

1 a −5, 4
  b 2, 3
  c −4, 7
  d −7, −6

2 a $\frac{1}{2}, 1$
   b $-\frac{2}{3}, \frac{1}{3}$
   c $-1, \frac{7}{2}$
   d $-5, -\frac{2}{3}$
3 a ± 13
   b $\pm \frac{11}{2}$
   c $\pm \frac{5}{4}$
   d $\pm \frac{3}{2}$
4 a i $(x+5)(x+2)$
      ii $\left(-\frac{7}{2}, -\frac{9}{4}\right)$
      iii minimum
      iv
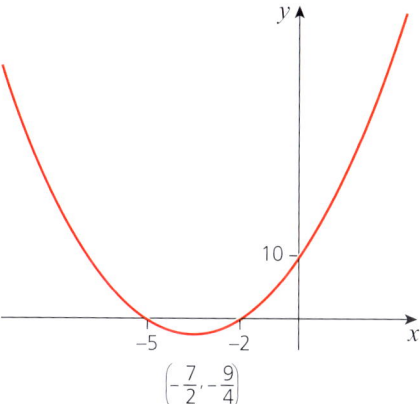
   b i $(8+x)(2-x)$
      ii $(-3, 25)$
      iii maximum
      iv
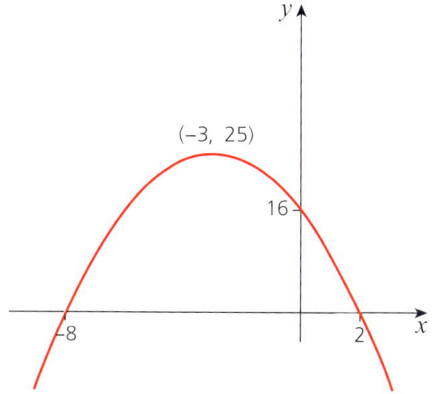
   c i $(5+x)(1-2x)$
      ii $\left(-\frac{9}{4}, \frac{121}{8}\right)$
      iii maximum
      iv
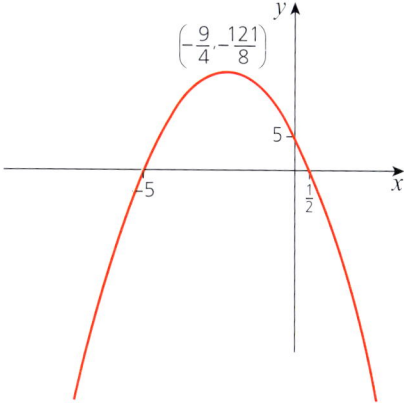
   d i $(2x+3)(x+4)$
      ii $\left(-\frac{11}{4}, -\frac{25}{8}\right)$
      iii minimum
      iv
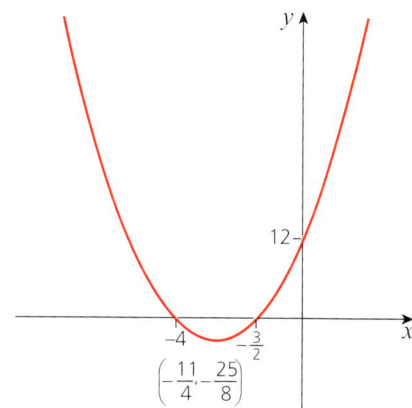

5 a $(x+2)^2 + 5$
   b $(x-5)^2 - 29$
   c $\left(x+\frac{5}{2}\right)^2 - \frac{53}{4}$
   d $\left(x-\frac{9}{2}\right)^2 - \frac{89}{4}$
6 a $2(x-3)^2 - 13$
   b $3(x+2)^2 + 8$
   c $4(x-1)^2 + 1$
   d $2\left(x+\frac{9}{4}\right)^2 - \frac{33}{8}$
7 a $-2 \pm \sqrt{13}$
   b $3.5 \pm \sqrt{14.25}$
   c $\frac{-3 \pm \sqrt{27}}{2} = -\frac{3}{2} \pm \sqrt{\frac{27}{4}}$
   d $-1.5 \pm \sqrt{7.25}$

8 a i $(-3, 6)$
  ii minimum
  iii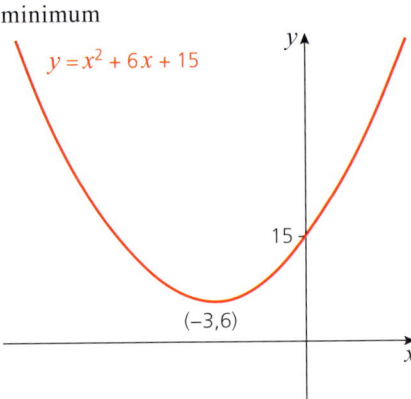

  iii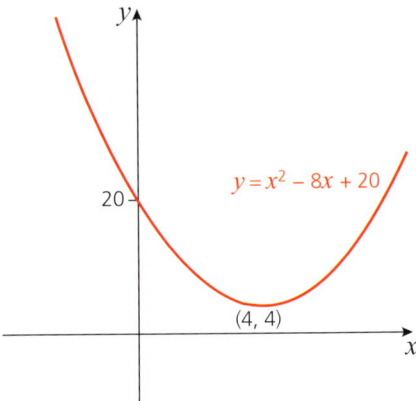

8 b i $(1, 9)$
  ii maximum
  iii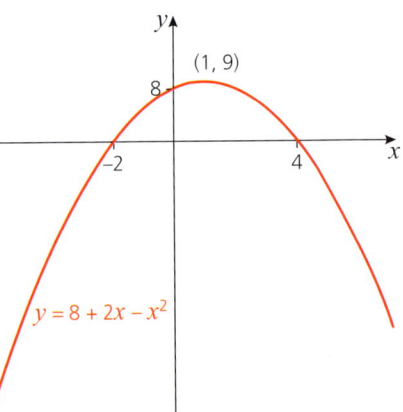

8 c i $\left(-\dfrac{1}{2}, -\dfrac{19}{2}\right)$
  ii minimum
  iii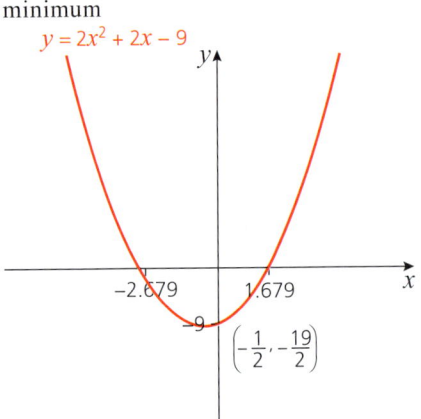

9 a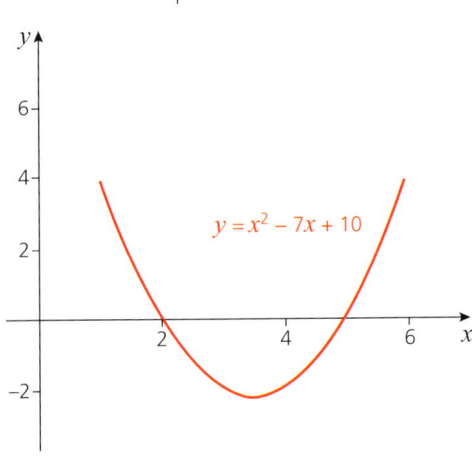

Range $\left[-\dfrac{9}{4}, 4\right]$

b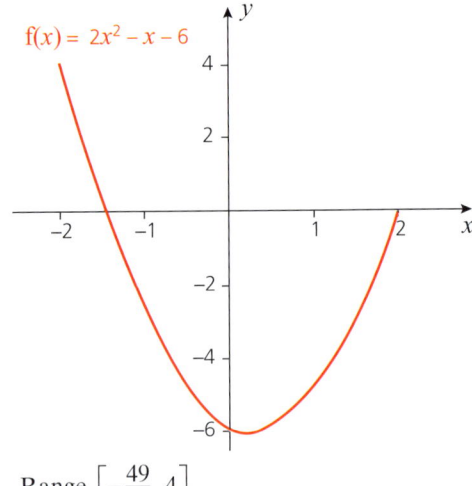

Range $\left[-\dfrac{49}{8}, 4\right]$

8 d i $(4, 4)$
  ii minimum

Real-world activity Page 28

1

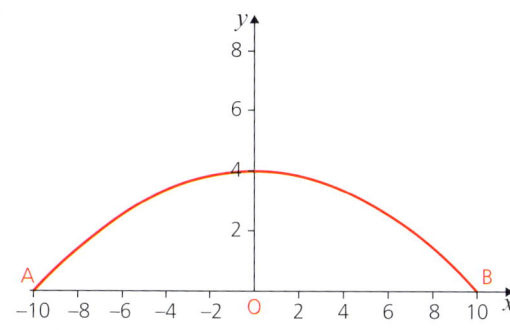

2 Height 4 m, span 20 m
3 $80y = 400 - x^2$

Discussion point Page 33

Lines parallel to the y-axis are straight vertical lines and as such a quadratic curve will always intersect them rather than touch them.

Exercise 2.2 Page 37

1
  a two real and different roots: $-2, -1$
  b two real and different roots: ± 3
  c no real roots
  d two real and different roots: $0, \frac{5}{2}$
  e two real and different roots: $-6, 3$
  f two equal roots: -5
  g two real and different roots: $-\frac{1}{3}, \frac{1}{5}$
  h two real and different roots: $-3, \frac{1}{3}$

2
  a i; ii $-2.32, 4.32$
  b i; ii $-1, 0$
  c i; ii $-2.68, 1.68$
  d i; ii $-2.27, 1.77$

3
  a $-2.43, 0.93$
  b no real roots
  c $-0.18, -2.82$
  d $-1.89, 2.39$

4
  a Two equal roots.
  b Two distinct roots.
  c No real roots.
  d Two distinct roots.
  e No real roots.
  f Two equal roots.

5
  a The line is a tangent to the curve at $(-3, -27)$.
  b Does not meet.
  c The line intersects the curve at $\left(\frac{2}{3}, \frac{2}{3}\right)$ and $\left(\frac{3}{2}, \frac{3}{2}\right)$.
  d The line intersects the curve at $(1.84, 6.68)$ and $(8.16, 19.32)$.
  e The line intersects the curve at $(-4, 12)$ and $(1, 2)$.
  f The line is a tangent to the curve at $(1.5, 18)$.
  g Does not meet.
  h Does not meet.

6
  a $x < 1$ or $x > 5$
  b $-4 \leq a \leq 1$
  c $-2 < y < 2$
  d $x \neq 2$
  e $-4 < a < 2$
  f $y < -1$ or $y > \frac{1}{3}$

Real-world activity Page 38

1 $y = 4 - \frac{4x^2}{49}$

$y = -4 + \frac{4x^2}{49}$

$y = 3 - \frac{x^2}{12}$

$y = -3 + \frac{x^2}{12}$

2 $y = 2 - \frac{8x^2}{49}$

$y = -2 + \frac{8x^2}{49}$

$y = \frac{3}{2} - \frac{x^2}{6}$

$y = -\frac{3}{2} + \frac{x^2}{6}$

Past-paper questions Page 39

1
  i $2\left(x - \frac{1}{4}\right)^2 + \frac{47}{8}$
  ii $\frac{47}{8}$ is the minimum value when $x = \frac{1}{4}$

2 $4 < k < 12$ or $k > 4$ and $k < 12$

3 $mx + 2 = x^2 + 12x + 18$
$x^2 + (12 - m)x + 16 = 0$
$(12 - m)^2 = 4 \times 16$
leading to $m = 4, 20$

Chapter 3 Factors of polynomials

Discussion point Page 40

23, 128. Keep adding 105 ($= 3 \times 5 \times 7$) to find subsequent possible answers.

Exercise 3.1 Page 43

1 $x^4 + 3x^3 - x^2 - 7x - 4$
2 $x^4 - 3x^3 + 5x^2 - x - 2$
3 $4x^4 - 8x^3 + 3x^2 + 10x - 5$
4 $x^4 - 4x^2 + 12x - 9$
5 $4x^4 - 12x^3 + 9x^2 - 16$
6 $x^4 - 6x^3 + 13x^2 - 12x + 4$
7 $x^2 - 2x - 1$
8 $x^2 - x - 1$

335

9 $x^3 - x^2 + x - 1$

10 $(x-2) - \dfrac{12}{(x+2)}$

Exercise 3.2 Page 49

1 a f(1) = 0 so factor
  b f(−1) = −2 so not a factor
  c f(1) = 0 so factor
  d f(−1) = −8 so not a factor

2 a f(−2) = f(−1) = f(3) = 0; $(x+2)(x+1)(x-3)$

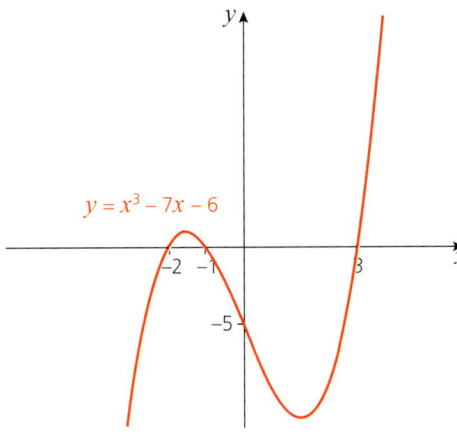

  b f(−3) = f(1) = f(2) = 0; $(x+3)(x-1)(x-2)$

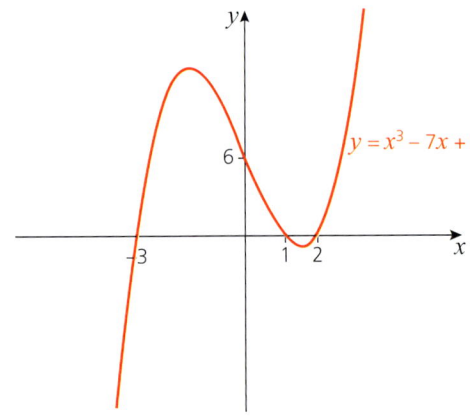

  c f(−5) = f(−1) = f(1) = 0 $(x+5)(x+1)(x-1)$

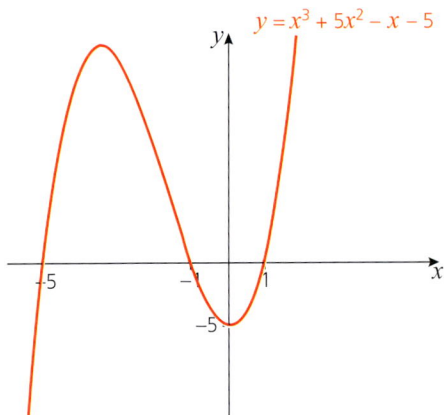

  d f(−1) = f(1) = f(5) = 0; $(x+1)(x-1)(x-5)$

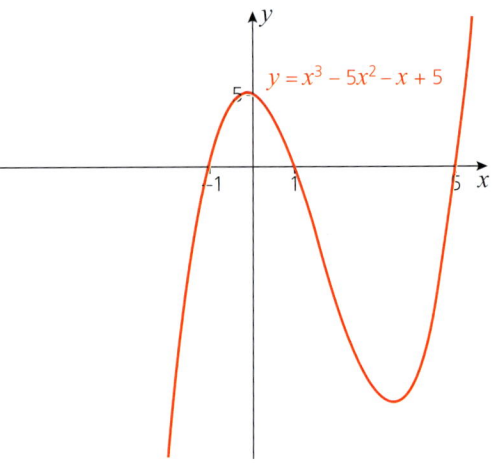

3 a $(x+1)(x^2+1)$
  b $(x-1)(x^2+1)$
  c $(x+2)(x^2+x+1)$
  d $(x-2)(x^2-x+1)$

4 3

5 3

6 a $a + b = 5, 3a + b = 27$
  b $a = 11, b = -6$

Exercise 3.3 Page 50

1 a f(2) = 6
  b f(−2) = −10
  c f(4) = 136
  d f(−4) = −144

2 $a = -8, b = 1$

3 $x = 1, -2, -3$

4 $a = 1, -3$

5 $a = -7, b = -6; x = -1, -2, 3$

6 a

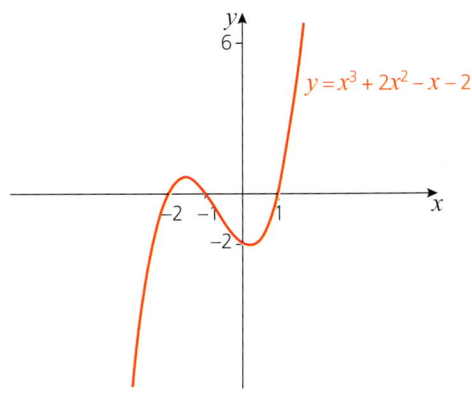

b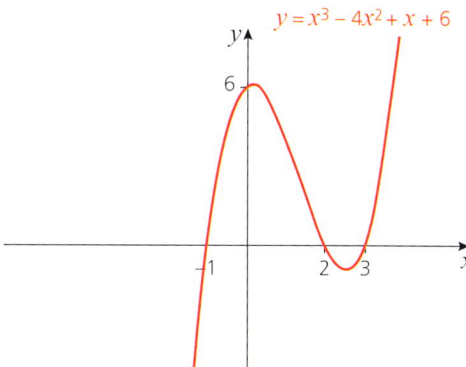

c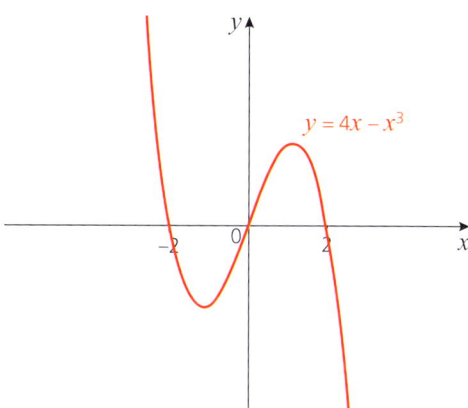

d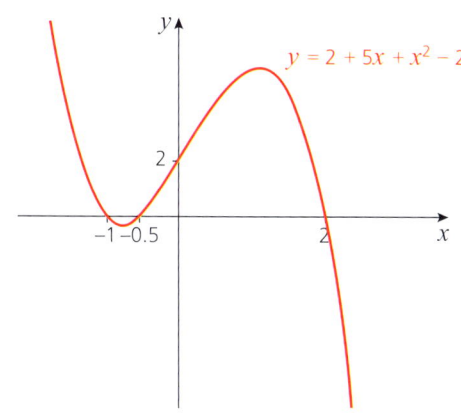

Review exercise 1

1. **a** $\frac{2}{3}$

   b $\frac{7x+2}{3x+6}$, $a = 7$, $b = 2$, $c = 3$, $d = 6$

   c $f^{-1}(x) = 1 - x^2$

   d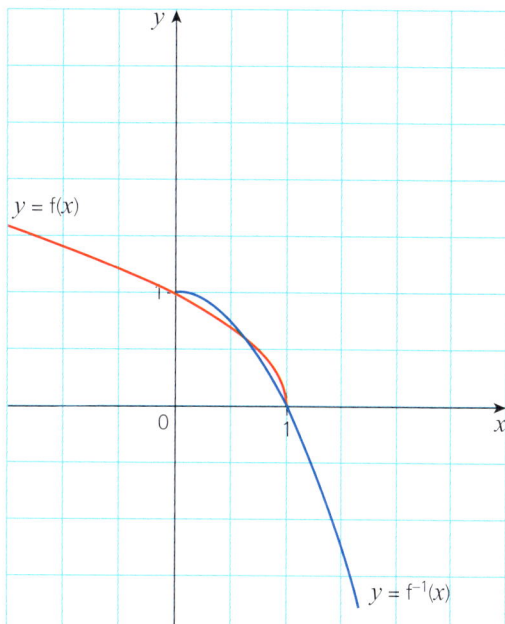

   Domain $x \geq 0$
   Range $f^{-1}(x) \leq 1$

2. **a** $f^{-1}(x) = \sqrt{x+1}$
   Domain $x \geq -1$
   Range $f^{-1}(x) \geq 0$

   b $f^{-1}(x)$ is a reflection of $f(x)$ in the line $y = x$.

   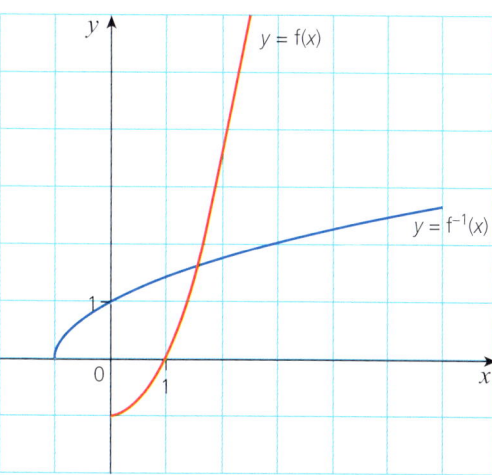

Past-paper questions Page 51

1. **i** $a = 14$
   ii $(2x - 1)(7x^2 - 4x + 2)$
2. **i** $a = -7$
   ii $f(-3) = -49$
   iii $f(x) = (2x - 1)(2x^2 + 3x - 2)$
   iv $f(x) = (2x - 1)(2x - 1)(x + 2)$
   Leading to $x = 0.5, -2$
3. **i** $a = -14$
   ii $(x + 2)(6x^2 - 17x + 20)$
   iii $6x^2 - 17x + 20 = 0$ has no real roots;
   $x = -2$

3 a i This function is not one–one; the graph has a turning point.

   ii $f^2(x) = \sqrt{2 + x^2}$

b i Any value of k greater than or equal to 0.

   ii $g^{-1}(x) = \sqrt{x^2 - 1}$

4 a $2(x + \frac{5}{4})^2 - \frac{49}{8}$, $p = 2$, $q = -\frac{5}{4}$, $r = -\frac{49}{8}$

b The least value is $-\frac{49}{8}$

This occurs when $x = -\frac{5}{4}$

5 a $-1 \leq x \leq \frac{3}{4}$

b The discriminant is independent of k and is less than 0. So, the equation has no real roots, whatever the value of k.

6 $k = 8$ or $k = 16$

7 $-1 < k < 2$

8 i Multiplying out the brackets gives $x^4 - 2x^3 - 3x^2 + 8x - 4$ as required.

   ii $p(x) = (x - 1)^2(x - 2)(x + 2)$

9 a $b = -6$ as required and $a = -5$

b $p = 3$, $q = 11$, $r = 6$

c $x = 1$, $x = -\frac{2}{3}$ or $x = -3$

10 a $a = -2$ as required and $b = 9$

b $q(x) = 2x^2 - 3x + 4$

c $q(x)$ has no real roots so there is only one real root, $x = -2$

Chapter 4 Equations, inequalities and graphs

Discussion point Page 55

You can describe it in words, as in 'both x and y must be between -0.5 and 0.5' or you can do it in symbols, for example $-0.5 < x < 0.5$ and $-0.5 < y < 0.5$. Another way with symbols is to use the modulus sign and write $|x < 0.5|$ and $|y < 0.5|$. You could even use set notation and write it as $|x < 0.5| \cup |y < 0.5|$.

Discussion point Page 56

A small family car such as a Volkswagen Golf has an average consumption of 5.3 litres per 100 kilometres, so the fuel economy is at worst 0.530 litres per kilometre which is remarkably similar to Vettel's Ferrari Formula 1 car.

$E = \frac{f}{d}$

$E = \frac{5.3}{100}$

$E = 0.530$

Find out the fuel consumption of other models of road vehicles and see how they compare.

Discussion point Page 58

Using three points helps to eliminate errors. A mistake in the calculation using only two points would give a completely different line.

Exercise 4.1 Page 62

1 a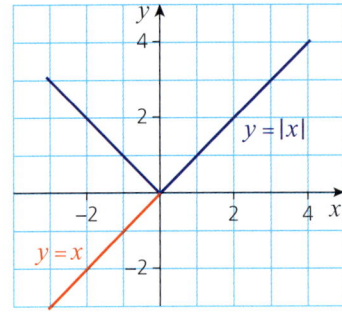

b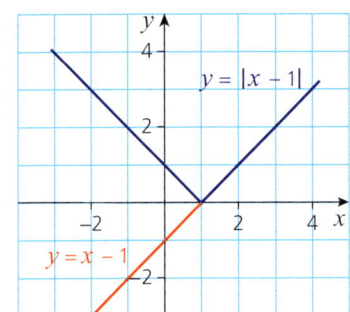

c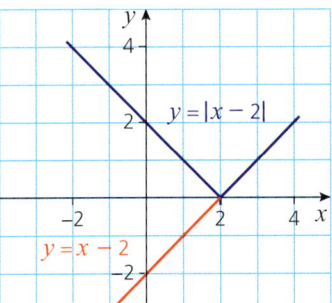

2 a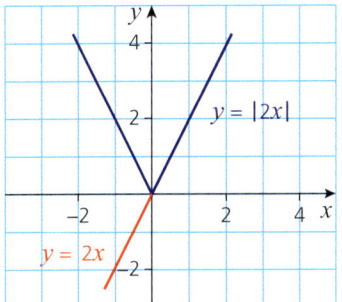

b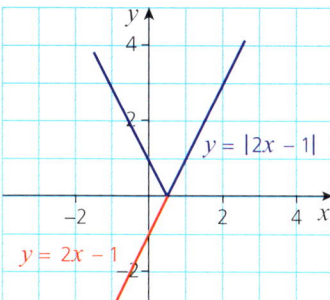

c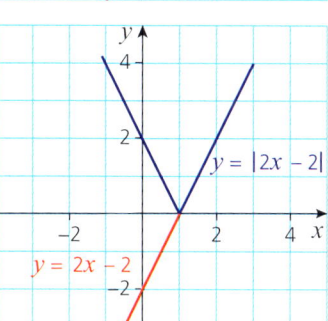

3 a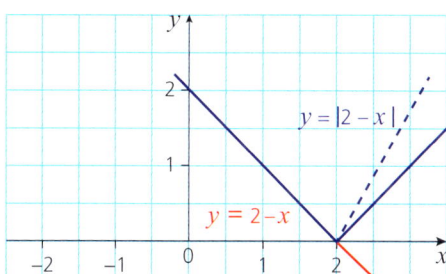

b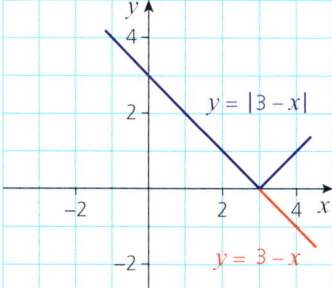

c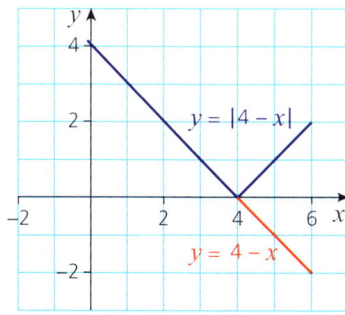

4 a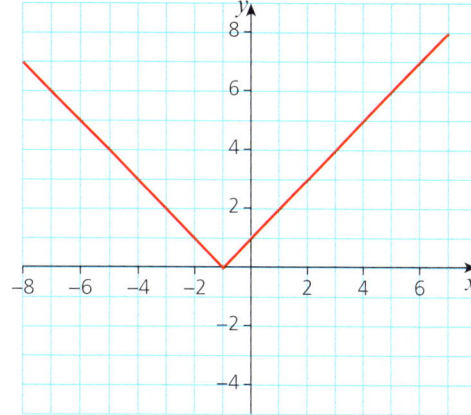

b $x = -6$ or $x = 4$
c $x = -6$ or $x = 4$

5 a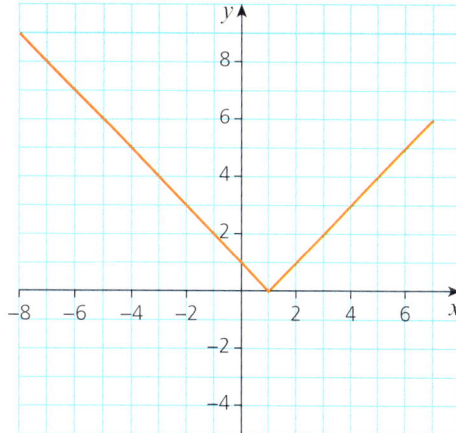

b $x = -4$ or $x = 6$
c $x = -4$ or $x = 6$

6 a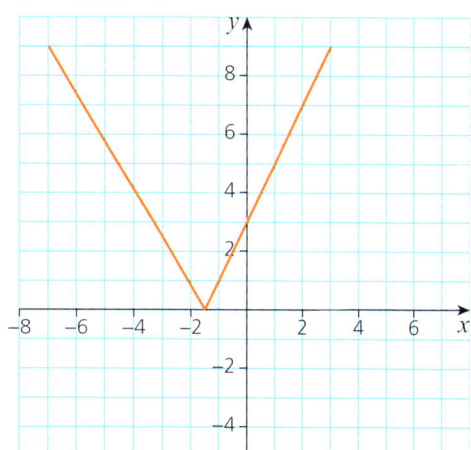

b $x = -5$ or $x = 2$
c $x = -5$ or $x = 2$

7 a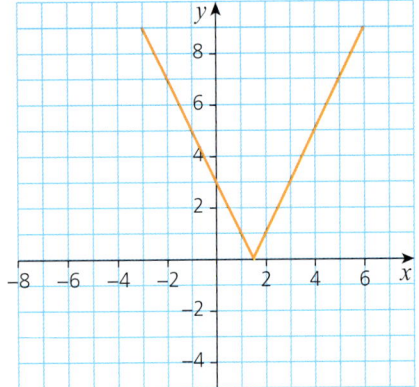
  b $x = -2$ or $x = 5$
  c $x = 2$ or $x = 5$

8 a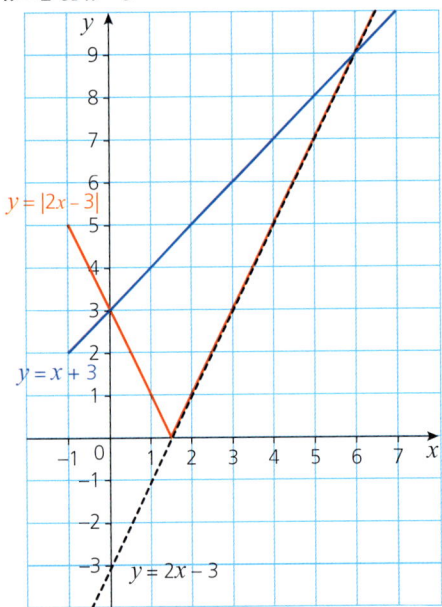
  b $x = 0$ or $x = 6$
  c $x = 0$ or $x = 6$

9 a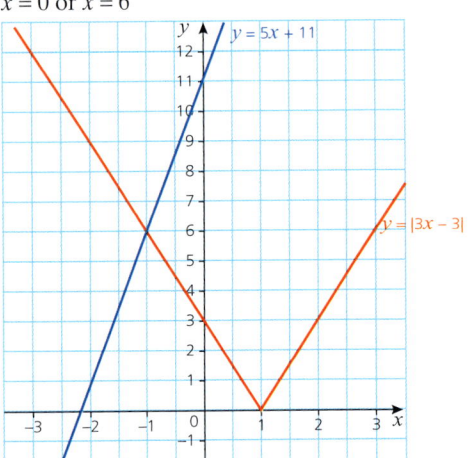
  b $x = -1$
  c $x = -1$

10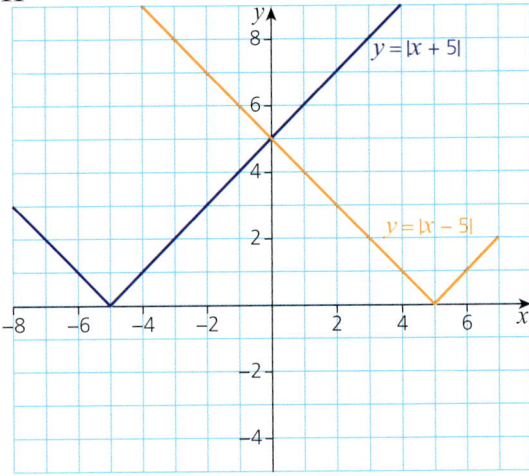

$x = 0$

11

$x = 0$

12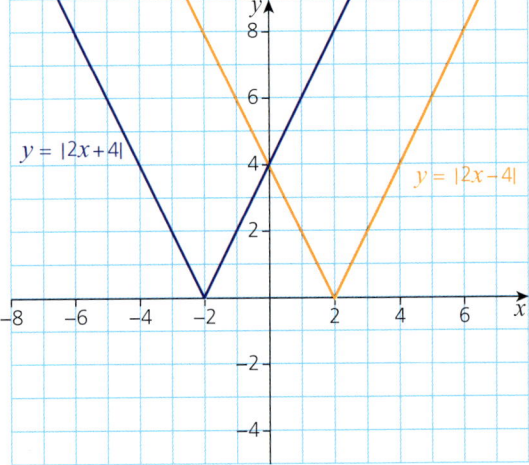

$x = 0$

13 a $x = -9, x = -5$ or $x = -1$
  b $x = -1, x = -5$ or $x = -9$
14 $x = 4.68$ (3 s.f.), $x = 7$, $x = 9$ or $x = 11.3$ (3 s.f.)
15 $x = -1.5$ or $x = 3$

Discussion point Page 66
$-(x + 7) = 4x$ is the same as $x + 7 = -4x$, which is already being considered.

Discussion point Page 67
It is easier to plot points with integer coordinates.

Exercise 4.2 Page 68
1 a $|x - 6| \leq 9$
  b $|x - 6| \leq 10$
  c $|x - 6| \leq 11$
2 a $-1 \leq x \leq 3$
  b $-1 \leq x \leq 5$
  c $-1 \leq x \leq 7$
3 a $-3 < x < 5$
  b $x < -3$ or $x > 5$
  c $-4 < x < 1$
  d $x < -4$ or $x > 1$
4 a

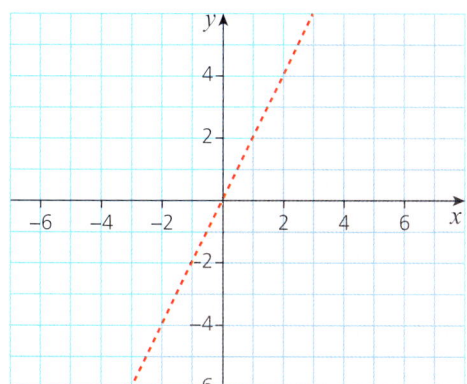

b

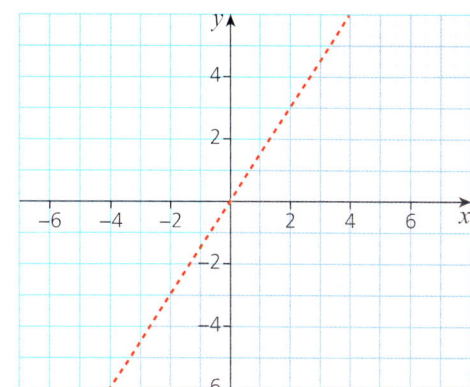

c

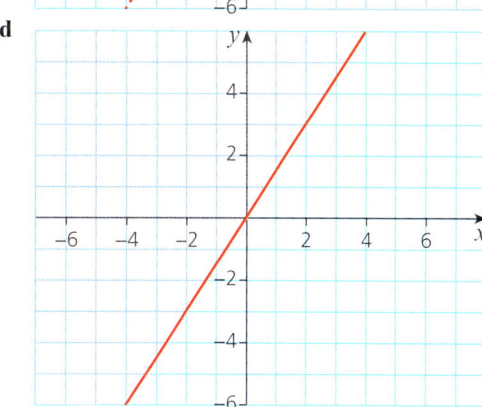

d

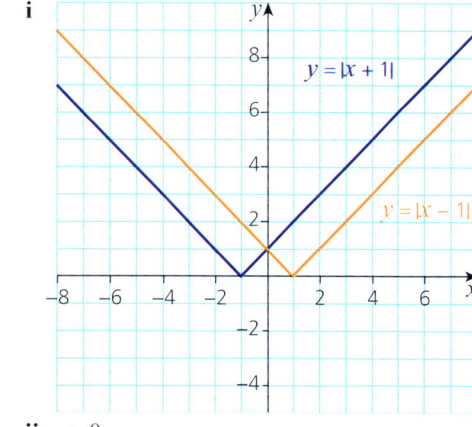

5 a i

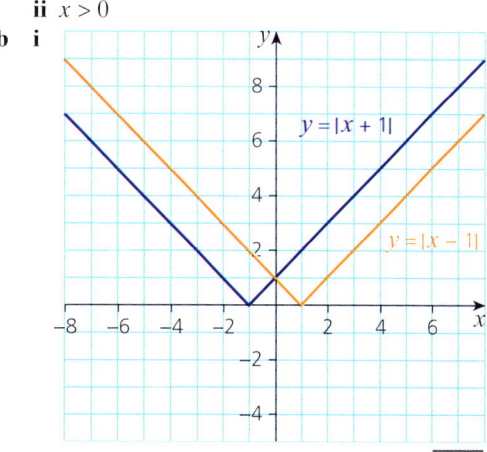

  ii $x > 0$
  b i

(see graph)

ii $x < 0$

c i
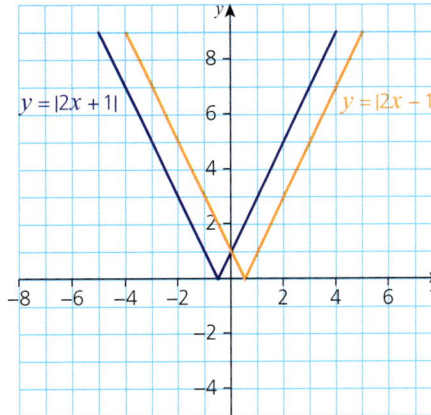

ii $x \geq 0$

d i
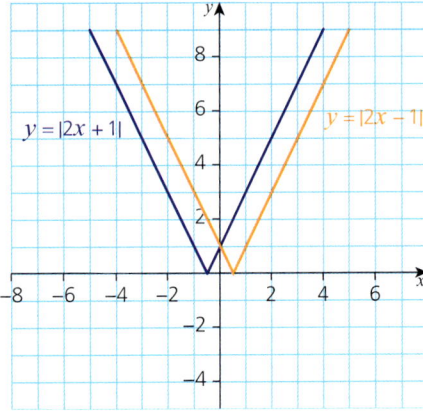

ii $x \leq 0$

6 a $y > x + 1$
   b $y \leq 2 - x$
   c $y \geq |x|$

7 a
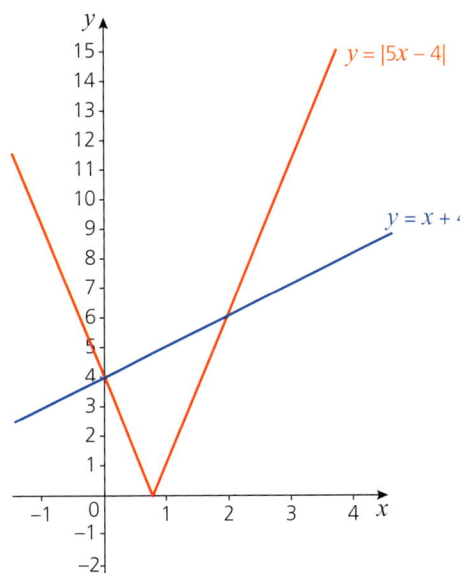

b $0 \leq x \leq 2$
c $0 \leq x \leq 2$

8 a $x > -0.4$ or $x < -2$
   b $3.3 \leq x \leq 3.7$
   c $-0.5 \leq x \leq 4.5$
   d $x \leq -6$ and $x \geq -\frac{4}{3}$
   e $-\frac{3}{8} \leq x \leq \frac{5}{6}$

9 a, b $-3 \leq x \leq -1$, $1 \leq x \leq 3$

10 $x < -2.89$ (3 s.f.), $-2 < x < 0.5$ or $x > 1.39$ (3 s.f.)

Discussion point Page 72

$3(x + 2)(x - 1)(x - 7) \geq -100 \Rightarrow -2.9 \leq x \leq 2.6$ or $x \geq 6.2$.

$3(x + 2)(x - 1)(x - 7) > -100 \Rightarrow -2.9 < x < 2.6$ or $x > 6.2$

$3(x + 2)(x - 1)(x - 7) < -100 \Rightarrow x < -2.9$ or $2.6 < x < 6.2$

Exercise 4.3 Page 72

1 a $x = 4$
   b $x = 4$
   c $x = 25$
   d no solutions

2 $x = -64, x = 1$

3 $x = 27, x = 1$; also $x = -27, x = -1$

4 a $x = 9, x = 16$
   b $x = 1$
   c $x = -125, x = 8$

5 a $x = 2, x = 1$; also $x = -2, x = -1$
   b $x^2 = -1$ or $x^2 = -4$, hence no solution as x^2 cannot be negative.
   c i $x = 8, x = 1$; also $x = -8, x = -1$
      ii no solution

6 a
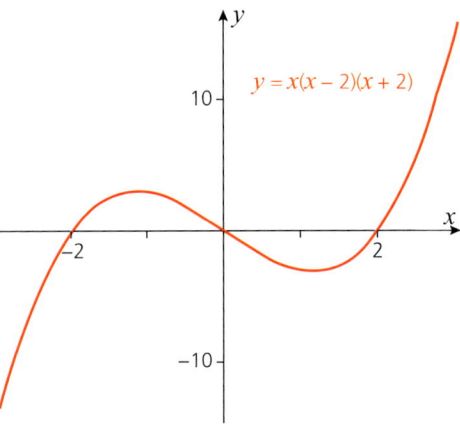

b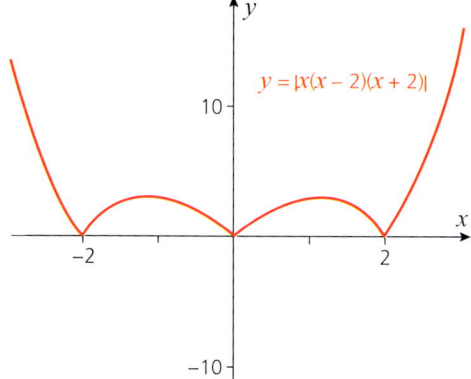

c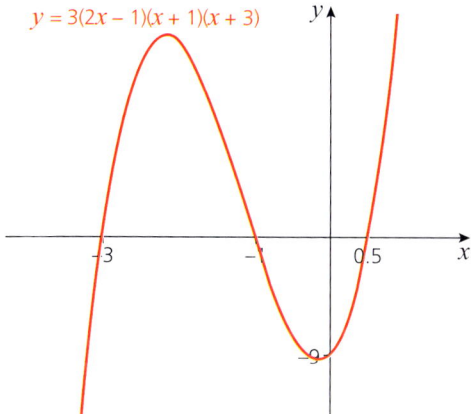

d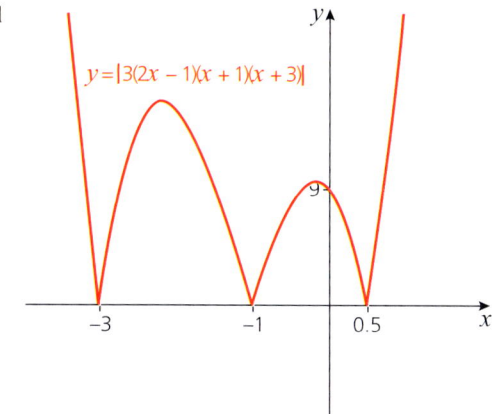

7 a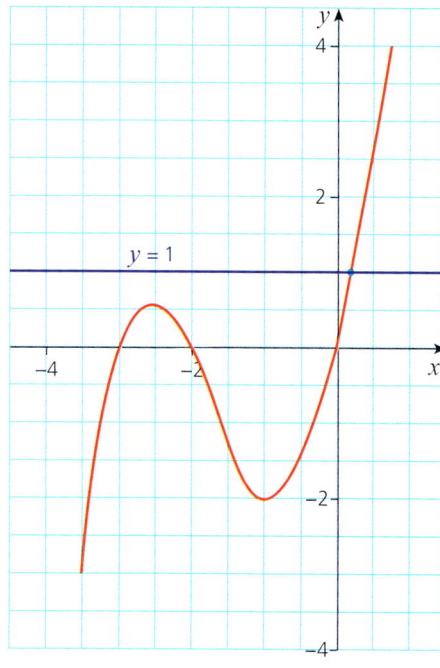

$x \geq 0.148$

b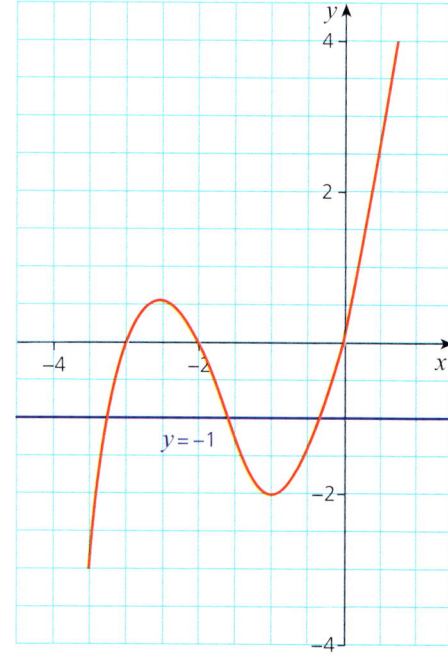

$x < -3.247$ or $-1.555 < x < -0.198$

c

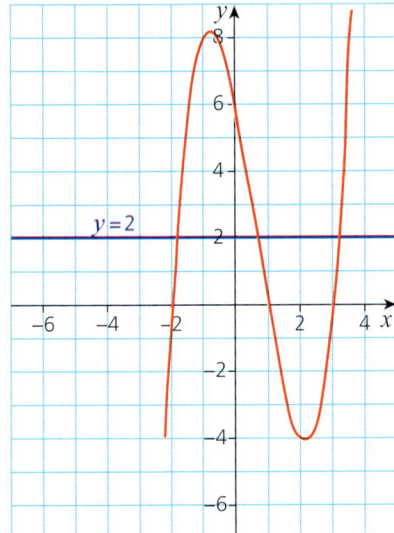

$-1.856 < x < 0.678$ or $x > 3.177$

d

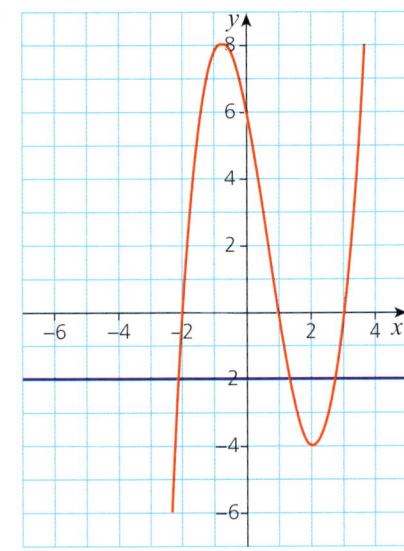

$x < -2.125$ or $1.363 < x < 2.762$

8 a $y = (x + 3)(x + 1)(x - 1)$
   b $y = (2x + 3)(2x + 1)(2x - 1)$
   c $y = (x + 2)(x - 2)^2$
9 a $y = |(x + 2)(x - 1)(x - 2)|$
   b $y = |(x + 3)(x + 2)(x - 1)|$
   c $y = |(x + 2)(2x - 1)(x - 2)|$
10 There is no vertical scale.

Past-paper questions Page 72

1 i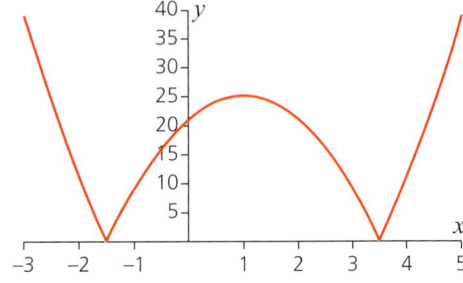

   ii Two

2 i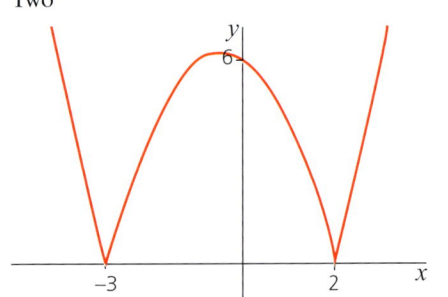

   $(-3, 0), (2, 0), (0, 6)$

   ii $\left(-\frac{1}{2}, \frac{25}{4}\right)$

   iii $6.25 < k \leq 14$

3 $9x^2 + 2x - 1 < (x + 1)^2$
   $8x^2 < 2$
   $-\frac{1}{2} < x < \frac{1}{2}$

Chapter 5 Simultaneous equations

Exercise 5.1 Page 82

1 a $x = 5, y = 7$

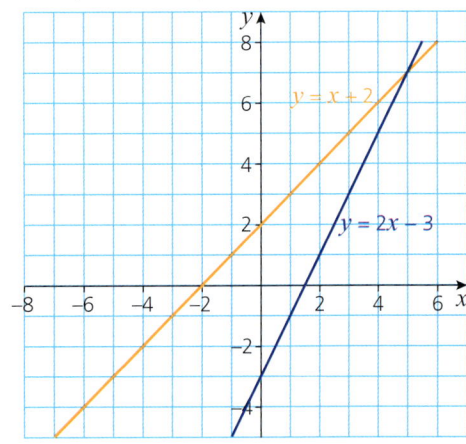

b $x = -1, y = 2$

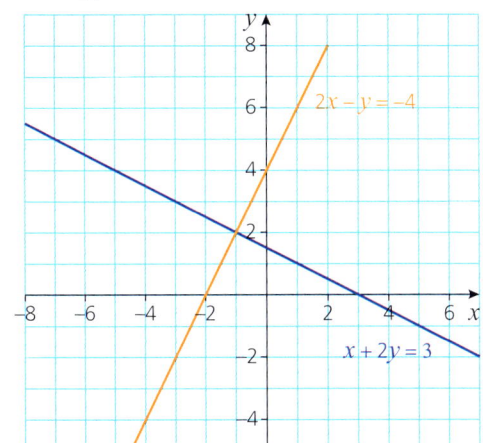

2 a $x = 3, y = 7$
  b $x = 7, y = 3$
3 a $x = -2, y = 2$
  b $x = 2, y = -2$
4 $x = 1, y = 1$
5 $x = -5, y = -2$
6 a $x = 3, y = 1$
  b $x = 3, y = 0.5$
7 a $x = 2, y = 3$
  b $x = 2, y = 1.5$
8 $x = 1, y = -2$
9 $x = -3, y = 4$
10 $3.40
11 $7
12 $340
13 $5

Exercise 5.2 Page 86

1 $x = 2, y = 3$ and $x = 2, y = -3$
2 $x = 2, y = 10$ and $x = 4, y = 20$

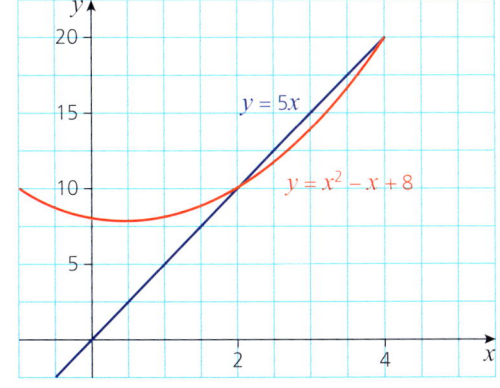

3 $x = -1, y = -4$ and $x = 4, y = 1$

4 $x = -1.5, y = 11$ and $x = 1.25, y = 0$
5 $x = \frac{2}{3}$ and $y = 3$
6 A $(-4, 3)$, B $(3, 4)$
7 $r = 15, x = 4$
8 a $x = 1, y = 2$
  b The line is a tangent to the curve.

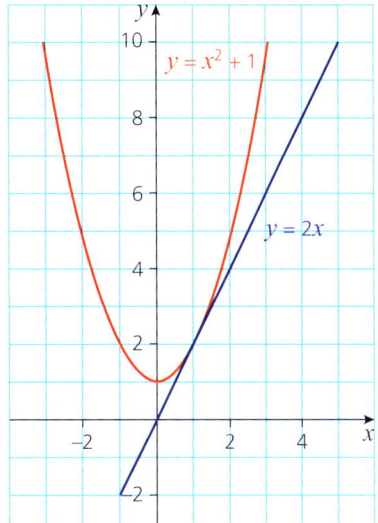

9 a There are no solutions.
  b The line does not intersect the curve.

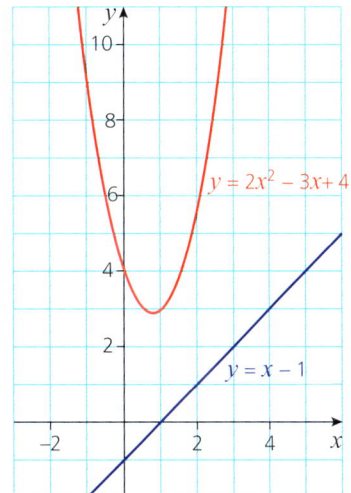

10 $x = 2, y = 1$ or $x = 3, y = 3$

Past-paper questions Page 85

1 $y = -0.5x + 3.75$
2 $3y + x - 2 = 0$
3 $k < -2, k > 8$

Chapter 6 Logarithmic and exponential functions

Discussion point Page 89
In 1700, about $2^{10} = 1024$. In 1000 about $2^{34} = 17\,000\,000\,000$ or 17 billion.
This is more than the present world population and much more than it was then. The calculation depends on the assumption that all your ancestors of any generation are different people and this is clearly not true; cousins, whether close or distant, must have married each other and so had some ancestors in common.

Discussion point Page 90
Answers will vary.

Discussion point Page 100
$5^{2 \times 1} - 5^1 - 20 = 25 - 5 - 20 = 0$, as required. So, $x = 1$ satisfies the original equation $5^{2x} - 5^x - 20 = 0$.

Exercise 6.1 Page 102

1. a 3
   b 0
   c 2
   d −2
2. a 4
   b 4
   c 3
   d −3
3. a 2
   b 6
   c −3
   d −6
4. a log15
   b log64
   c log4
   d log5
   e log72
   f $\log \frac{81}{64}$
   g $\log \frac{1}{8}$
5. a log 108
   b log 3
   c log 4500
6. a $\log_2 5$
   b $\log_9 100$
   c $\frac{3}{\lg e}$
7. a ln 6
   b ln 49
   c ln 192
8. a $3 \log x$
   b $6 \log x$
   c $\frac{3}{2} \log x$
9. a 9.3 cm = 93 mm
   b 517.06 cm² = 51 706 mm²
10. a Stretch in the y-direction scale factor 3.

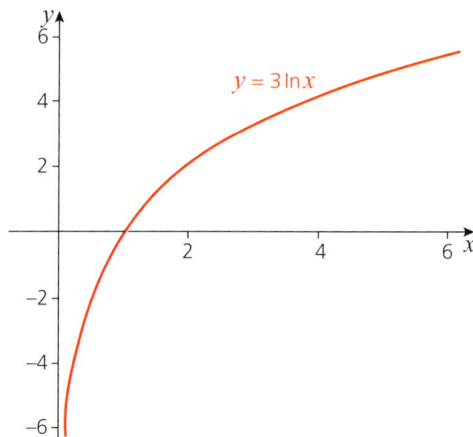

   b Translation

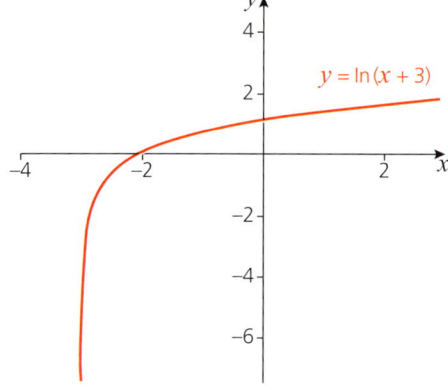

c Stretch in the y-direction scale factor 3 and stretch in the x-direction scale factor $\frac{1}{2}$.

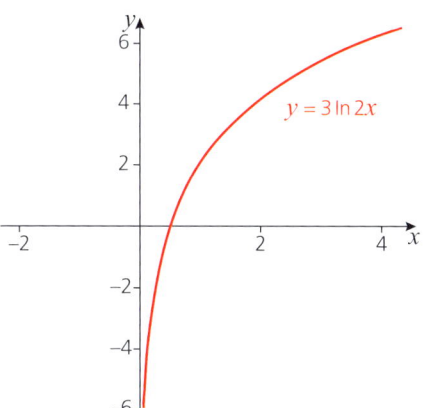

d Stretch in the y-direction scale factor 3 and translation $\begin{pmatrix} 0 \\ 2 \end{pmatrix}$.

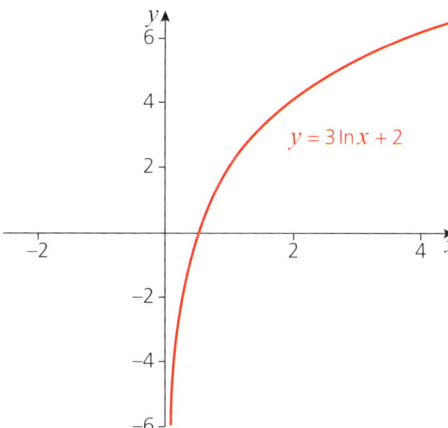

e Stretch in the y-direction scale factor −3 and translation $\begin{pmatrix} -1 \\ 0 \end{pmatrix}$.

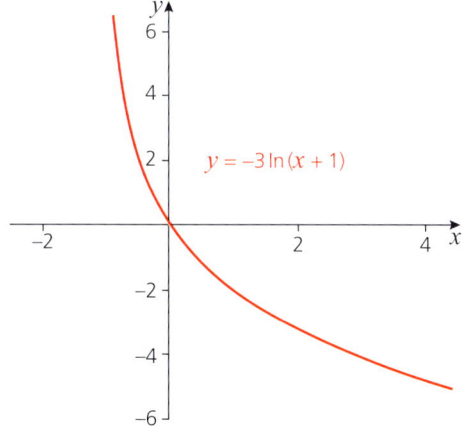

f Translation $\begin{pmatrix} -2 \\ 0 \end{pmatrix}$ followed by stretch in the y-direction scale factor $\frac{1}{2}$.

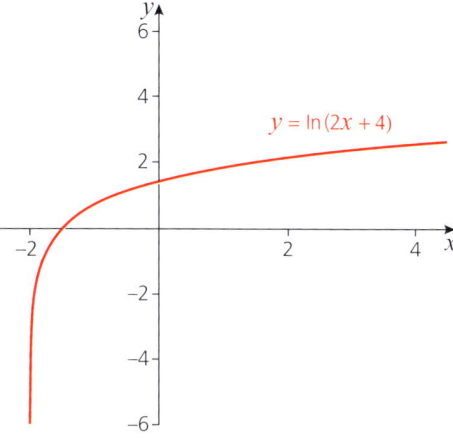

11 a ii $y = \log(x - 1)$
b iv $y = 3\ln x$
c v $y = \log(2 - x)$
d vi $y = \ln(x + 2)$
e i $y = \log(x + 1)$
f iii $y = -\ln x$

12 a $\frac{1}{2}$
b $\frac{2}{9}$
c $\frac{1}{3}\ln 2, 0$

13 $c = 0.993$
14 $x = 3.15$
15 a 9.4 years (10 years)
b 0.198% per month
c i 113 months
ii 3435 days
16 a $\frac{1}{3}$
b $x = \log_{10}(-3)$ which is not a valid solution. So, the equation has no solution.
c 1000 or 10 000
d 0.0001 or 100

Exercise 6.2 Page 111

1 a

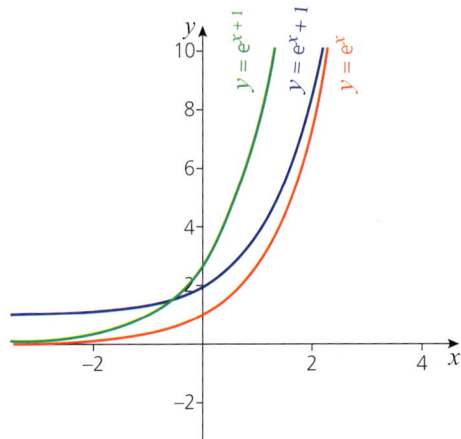

   ii $y = e^x$ at $(0, 1)$, $y = e^x + 1$ at $(0, 2)$ and $y = e^{x+1}$ at $(0, e)$

b

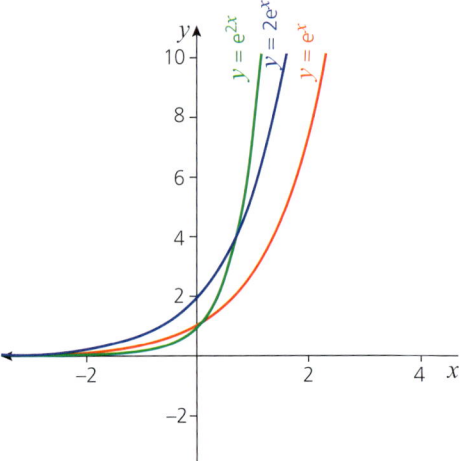

   ii $y = e^x$ at $(0, 1)$, $y = e^{2x}$ at $(0, 1)$ and $y = 2e^x$ at $(0, 2)$

c

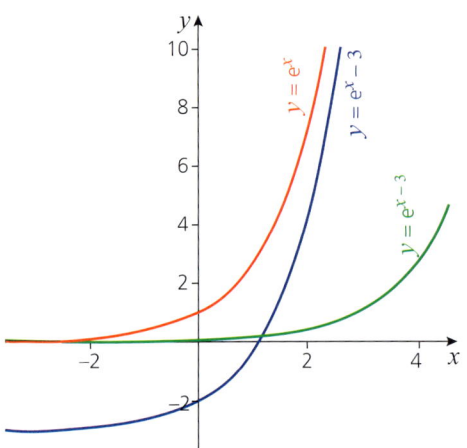

   ii $y = e^x$ at $(0, 1)$, $y = e^x - 3$ at $(\ln 3, 0)$ and $(0, -2)$ and $y = e^{x-3}$ at $(0, e^{-3})$

2

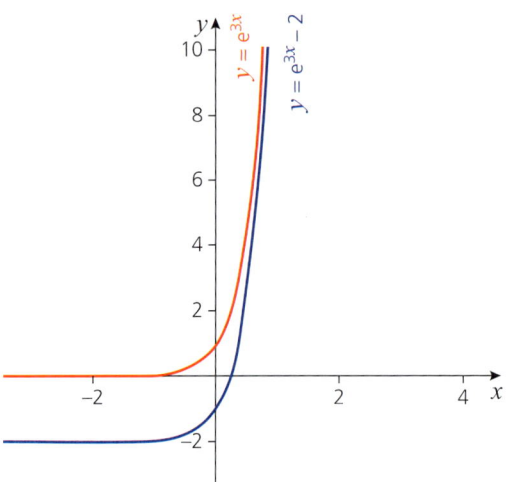

3

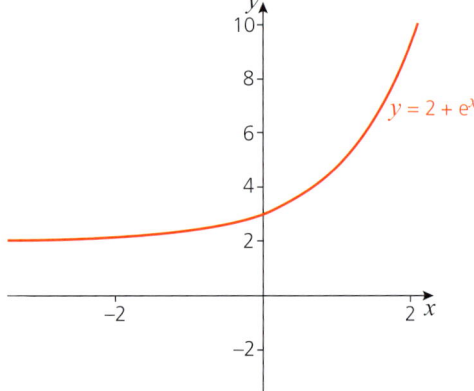

4 a $(0, 3)$

b (0, 1)

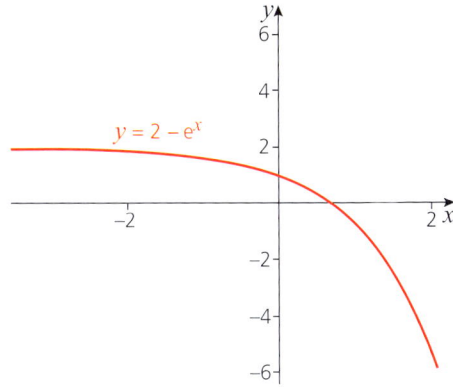

c (0, 3)

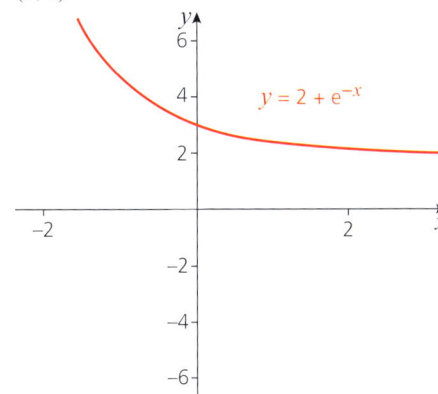

d (0, 1)

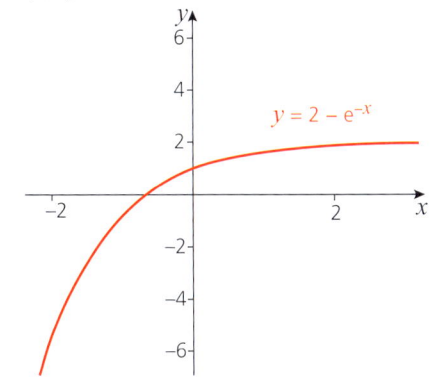

5 **a** 8.55
  b 3.23
  c 0.303
  d 4.30
6 **a** $4377
  b 23 years
7 **a** 5000 m
  b 42.59 seconds
8 **a** ii $y = e^x + 2$
  b iv $y = 2 - e^{-x}$

  c i $y = e^{2x}$
  d vi $y = e^{-2x} - 1$
  e v $y = 3e^{-x} - 5$
  f iii $y = 2 - e^x$
9 **a**

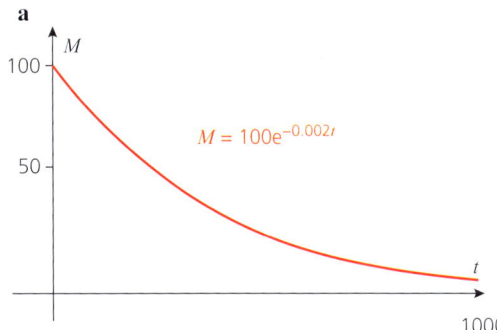

  b 347 days
10 **a** $17.82
  b 5 years
11 **a** 1.0986 (4 d.p.)
  b 0.5493 (4 d.p.)
12 **a** −0.6309 or 0.6309
  b 0.65
  c 2, 0
  d 0.4055 or 1.6094
  e 0.5108 or 0.6931
  f 0.0451 or 11.0385

Past-paper questions Page 114

1 **i** $\log_a p + \log_a q = 9$
       $2\log_a p + \log_a q = 15$
       $\log_a p = 6$ and $\log_a q = 3$
  ii $\log_p a + \log_q a = \dfrac{1}{\log_a p} + \dfrac{1}{\log_a q} = 0.5$
2   $a = b^2$, $2a - b = 3$
    $2b^2 - b - 3 = 0$ or $4a^2 - 13a + 9 = 0$
    leading to $a = \dfrac{9}{4}$, $b = \dfrac{3}{2}$
3 **i** $B = 900$
  ii $B = 500 + 400e^2 = 3456$
  iv $10000 = 500 + 400e^{0.2t}$
       $e^{0.2t} = 23.75$
       $0.2t = \ln 23.75$
       $t = 16$ (days)

Review exercise 2

1 **a** $x = -\dfrac{7}{2}$ or $x = \dfrac{17}{2}$
  b $x \leq -3$ or $x \geq 1$

2 a

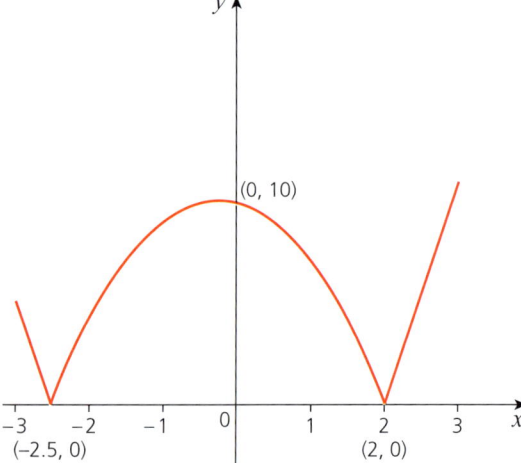

b 2

3 $-\dfrac{7}{2} \le x \le 1$

4 a $p(x) = (x-2)(x+2)(x-1)$
b i, ii

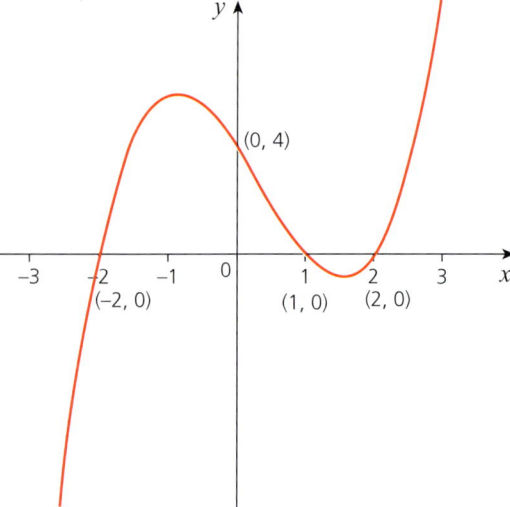

c $-2 \le x \le 1$ or $x \ge 2$

5 $x = 4, y = 8$ or $x = -2, y = 2$
6 $k < 1$ or $k > 4$
7 $x = 4, y = 2$ or $x = 9, y = 3$
8 i $12 000
 ii 2(.0273...) years

9 a

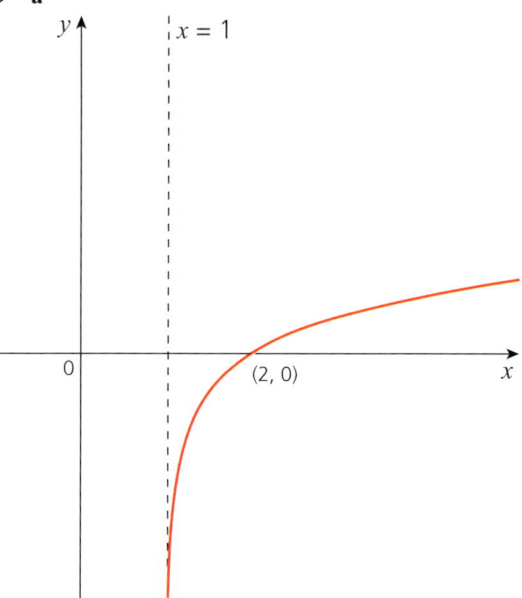

b

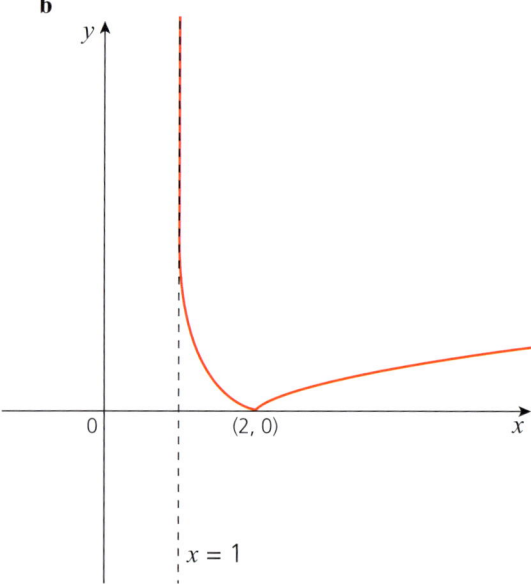

10 $x = 0.97$ (2 d.p.)
11 $\log_5 ab$

Chapter 7 Straight line graphs

Discussion point Page 118

The graph tells us that the natural length of the rope, i.e. when there is no load attached, is 15 m. It shows a proportional relationship between the rope and the load; the heavier the load, the longer the rope becomes as it is stretched. It also tells us that for every extra 40 kg of load it stretches another 5 metres, so for every extra 1 kilogram it stretches 0.125 metres or 12.5 cm.

Activity Page 123

3 gradient of AB $m_1 = \dfrac{q}{p}$

gradient of BC $m_2 = -\dfrac{p}{q}$

Exercise 7.1 Page 125

1 **a** **i** 2 **ii** $-\dfrac{1}{2}$ **iii** $4\sqrt{5}$ **iv** (6, 7)

  b **i** $-\dfrac{11}{5}$ **ii** $\dfrac{5}{11}$ **iii** $\sqrt{146}$ **iv** (7.5, −2.5)

  c **i** $\dfrac{15}{2}$ **ii** $-\dfrac{2}{15}$ **iii** $\sqrt{229}$ **iv** (7, 7.5)

  d **i** $-\dfrac{1}{5}$ **ii** 5 **iii** $\sqrt{26}$ **iv** (−0.5, −6.5)

3 **a** $6 + 3\sqrt{10} + 3\sqrt{2}$
  b Area = 9

4 **a** All sides $\sqrt{173}$
  b Midpoints both $\left(\dfrac{7}{2}, \dfrac{1}{2}\right)$
  c Gradient $PQ = -\dfrac{2}{13}$ gradient $PS = -\dfrac{13}{2}$ so not perpendicular; rhombus

5 **a**

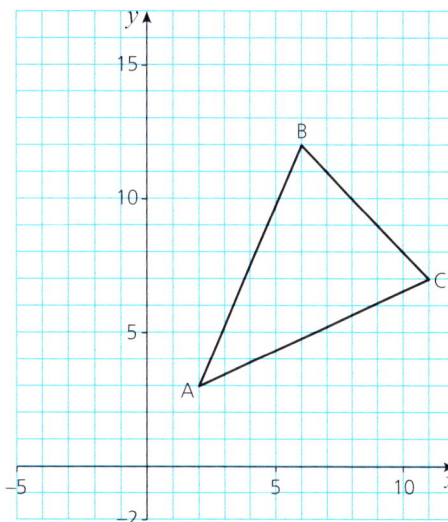

  b two equal sides are AB and AC, length $\sqrt{97}$
  c midpoint BC $\left(\dfrac{17}{2}, \dfrac{19}{2}\right)$
  d $\dfrac{65}{2}$

6 **b** (7, 4)
7 **a** 2
  b 1 : 2
8 **a**

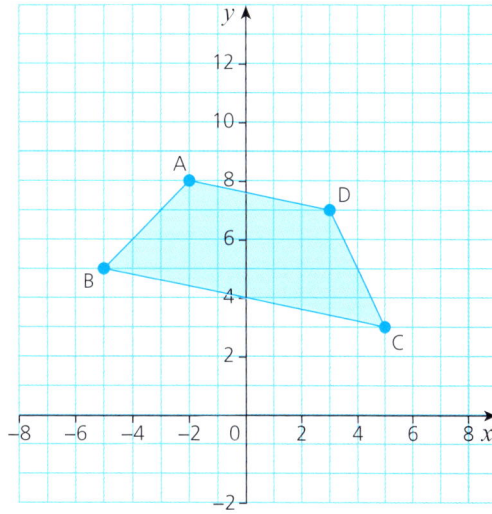

  b gradient BC = gradient $AD = -\dfrac{1}{5}$ so one pair of parallel sides, so trapezium
  c (8, 6)

9 **a** **i** $y = 2x - 13$ **ii** $2y + x + 1 = 0$
  b **i** $x + 3y = 17$ **ii** $y = 3x + 19$
  c **i** $2x = 3y + 16$ **ii** $2y + 3x + 15 = 0$

10 **a** $2y = 3x - 13$
  b $y = -4x - 8$
  c $9y + x = 18$

351

11 a $2y = x - 4$
    b $(5, 0.5)$
    c $y = -2x + 10.5$
    d $(0, 10.5)$ $(5.25, 0)$

Exercise 7.2 Page 131

1 a iii   b ii   c i   d iv
2 a $\ln y = \ln k + x \ln a$; plot $\ln y$ against x, gradient $\ln a$ and intercept $\ln k$
   b $\ln y = \ln k + a \ln x$; plot $\ln y$ against $\ln x$, gradient a and intercept $\ln k$
   c $\ln y = \ln a + x \ln k$; plot $\ln y$ against x, gradient $\ln k$ and intercept $\ln a$
   d $\ln y = \ln a + k \ln x$; plot $\ln y$ against $\ln x$, gradient k and intercept $\ln a$

3 b

t	1	2	3	4	5	6
$\ln A$	0.5878	0.9555	1.2809	1.6292	1.9741	2.3321

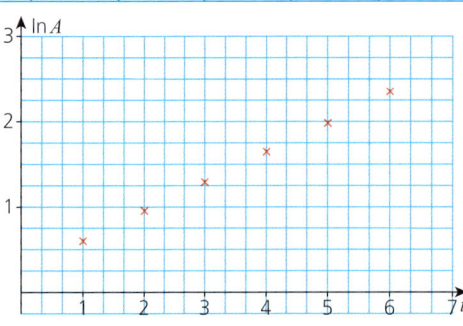

Points are close to a straight line.
   c $k = 1.25$, $b = 1.44$ (values may vary slightly)
   d (values may vary slightly)
      i 4.3 days   ii 6.45 cm²

4 b

$\ln a$	0.6931	1.3863	1.7918	2.0794	2.3026	2.4849
$\ln b$	2.2824	2.4932	2.6174	2.7014	2.7726	2.8273

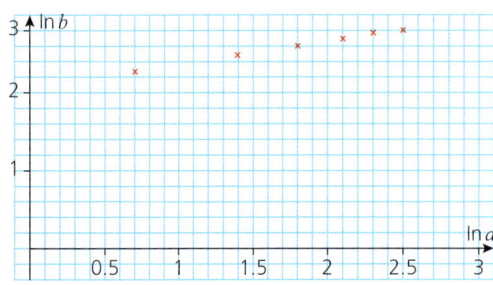

Points are close to a straight line.
   c $n = 0.304$, $P = 7.938$ (values may vary slightly)

5 a $a = 3$, $n = 2.5$

$\ln x$	0.2624	0.4700	0.6419	0.7885	0.9163	1.0296
$\ln y$	1.7579	2.1041	2.7014	3.0681	3.3878	3.6738

   b The incorrect value is the second one. Using the answers given for part **a**, the correct value should be 9.7.
6 a $\ln P = \ln k + t \ln a$ corresponds to the equation for a straight-line graph: $y = mx + c$
   b $a = 1.2$, $k = 3.0$

t	1	2	3	4	5
$\ln P$	1.2809	1.4586	1.6487	1.8245	2.0149

   c 114–115 thousand people; not reliable.

7 a

p^3	1.331	1.728	2.197	2.744	3.375	4.096
q^2	6.3001	7.5076	8.8804	10.5625	12.3904	14.5924

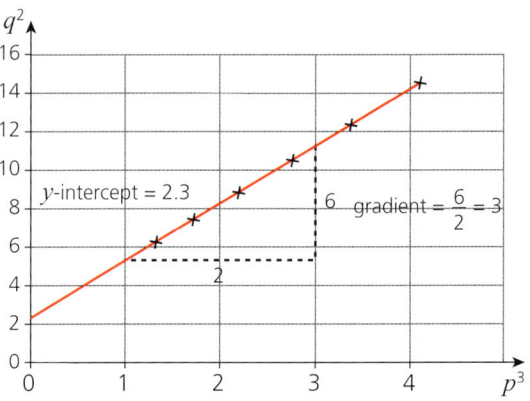

$A = 3.0$, $B = 2.3$ (values may vary slightly)
   b Data forms a straight line so the model is a good fit for the given data.

8 a

t^2	0.25	1	2.25	4	6.25	9
e^{2y}	3.00	6.05	11.02	18.17	27.11	38.09

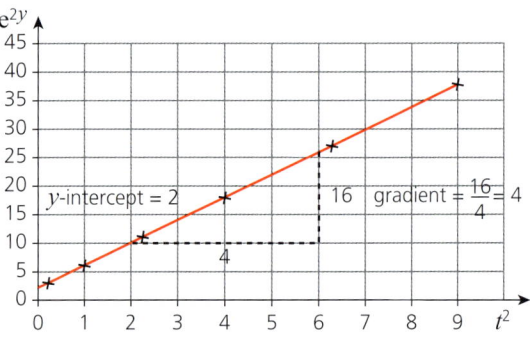

   b $A = 4.0$, $B = 2.0$
   c 2.31 cm
   d Result not valid as 5 hours is outside the range of the given data.

9 a $A = 0.8$, $B = 1.5$

$\ln t$	0.41	0.92	1.25	1.50	1.70	1.87
y^3	1.82	2.25	2.52	2.69	2.86	2.99

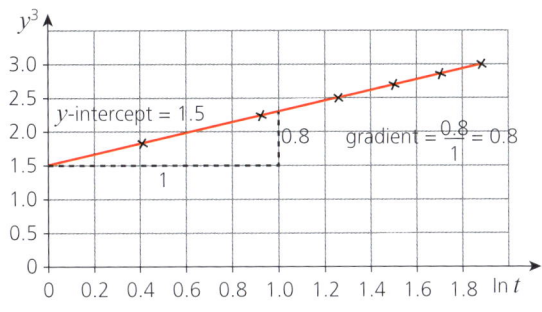

b 10 minutes 25.2 seconds

Past-paper questions Page 133

1 **i** $(3, 5)$
   ii $y = 2x - 1$
   iii 15 square units
2 **i** $D(3, 8)$, $E(5.4, 9.2)$
   ii Area $= \frac{1}{2}(13 + 3) \times 4 = 32$
3 **i** $\sqrt{20}$ or 4.47
   ii Gradient $AB = \frac{1}{2}$, $\perp$ gradient $= -2$
         $\perp$ line $y - 4 = -2(x - 1)$
         $y = -2x + 6$
   iii $(3, 0)$ $(-1, 8)$

Chapter 8 Coordinate geometry of the circle

Discussion point Page 136

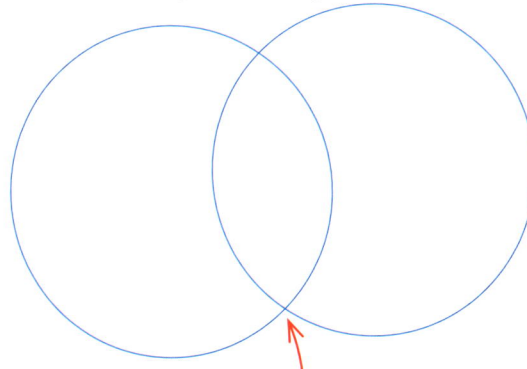

Two points of intersection so two possible locations

With two satellites you would get two possible locations. There would be two circles with known radii which would have two distinct points of intersection.

In reality, the method of trilateration would use equations of spheres rather than circles. Therefore, a minimum of four satellites would be needed to find your precise location.

Exercise 8.1 Page 141

1 **a** $(x - 2)^2 + (y - 7)^2 = 9$
   b $(x - 6)^2 + y^2 = 64$
   c $(x + 2)^2 + (y - 5)^2 = 1$
   d $(x - 3)^2 + (y + 10)^2 = 25$
   e $(x + 5)^2 + (y + 9)^2 = 4$
2 **i**   Radius 4, centre $(0, 0)$
   ii Radius 8, centre $(1, 0)$
   iii Radius 1, centre $(-6, 5)$
   iv Radius 5, centre $(1, -1)$
   v   Radius 6, centre $(-4, -4)$
3 **a** centre $(0, 0)$, radius 6

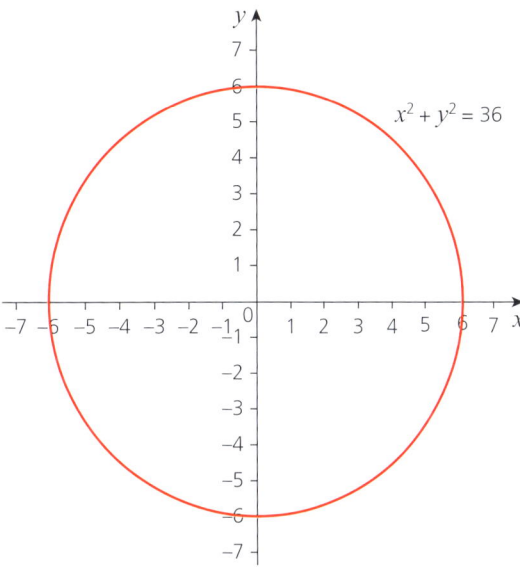

   b centre $(0, -1)$, radius 5

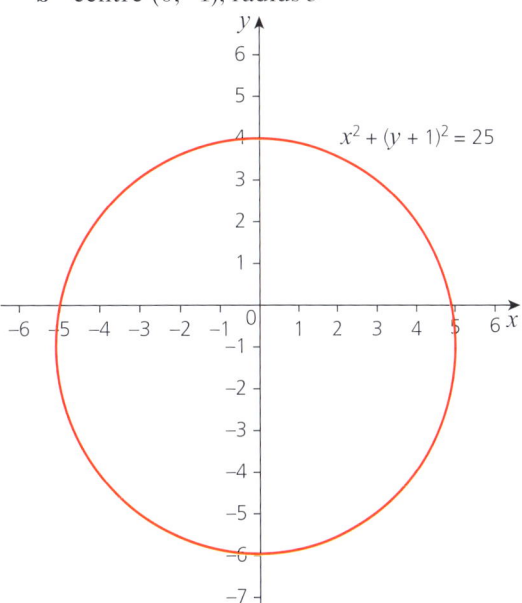

c centre (4, 7), radius 3

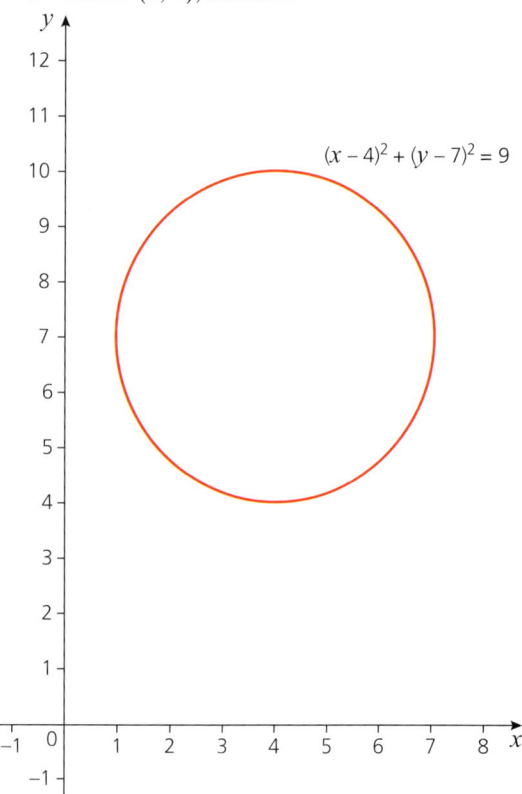

4 $(x-1)^2 + (y+2)^2 = 25$
5 $(x+3)^2 + (y+6)^2 = 289$
6 a (1, 3)
  b 5
  c $(x-1)^2 + (y-3)^2 = 25$
  d $(1-1)^2 + (-2-3)^2 = 0 + 25 = 25$ as required.
7 a $a = 3$, $b = -7$ and $r = 2$
  b Radius is 2 and centre is (3, −7)
8 a Radius is 9 and centre is (−5, 0)
  b Radius is $\sqrt{5}$ and centre is (−1, −1)
  c Radius is $2\sqrt{2}$ and centre is (5, 8)
9 $(x+2)^2 + (y-12)^2 = (\sqrt{29})^2$
10 a, b Centre is $(-g, -f)$ and radius is $\sqrt{g^2 + f^2 - c}$

Discussion point Page 144

Making y the subject would give $y = -\frac{1}{2}x - 3$. The equation with x as the subject is slightly easier to work with because it only contains integer values.

Exercise 8.2 Page 146

1 a The line intersects the circle at (2, 4) and (−4, −2).
  b No points of intersection
  c The line is a tangent to the circle at the point (6, −2).
  d The line intersects the circle at (0, −9) and (−12, −15).
  e No points of intersection
  f The line is a tangent to the circle at the point (4, 14).
2 No points of intersection
3 One point of intersection so must be a tangent to the circle.
4 a The line intersects the circle at (−2, −1) and (0, 5).
  b (−1, 2)
  c The centre of the circle is (−1, 2); the midpoint of the line is the centre of the circle so the line must be a diameter of the circle.
5 a

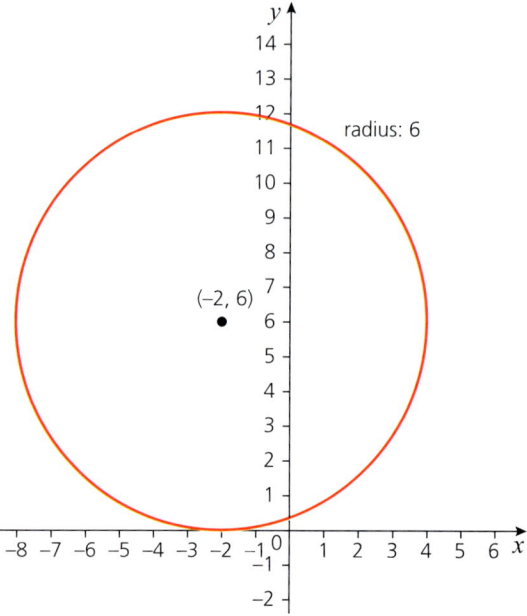

  b $y = 0$, $y = 12$
  c $x = -8$, $x = 4$
6 a $\frac{1}{3}$
  b −3
  c $y = -3x - 2$
7 a $y = 2x - 7$
  b $y = x + 1$
  c $y = -\frac{1}{2}x + 1$
  d $y = 3x + 4$
  e $y = -\frac{3}{4}x - \frac{9}{4}$
8 a $y = -x + 3$
  b $(x-5)^2 + (y-6)^2 = 32$
  c $\left(\frac{7}{3}, \frac{2}{3}\right)$

Discussion point Page 150

Substituting $y = 2x - 4$ into $x^2 + y^2 - 24x + 44 = 0$ gives

$x^2 + (2x - 4)^2 - 24x + 44 = 0$

$\Rightarrow x^2 + 4x^2 - 16x + 16 - 24x + 44 = 0$

$\Rightarrow 5x^2 - 40x + 60 = 0$

$\Rightarrow x^2 - 8x + 12 = 0$

$\Rightarrow (x - 2)(x - 6) = 0$

So $x = 2$ or $x = 6$
When $x = 2$, $y = 2 \times 2 - 4 = 0$

When $x = 6$, $y = 2 \times 6 - 4 = 8$

Therefore, the two circles intersect at the points $(2, 0)$ and $(6, 8)$.

Exercise 8.3 Page 153

1. **a** No points of intersection
   b The circles intersect at $(5, 11)$ and $(5, -5)$.
   c The circles touch at $(2, 4)$.
   d The circles intersect at $(7, -6)$ and $(8, 1)$.
2. **a** The circles intersect at $(6, -1)$ and $(10, 7)$.
   b $y = 2x - 13$
3. **a** $B(-2, 1)$.
   b $y = x + 3$
   c $(3, 6)$
   d The line joining the centres is the perpendicular bisector of the chord AB.
4. **a** The circles touch at $(6, 12)$.
   b 10
   c The circle $(x - 3)^2 + (y - 8)^2 = 25$ must lie inside the circle $(x + 3)^2 + y^2 = 225$. The distance between the centres, 10, is equal to the difference of the radii. So, the circles will touch internally.
5. **a** No points of intersection
   b $x^2 + (y - 2)^2 = 64$ has centre $(0, 2)$ and radius 8
   $x^2 + (y - 4)^2 = 16$ has centre $(0, 4)$ and radius 4
   c The circle $x^2 + (y - 4)^2 = 16$ lies inside the circle $x^2 + (y - 2)^2 = 64$.

Practice questions Page 154

1. **a** $(x - 2)^2 + (y - 3)^2 = 5^2$
   b $(-2, 6)$
2. **a** $(3, 3)$ does lie on the circle.
   b A is $(-1, -5)$, B is $(7, -5)$.
   c Area $= \frac{1}{2} \times 8 \times 8 = 32$ units^2
3. **a** Discriminant is less than 0 so there is no solution to the equation. Hence, no points of intersection.
   b **i** 13
   ii $r_1 + r_2 = 12 < 13$; hence, one circle does not lie inside the other.

Chapter 9 Circular measure

Discussion point Page 157

Approximately 26 centimetres per second

Exercise 9.1 Page 162

1. **a** $\frac{2\pi}{3}$
   b 3π
   c $\frac{11\pi}{90}$
   d $\frac{5\pi}{6}$
   e $\frac{5\pi}{24}$
2. **a** $120°$
   b $100°$
   c $171°$
   d $25.7°$
   e $67.5°$

3.

Radius, r (cm)	Angle at centre in degrees	Angle at centre in radians	Arc length, s (cm)	Area, A (cm^2)
8	$120°$	$\frac{2\pi}{3}$	$\frac{16\pi}{3}$	$\frac{64\pi}{3}$
10	$28.6°$	0.5	5	25.0
5.73	$60°$	$\frac{\pi}{3}$	6	17.2
6	$38.2°$	$\frac{2}{3}$	4	12
5.53	$75°$	$\frac{5\pi}{12}$	7.24	20

4.

Radius, r (cm)	Angle at centre, θ (radians)	Arc length, s (cm)	Area, A (cm^2)
10	$\frac{\pi}{3}$	$\frac{10\pi}{3}$	$\frac{50\pi}{3}$
12	2	24	144
20.37	$\frac{\pi}{4}$	16	162.97
5	2	10	25
6.51	$\frac{3\pi}{5}$	12.28	40

5. **a** $\frac{32\pi}{3}$ cm^2
   b $8 + \frac{16\pi}{3}$ cm
6. **a** 15π cm^2
   b 9 cm^2
   c $15\pi - 9$ cm^2

7 a 1.855
b 67.36 cm²

8 a $\frac{3\pi}{4}$
b 24π cm²
c $16\sqrt{2}$ cm²
d $24\pi - 16\sqrt{2} = 52.77$ cm²

Past-paper questions Page 163

1 i Area $= \frac{1}{2} 18^2 \sin 1.5 - \frac{1}{2} 10^2 (1.5)$
$= 161.594 - 75$
$= 86.6$ cm²

ii $AC = 15$ or 10×1.5
$BD = \sqrt{18^2 + 18^2 - (2 \times 18 \times 18 \cos 1.5)}$
$= 24.5$
Perimeter $= 15 + 24.5 + 16$
$= 55.5$ cm

2 i $\sin\frac{\theta}{2} = \frac{6}{8}, \frac{\theta}{2} = 0.8481$
$\theta = 1.696$

ii Arc length $= (2\pi - 1.696) \times 8 = 36.697$
Perimeter $= 12 + (2\pi - 1.696) \times 8$
$= 48.7$

iii Area $= \frac{8^2}{2}(2\pi - 1.696) + \frac{8^2}{2}\sin 1.696$
$= 178.5$

3 i Area of sector $= \frac{1}{2} \times x^2 \times 0.8$
Area of triangle $=$
$\frac{1}{2} \times 5\cos 0.8 \times 5\sin 0.8 = 6.247$ cm²
$0.08 x^2 = 6.247$
$x = 8.837$

ii Perimeter $= 19.85$ cm (2 d.p.)
iii Area $PQSR = 25$ cm²

Chapter 10 Trigonometry

Discussion point Page 166

You can take measurements from the picture. Then use them to make a scale drawing of the cross-section of the pyramid that goes through its vertex, and measure the angle with a protractor.

This is not a very accurate method. The alternative is to use trigonometry to calculate the answer.

Exercise 10.1 Page 172

1 a $29^2 = 841 = 400 + 441 = 20^2 + 21^2$
b $\sin Q = \frac{20}{29}$
$\cos Q = \frac{21}{29}$
$\tan Q = \frac{20}{21}$

c i $\left(\frac{20}{29}\right)^2 + \left(\frac{21}{29}\right)^2 = \frac{400 + 441}{841} = 1$
ii $\tan Q = \frac{20}{29} \div \frac{21}{29} = \frac{20}{21}$

5 a 6 cm
6 a $2\sqrt{2}$ cm
7 a $4d$
8 a 2 cm

Exercise 10.2 Page 174

1 a i $\frac{1}{2}$ **ii** $\frac{\sqrt{3}}{2}$ **iii** $\frac{1}{\sqrt{3}}$
b i 2 **ii** $\frac{2}{\sqrt{3}}$ **iii** $\sqrt{3}$

2 a i $\frac{1}{\sqrt{2}}$ **ii** $\frac{1}{\sqrt{2}}$ **iii** 1
b i $\sqrt{2}$ **ii** $\sqrt{2}$ **iii** 1

3 a i $\frac{\sqrt{3}}{2}$ **ii** $\frac{1}{2}$ **iii** $\sqrt{3}$
b i $\frac{2}{\sqrt{3}}$ **ii** 2 **iii** $\frac{1}{\sqrt{3}}$

4 a $B = 60°, C = 30°$
b $\sqrt{3}$

5 a $B = 30°, C = 60°$
b $AB = 2\sqrt{3}$ units, $BC = 4$ units

6 $\frac{4}{\sqrt{7}}, \frac{\sqrt{7}}{3}$

7 a $45°; 45°$
b $\sec L = \sqrt{2}$, $\csc L = \sqrt{2}$, $\tan L = 1$

8 b $14.0°$

9 a $0 < \alpha < 90$
$0 < \alpha < \frac{\pi}{2}$
b no
c no

Discussion point Page 178

1 The graphs continue the same wave patterns to the left.

2
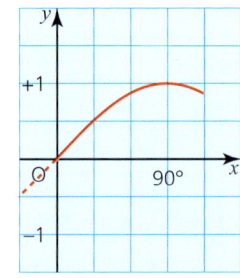

Starting with $y = \sin x$ for $0°$ to $90°$:
- reflect it in the line $x = 90°$ to obtain the part of the curve from $90°$ to $180°$.
- rotate the curve from $0°$ to $180°$ through $180°$ about the point $(180°, 0)$ to obtain the part of the curve between $180°$ and $360°$. You now have the complete curve of $y = \sin x$ from $0°$ to $360°$.

To obtain the curve of $y = \cos x$, translate the curve for $y = \sin x$ from 0° to 360° by −90° in the horizontal direction. Then translate the part of the curve between −90° and 0° by +360° in the horizontal direction.

Discussion point Page 180
The curves do not change.

Exercise 10.3 Page 182

1 a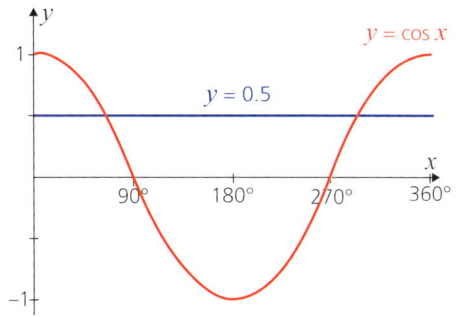

   b 60° or 300°
   c $(60 + 360n)°$ or $(300 + 360n)°$
   d $-\dfrac{1}{2}$

2 a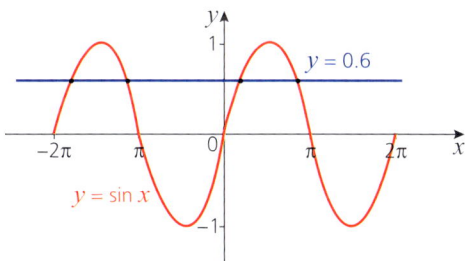

   b $-1.2\pi, -1.8\pi, 0.2\pi, 0.8\pi$
   c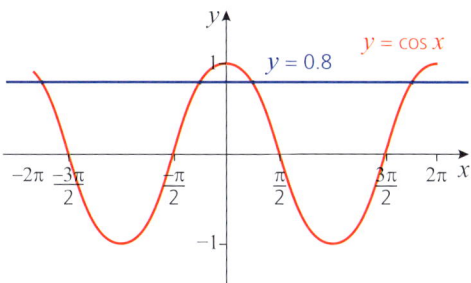

   d $-0.2\pi, -1.8\pi, 0.2\pi, 1.8\pi$
   e For acute angles, and for other angles where both sin and cos are positive, $\sin^{-1} 0.6 = \cos^{-1} 0.8$. For other angles, one or other is negative.

3 a $\dfrac{\pi}{3}, \dfrac{4\pi}{3}$ (or 60°, 240°)
   b $\dfrac{\pi}{6}, \dfrac{5\pi}{6}$ (or 30°, 150°)
   c $\dfrac{5\pi}{6}, \dfrac{7\pi}{6}$ (or 150°, 210°)
   d $\dfrac{\pi}{6}, \dfrac{7\pi}{6}$ (or 30°, 210°)
   e 2.35 rad, 3.94 rad (or 134.4°, 225.6°)
   f 1.27 rad, 5.02 rad (or 72.5°, 287.5°)
   g 3.48 rad, 5.94 rad (or 199.5°, 340.5°)
   h 270°

4 a $\dfrac{1}{\sqrt{2}}$
   b $\dfrac{1}{2}$
   c 1
   d $\dfrac{\sqrt{3}}{2}$
   e $-\dfrac{\sqrt{3}}{2}$
   f 0
   g $\dfrac{1}{\sqrt{2}}$
   h $\dfrac{1}{\sqrt{2}}$
   i 1

5 a 150°
   b −160.3°
   c 87.4°

6 a Graph has a line of symmetry at $x = 90°$.

   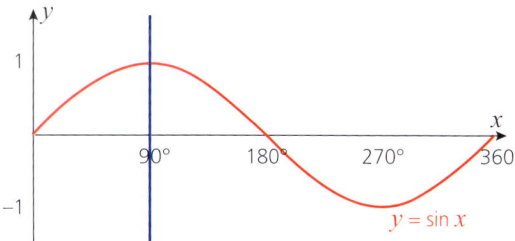

   b i False   ii True   iii False   iv True

7 a $0° < \alpha < 90°$
   b No. If $\cos\alpha$ and $\sin\alpha$ are both positive, $\tan\alpha$ is positive.
   c No. Values of α do exist where $\sin\alpha$ and $\cos\alpha$ are both negative (such as when $180° < \alpha < 270°$); however, in these cases, $\tan\alpha$ would always be positive.

8 a 80°, 320°
   b 20°, 200°
   c 55°, 325°
   d 30°, 120°, 210°, 300°
   e 60°, 300°
   f 75°, 105°, 255°, 285°
   g 10°, 50°, 130°, 170°, 250°, 290°
   h 0°, 90°, 180°, 270°, 360°
   i 15°, 75°, 135°, 195°, 255°, 315°

9 a −6.18 rad, −3.24 rad, 0.1 rad, 3.04 rad
  b $\frac{-5\pi}{3}, \frac{-\pi}{3}, \frac{\pi}{3}, \frac{5\pi}{3}$
  c −4.25 rad, −1.11 rad, 2.03 rad, 5.18 rad
  d −2.73 rad, −0.412 rad, 3.55 rad, 5.87 rad
  e $-2\pi, \frac{-3\pi}{2}, -\pi, 0, \frac{\pi}{2}, \pi, 2\pi$

Discussion point Page 185

Stretching the point (90°, 1) by a scale factor of 2 parallel to the y-axis gives the point (90°, 2). Translating this through 3 units in the y-direction gives (90°, 5). If the translation is done before the stretch, the point (90°, 1) moves through (90°, 4) to (90°, 8), which is incorrect.

Exercise 10.4 Page 190

1 i a
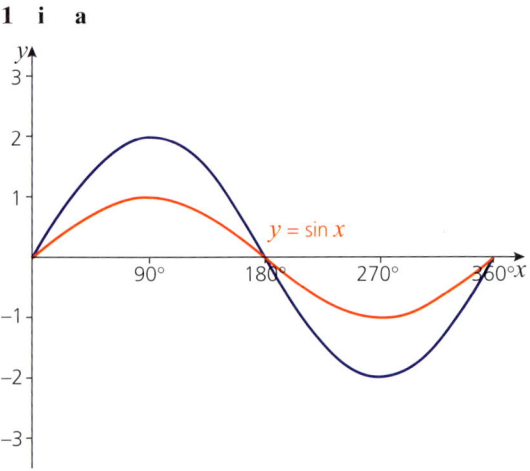
  b amplitude 2, period 360°
  ii a
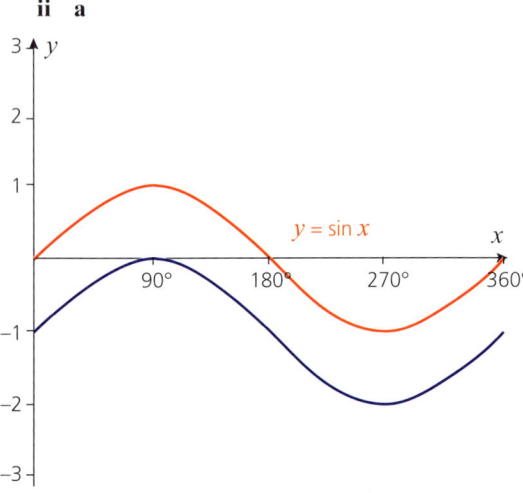
  b amplitude 1, period 360°

  iii a
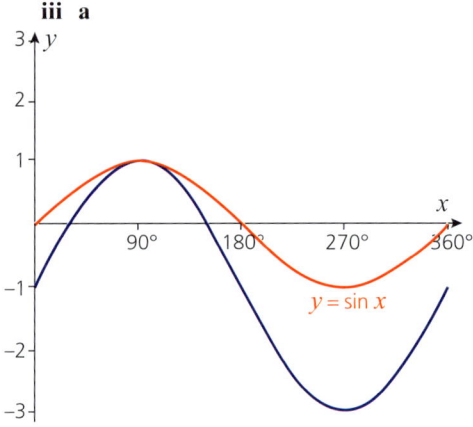
  b amplitude 2, period 360°
  iv a
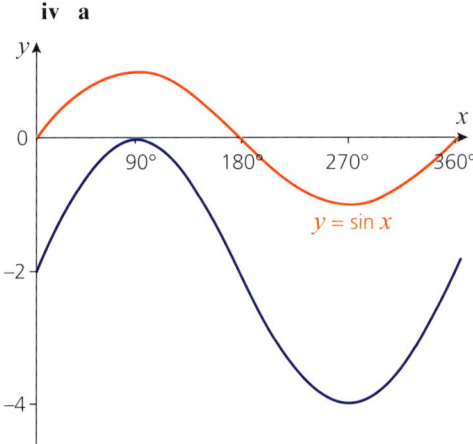
  b amplitude 2, period 360°
  c The order of the transformations matters.

2 i a and b
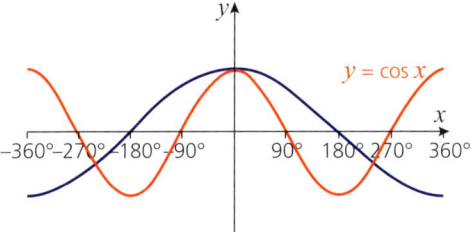
  c amplitude 1, period 720°
  ii a and b
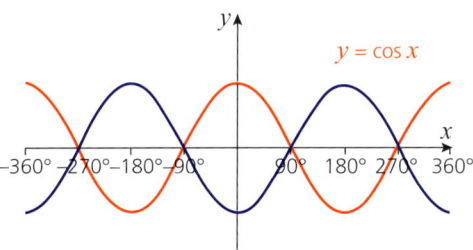
  c amplitude 1, period 360°

iii **a** and **b**

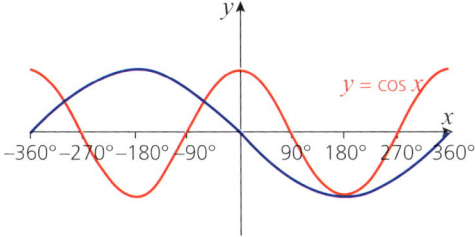

c amplitude 1, period 720°

iv **a** and **b**

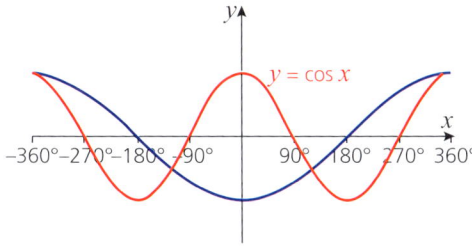

c amplitude 1, period 720°
d The order of the transformations matters.

3 i a and **b**

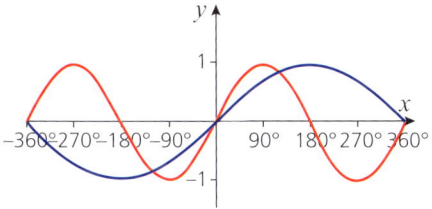

c amplitude 1, period 720°

ii **a** and **b**

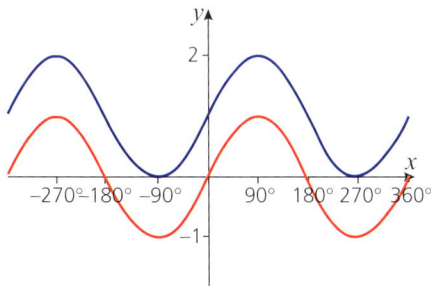

c amplitude 1, period 360°

iii **a** and **b**

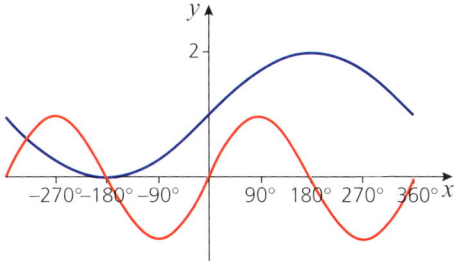

c amplitude 1, period 720°

iv **a** and **b**

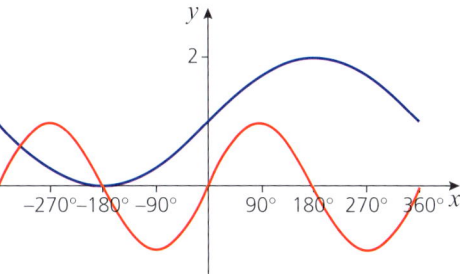

c amplitude 1, period 720°
d The order does not matter when the transformations are in different directions.

4 a Stretch in the y-direction scale factor 3 and stretch in the x-direction scale factor $\frac{1}{2}$ (either order).
b Stretch in the y-direction scale factor 2 followed by a translation of one unit vertically upwards.
c Stretch in the y-direction scale factor 2 and translate to the right by 180° (either order).
d Translation to the left $\frac{\pi}{2}$ followed by a stretch in the y-direction scale factor 3 and a translation 3 units vertically upwards.

5 A stretch, scale factor 2, parallel to y-axis and a translation of 1 unit vertically upwards.

6 A stretch, scale factor $\frac{1}{2}$, parallel to the x-axis and a stretch, scale factor 3, parallel to the y-axis

7 Stretch in the x-direction scale factor $\frac{1}{2}$ followed by a stretch in the y-direction scale factor 3 and a translation vertically downwards by 2. (The stretch in the x-direction can be at any stage.)

8 a Stretch in the *x*-direction with scale factor $\frac{1}{2}$ and a stretch in the *y*-direction with scale factor 4.

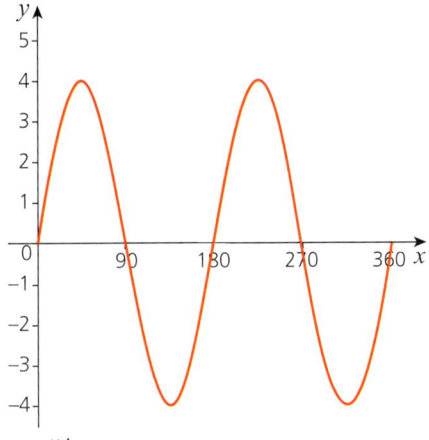

b i

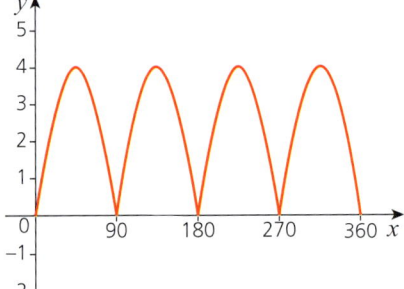

ii

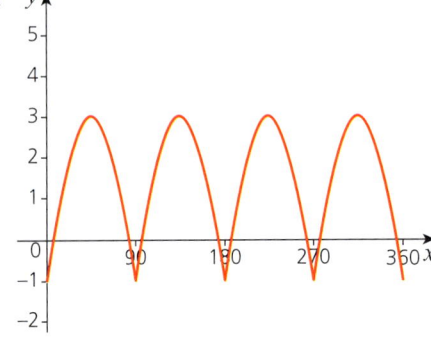

iii

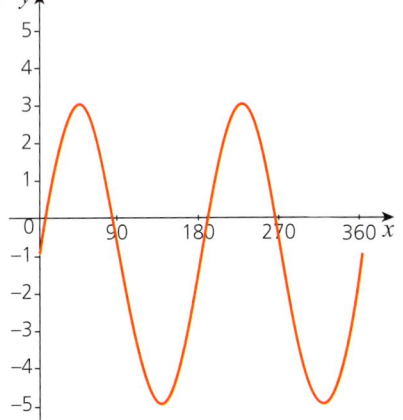

iv

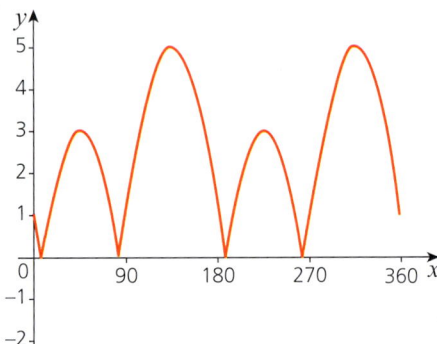

9 a

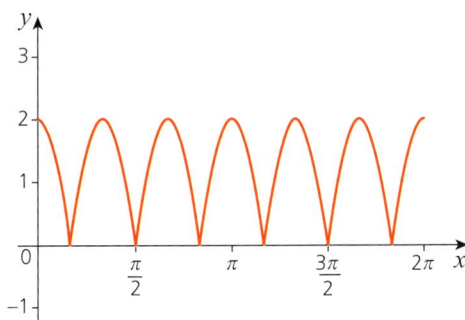

b

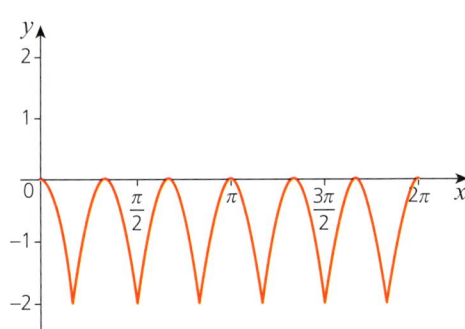

c

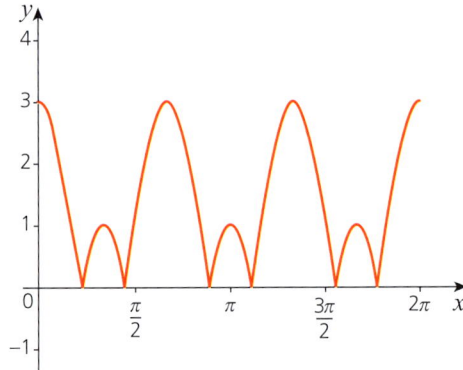

Exercise 10.5 Page 199

5. 54.74°; 125.26°
7. $\frac{\pi}{3}; \frac{2\pi}{3}; \frac{4\pi}{3}$ and $\frac{5\pi}{3}$.
9. **a** $-2\pi; 0; 2\pi$
   b There are no solutions.
10. **a** $\theta = -270°, -90°$ or $90°, 270°$
    b $\theta = -182.84°$ or $357.16°$
    c $\theta = 0.99, 1.63, 4.13$ or 4.77
11. **b** $x = 0°; 360°; 156.44°; 203.56°$
12. **b** $0; \pi; 2\pi; \frac{\pi}{6}; \frac{5\pi}{6}$

Past-paper questions Page 199

1. **a** $\sin^2 x = \frac{1}{4}$

   $\sin x = (\pm)\frac{1}{2}$

   $x = 30°, 150°, 210°, 330°$

   b $(\sec^2 3y - 1) - 2\sec 3y - 2 = 0$

   $\sec^2 3y - 2\sec 3y - 3 = 0$

   $(\sec 3y + 1)(\sec 3y - 3) = 0$

   leading to $\cos 3y = -1$, $\cos 3y = \frac{1}{3}$

   $3y = 180°, 540°; 3y = 70.5°, 289.5°, 430.5°$

   $y = 60°, 180°, 23.5°, 96.5°, 143.5°$

   c $z - \frac{\pi}{3} = \frac{\pi}{3}, \frac{4\pi}{3}$

   $z = \frac{2\pi}{3}, \frac{5\pi}{3}$

2. **a** **i** $A = 3, B = 2$
       ii $C = 4$
   b Period = 120° or $\frac{2\pi}{3}$

   Amplitude = 5

3. LHS $= \frac{1 + \cos\theta}{(1 - \cos\theta)(1 + \cos\theta)} + \frac{1 - \cos\theta}{(1 + \cos\theta)(1 - \cos\theta)}$

   $= \frac{2}{1 - \cos^2\theta}$

   $= \frac{2}{\sin^2\theta}$

   $= 2\csc^2\theta = $ RHS

Review exercise 3

1. $y = -\frac{3}{2}x + \frac{39}{4}$
2. **a** $\ln y = n \ln x + \ln a$
   b $n = -0.6, a = 276$ (3 s.f.) *(values may vary slightly)*
3. **a** $\left(\frac{11}{3}, \frac{14}{3}\right)$
   b P is $(-3, -2)$ and Q is $(2, 3)$
   c $33\frac{1}{3}$ units^2
4. **a** $(-3, 0)$ and $(1, 4)$
   b $y = x + 3$
5. **a** The distance between the point $(2, 8)$ and the centre $(-1, 4)$ is less than the radius. So, the point $(2, 8)$ must lie inside the circle.
   b $r_2 - r_1 = d$, hence, the circles must touch at one point (they touch internally).
6. **i** $\theta = \frac{5}{8}$ radians
   ii 5 cm
   iii 13.3 cm^2 (3 s.f.)
7. **i** $a = -2, b = 4, c = 6$
   ii

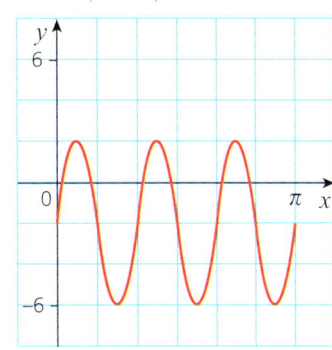

8. **a** LHS $= \frac{1 - \cos^2 A}{\sin A \cos A} = \frac{\sin^2 A}{\sin A \cos A} = \frac{\sin A}{\cos A}$

   $= \tan A$

   b $x = 22.5°$ or $x = 112.5°$

Chapter 11 Permutations and combinations

Discussion point Page 206

There are $6 \times 6 \times 6 \times 6 = 1296$ possibilities.

Suppose it takes 10 seconds to check each combination, so you check six combinations in 1 minute.

It will take $1296 \div (6 \times 60) = 3.6$ hours or 3 hours 36 minutes.

Exercise 11.1 Page 208

1. **a** 5040
   b 72
   c $\frac{12}{7}$
2. **a** $\frac{1}{n+1}$
   b $n - 2$
3. **a** $(n+1)(n+2)$
   b $n(n+1)$
4. **a** $\frac{9! \times 3!}{(6!)^2}$
   b $\frac{15!}{13! \times 5!}$
   c $\frac{(n+2)!}{n! \times 4!}$

5 a $8 \times 6!$
   b $(n+1)(n-1)!$
6 $\dfrac{7!}{5!}$
7 24
8 5040
9 120
10 120
11 15!
12 a 6
   b $8! = 40320$
   c $9! = 362880$
   d $6! = 720$
   e $7! = 5040$
   f $8! = 40320$
13 a $8! = 40320$
   b $7! = 5040$
   c 10080
   d 30240

Discussion point Page 210

No, it does not matter in which order the machine picked them since this is a combination, not a permutation.

Discussion point (2) Page 211

As the amount of numbers to choose from increases, the probability decreases because there are more possible combinations.

Discussion point (3) Page 211

$${}^nC_r = \dfrac{n!}{r!(n-r)!}, \quad {}^nP_r = \dfrac{n!}{(n-r)!}$$

$${}^nC_r = \dfrac{1}{r!} \times \dfrac{n!}{(n-r)!} = \dfrac{1}{r!} \times {}^nP_r = \dfrac{{}^nP_r}{r!}$$

Discussion point Page 212

$${}^nC_r = \dfrac{n!}{r!(n-r)!}$$

$$\Rightarrow {}^nC_0 = \dfrac{n!}{0!(n-0)!} = \dfrac{n!}{1 \times n!} = 1$$

Similarly for ${}^nC_n$.

Exercise 11.2 Page 212

1 a i 210   ii 3024   iii 1 814 400
   b i 35   ii 126   iii 45
2 2730
3 1287
4 576
5 a 352 716
   b 24 000

6 a 210
   b i $\dfrac{1}{42}$ ii $\dfrac{10}{21}$
7 a 210
   b i $\dfrac{1}{14}$ ii $\dfrac{1}{42}$
8 a 8008
   b 2940
   c 7
   d 84
9 a 1001
   b 420
   c 336
10 32
11 9080
12 a 495
   b 360
13 a 21 772 800
   b 126
14 a 56 756 700
   b 3 870 720
   c 192
15 a 79 833 600
   b 3 628 800

Past-paper questions Page 214

1 i $\dfrac{14 \times 13 \times 12 \times 11 \times 10 \times 9}{6 \times 5 \times 4 \times 3 \times 2 \times 1}$ or $\dfrac{14!}{8! \times 6!}$
       3003
   ii Either 5 students + 1 teacher or 4 students + 2 teachers.
       56×6 or 70×15
       1386
   iii 30
2 a i   360
      ii 120
   b i   924
      ii 28
      iii $924 - ({}^8C_3 \times {}^4C_3) - ({}^8C_2 \times {}^4C_4)$
          (i.e. 924 – 3M 3W – 2M 4W)
          $924 - 224 - 28 = 672$
3 a i   15120
      ii 210

Chapter 12 Series

Discussion point Page 216

50×2^{63} mg = 4.6×10^{11} tonnes. In 2017 the world production of wheat was about 7.5×10^8 tonnes. King Shirham made an agreement that he could not possibly fulfil!

Real-world activity Page 220

1. $\dfrac{n-1}{2}$
2. Multiply the number in column 2 by $\dfrac{n-2}{3}$

Exercise 12.1 Page 223

1. **a** $1 + 4x + 6x^2 + 4x^3 + x^4$
   b $1 + 8x + 24x^2 + 32x^3 + 16x^4$
   c $1 + 12x + 54x^2 + 108x^3 + 81x^4$
2. **a** $16 + 32x + 24x^2 + 8x^3 + x^4$
   b $81 + 108x + 54x^2 + 12x^3 + x^4$
   c $256 + 256x + 96x^2 + 16x^3 + x^4$
3. **a** $x^4 + 4x^3y + 6x^2y^2 + 4xy^3 + y^4$
   b $x^4 + 8x^3y + 24x^2y^2 + 32xy^3 + 16y^4$
   c $x^4 + 12x^3y + 54x^2y^2 + 108xy^3 + 81y^4$
4. **a** 10
   b 21
   c 35
   d 21
   e 1
   f 286
5. **a** 15
   b 21
   c 28
6. $243 + 405kx + 270k^2x^2$
7. $729x^6 - 4374x^4 + 10\,935x^2$
8. **a** $6t + 2t^3$
9. **a** $-5, -5$
   b $5, -5$
10. **a** $1 - 12x + 60x^2 - 160x^3$
    b $64 - 576x + 2160x^2 - 4320x^3$
    c $729 - 5832x + 19\,440x^2 - 34\,560x^3$
11. **a** $x^{10} + 5x^7 + 10x^4 + 10x$
    b $x^{10} - 5x^7 + 10x^4 - 10x$
    c $x^{15} + 5x^{11} + 10x^7 + 10x^3$
    d $x^{15} - 5x^{11} + 10x^7 - 10x^3$
12. **a** 11520
    b 792
13. **a** 7505784
    b −25200000
14. $n = 5, a = 3$

Exercise 12.2 Page 229

1. **a** yes: 2, 40
   b no
   c no
   d yes: 4, 29
   e yes: −4, −12
2. **a** 21
   b 19
3. **a** $d = 6$
   b 16 terms
4. **a** 5
   b 2055
5. **a** 16, 20, 24
   b 456
6. **a** 15
   b 1140
7. **a** $a = 8, d = 6$
   b 10
8. **a** 2
   b 480
9. **a** 10000
   b 10200
   c 20200
   d Answer c is the total of the other two (a and b) answers.
10. **a** 66000
    b Sum is negative if there are more than 31 terms. (The sum of the first 31 values is 0.).
11. **a** $7n - 9$, 17th
    b $\dfrac{7}{2}n^2 - \dfrac{11}{2}n$, 25
12. **a** $54\,000
    b 10
13. **a** 47 days
    b 126.9 km
14. **a** 16
    b 15 cm
15. **a** $a = -25$ and $d = 15$
    b $n = 16$
16. **a** 3
    b 105
    c $2n^2 + n$
    d $6n^2 + n$

Discussion point Page 233

Approximately 3.7×10^{11} tonnes of wheat.
China produces approximately 130 million tonnes of wheat per year. Dividing 130 million by 365 days gives approximately 356 000 tonnes per day. This might be an amount of wheat that you would expect there to be in China at any one time. However, there may be more or less; it depends on how quickly it is used and when it is harvested.

Discussion point Page 237

Something else.

Exercise 12.3 Page 237

1. **a** yes: 2, 192
   b no
   c yes: −1, 10
   d yes: 1, 1
   e no
   f yes: $\dfrac{1}{2}, \dfrac{5}{32}$
   g no

2 a 320
  b 635
3 a 4
  b 49152
4 a 16
  b 8th
5 a 10
  b 7161
6 a 9
  b 199.609375
7 a 2
  b 4.5
  c 4603.5
8 a 0.5
  b 16
9 a 0.1
  b $\frac{8}{9}$
  c $\frac{8}{11}$
10 a 0.7
   b 14th
   c $\frac{1000}{3}$
   d 13
11 a $\frac{1}{3}$
   b 5
12 a $\frac{1}{3}$
   b $216 \times \frac{1}{3}^{(n-1)}$
   c $324 - 324\left(\frac{1}{3}\right)^n$
   d 324
   e 12 terms
13 a 10, 5, 2.5, 1.25, 0.625
   b 0, 5, 7.5, 8.75, 9.375
   c The first sequence is geometric with common ratio 0.5
14 a 60
   b 160.51°
15 a $15 \times \left(\frac{5}{8}\right)^{n-1}$
   b 79.27 m
18 a 4 and −4
   b when $r = -4$, $a = -3$ and the sum of the first ten terms = 629145
     when $r = 4$, $a = 3$ and the sum of the first ten terms = 1048575

Past-paper questions Page 239

1 $120(p^7q^3)$ and $45(p^8q^2)$
  $120p^7q^3 = 270p^8q^2$
  $252p^5q^5 = 252$
  $pq = 1$ and $4q = 9p$
  leading to $p = \frac{2}{3}, q = \frac{3}{2}$

2 i −27.5
  ii 38.5
3 i $64 + 192x + 240x^2 + 160x^3$
  ii 64

Chapter 13 Vectors in two dimensions

Discussion point Page 241

They tell you both the speed and the direction of the wind.

Exercise 13.1 Page 244

1 a $\begin{pmatrix} -4 \\ -2 \end{pmatrix}$

  b $\begin{pmatrix} 6 \\ -4 \end{pmatrix}$

  c $\begin{pmatrix} -3 \\ 0 \end{pmatrix}$

  d $\begin{pmatrix} -2 \\ -4 \end{pmatrix}$

2 a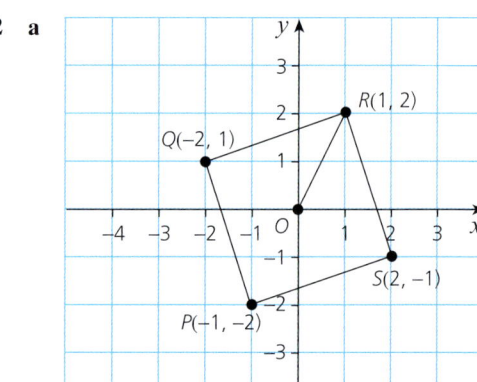

  b i $\begin{pmatrix} 1 \\ 2 \end{pmatrix}$

    ii $\begin{pmatrix} -1 \\ -2 \end{pmatrix}$

  c i $\begin{pmatrix} 2 \\ 4 \end{pmatrix}$

    ii $\begin{pmatrix} 4 \\ -2 \end{pmatrix}$

  d i–iv All four lengths = $\sqrt{10}$
  e PQRS is a square because all side lengths are equal and all angles are equal to 90°.

3 a, b, c, d

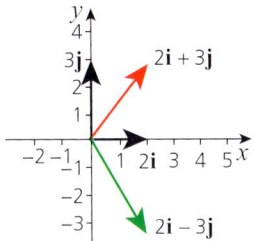

4 i

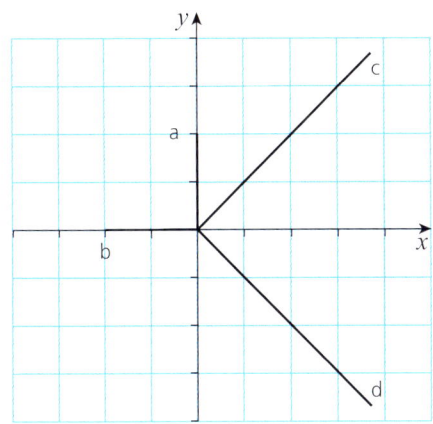

ii
a 4
b 3
c $\sqrt{74}$
d $\sqrt{74}$

5 a

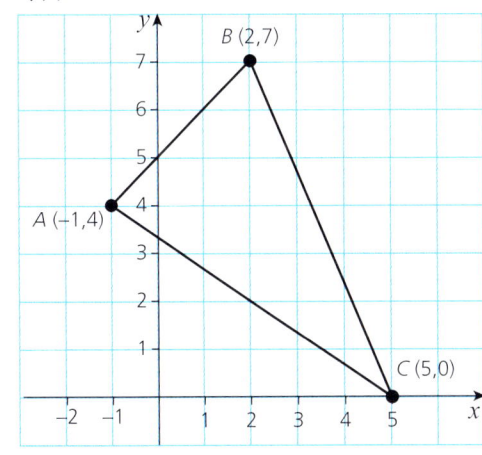

$\overrightarrow{AB} = \begin{pmatrix} 3 \\ 3 \end{pmatrix}$

$\overrightarrow{BC} = \begin{pmatrix} 3 \\ -7 \end{pmatrix}$

$\overrightarrow{AC} = \begin{pmatrix} 6 \\ -4 \end{pmatrix}$

b BC is the longest side of the triangle.

6 a

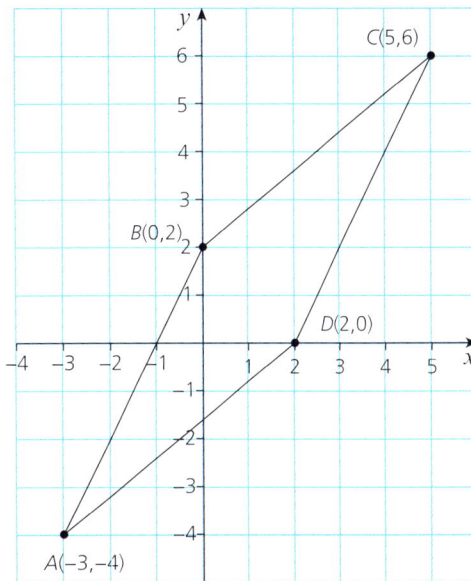

b $OA = -3\mathbf{i} - 4\mathbf{j}$
$OB = 2\mathbf{j}$
$OC = 5\mathbf{i} + 6\mathbf{j}$
$OD = 2\mathbf{i}$

c $AB = 3\mathbf{i} + 6\mathbf{j}$
$DC = 3\mathbf{i} + 6\mathbf{j}$
$BC = 5\mathbf{i} + 4\mathbf{j}$
$AD = 5\mathbf{i} + 4\mathbf{j}$

d $ABCD$ is a parallelogram.

7 a

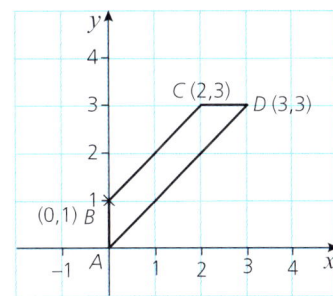

b $\begin{pmatrix} 3 \\ 3 \end{pmatrix}$

c isosceles trapezoid

Exercise 13.2 Page 250

1 a $\begin{pmatrix} 3 \\ 8 \end{pmatrix}$

b $\begin{pmatrix} 2 \\ 2 \end{pmatrix}$

c $\begin{pmatrix} 5 \\ 5 \end{pmatrix}$

2 a $-\mathbf{i}+5\mathbf{j}$
   b $13\mathbf{j}$
3 a $4\mathbf{i}+6\mathbf{j}$
   b $6\mathbf{i}-4\mathbf{j}$
   c $12\mathbf{j}$
   d $5\mathbf{i}+20\mathbf{j}$
   e $-4\mathbf{i}+25\mathbf{j}$
   f $\mathbf{i}+26\mathbf{j}$
   g $24\mathbf{i}-45\mathbf{j}$

4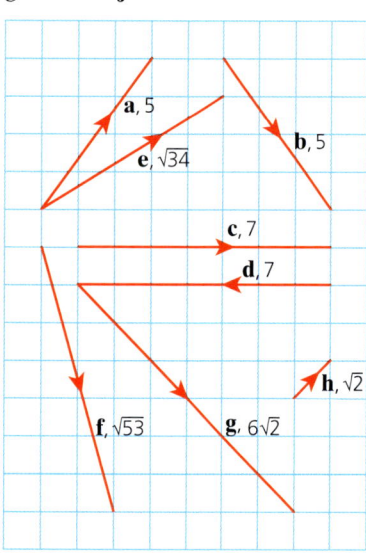

5 a i $2\mathbf{i}+3\mathbf{j}$
      ii $\begin{pmatrix}2\\3\end{pmatrix}$
   b i $\mathbf{i}-7\mathbf{j}$
      ii $\begin{pmatrix}1\\-7\end{pmatrix}$
   c i $3\mathbf{i}+5\mathbf{j}$
      ii $\begin{pmatrix}3\\5\end{pmatrix}$
   d i $10\mathbf{i}-10\mathbf{j}$
      ii $\begin{pmatrix}10\\-10\end{pmatrix}$
   e i $4\mathbf{i}-2\mathbf{j}$
      ii $\begin{pmatrix}4\\-2\end{pmatrix}$
   f i $4\mathbf{i}+2\mathbf{j}$
      ii $\begin{pmatrix}4\\2\end{pmatrix}$

6 a $\mathbf{b}-\mathbf{a}$
   b $\mathbf{a}-\mathbf{b}$
   c $\mathbf{a}$
   d $-\mathbf{a}$
   e $\mathbf{b}$
   f $-\mathbf{b}$
   g $-\mathbf{a}+\mathbf{b}-\mathbf{a}=\mathbf{b}-2\mathbf{a}$
   h $\mathbf{a}-\mathbf{b}+\mathbf{a}=2\mathbf{a}-\mathbf{b}$

7 a $AD=\begin{pmatrix}3\\2\end{pmatrix}$
      $DC=\begin{pmatrix}2\\3\end{pmatrix}$
   b $AC=\begin{pmatrix}5\\5\end{pmatrix}$
      $BD=\begin{pmatrix}1\\-1\end{pmatrix}$
   c $\left(\frac{3}{2},\frac{1}{2}\right)$
   d Gradient of $AC=\frac{5}{5}=1$
      Gradient of $BD=\frac{1}{-1}=-1$
      They are perpendicular because $1\times-1=-1$.

8 a $\frac{3}{5}\mathbf{i}-\frac{4}{5}\mathbf{j}$
   b $\left(\frac{5}{\sqrt{74}}\right)\mathbf{i}+\left(\frac{7}{\sqrt{74}}\right)\mathbf{j}$
   c $\begin{pmatrix}\frac{5}{13}\\\frac{12}{13}\end{pmatrix}$
   d $\begin{pmatrix}\frac{1}{\sqrt{10}}\\\frac{-3}{\sqrt{10}}\end{pmatrix}$
   e $\mathbf{i}$
   f $\begin{pmatrix}\frac{1}{\sqrt{2}}\\\frac{1}{\sqrt{2}}\end{pmatrix}$

9 a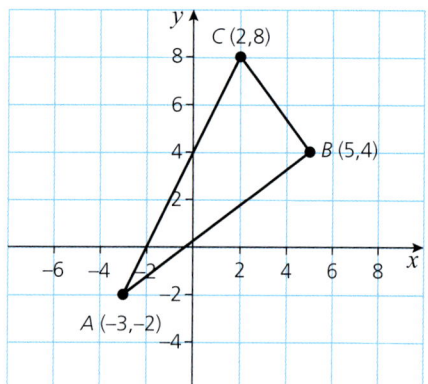

b $\overrightarrow{AB} = 8\mathbf{i} + 6\mathbf{j}$
$\overrightarrow{BC} = -3\mathbf{i} + 4\mathbf{j}$
$\overrightarrow{CA} = -5\mathbf{i} - 10\mathbf{j}$
c $|AB| = 10$
$|BC| = 5$
$|CA| = 5\sqrt{5}$
d Scalene triangle.

10 a i $\overrightarrow{OC} = 4\mathbf{i}$
ii $\overrightarrow{AB} = 4\mathbf{i} + 4\mathbf{j}$
iii $\overrightarrow{BC} = 4\mathbf{i} - 4\mathbf{j}$
iv $\overrightarrow{AD} = 4\mathbf{i} - 8\mathbf{j}$
v $\overrightarrow{CD} = -4\mathbf{i} - 8\mathbf{j}$
vi $\overrightarrow{AC} = 8\mathbf{i}$
b $|OC| = 4$
$|AB| = 4\sqrt{2}$
$|BC| = 4\sqrt{2}$
$|AD| = 4\sqrt{5}$
$|CD| = 4\sqrt{5}$
$|AC| = 8$
c AOB, BOC and ABC are all isosceles triangles and right triangles.

11 a

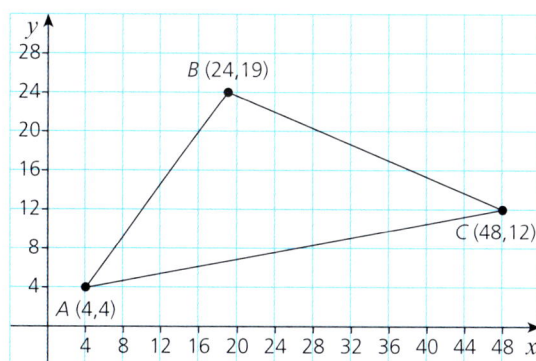

b $\overrightarrow{AB} = \begin{pmatrix} 20 \\ 15 \end{pmatrix}$
$\overrightarrow{BC} = \begin{pmatrix} 24 \\ -7 \end{pmatrix}$
$\overrightarrow{AC} = \begin{pmatrix} 44 \\ 8 \end{pmatrix}$
c $|AB| = 25$
$|BC| = 25$
$|AC| = 20\sqrt{5}$
d Isosceles triangle.

12 i Salman walks faster by 0.21 km per hour
ii 13.6 km (3 s.f.)

13 direction = 217°
distance = 379.47 km
time = 1 hour 54 minutes

Past-paper questions Page 249

1 ii $40\mathbf{i} + 96\mathbf{j}$
iii $(40 + 10t)\mathbf{i} + (96 + 24t)\mathbf{j}$
iv $(120 - 22t)\mathbf{i} + (81 + 30t)\mathbf{j}$
v 1830 hours, $65\mathbf{i} + 156\mathbf{j}$

2 i $\overrightarrow{AB} = \overrightarrow{OB} - \overrightarrow{OA}$
$= 9\mathbf{i} + 45\mathbf{j}$
ii $\overrightarrow{OC} = \overrightarrow{OA} + \frac{1}{3}\overrightarrow{AB}$
$\overrightarrow{OC} = 5\mathbf{i} + 12\mathbf{j}$
$|OC| = \sqrt{5^2 + 12^2}$
$= 13$
iii $\overrightarrow{OD} = \frac{4}{3}\mathbf{i} - 2\mathbf{j}$

3 $\overrightarrow{OC} = 5\mathbf{i} + 12\mathbf{j}$
$|\overrightarrow{OC}| = \sqrt{5^2 + 12^2} = 13$

Review exercise 4

1 a 3024
b 42
2 a 120
b 96
3 a 43758
b 66
c 43692
4 a 480700
b 450970
c 42
5 a $6561 + 17496x + 20412x^2 + 13608x^3$
b 9720
6 a First term is 18 and common difference is −5.
b $n = 47$
7 a i $r = -\frac{3}{5} \Rightarrow a = 25$
$r = \frac{3}{2} \Rightarrow a = 4$
ii 15.625
b 119382
8 $\overrightarrow{OC} = 5.2\mathbf{i} + 6.2\mathbf{j}$
$|OC| = 8.09$ (3 s.f.)
9 i $\mathbf{c} - \mathbf{a}$
ii $\frac{1}{3}\mathbf{a} + \frac{2}{3}\mathbf{c}$
iii $\frac{3}{5}\mathbf{b}$
iv $9\mathbf{b}$
v $-\frac{4}{9}\mathbf{a} + \frac{10}{9}\mathbf{c}$
10 a 25 kmh⁻¹
b 164°
c $7(4+t)\mathbf{i} - 24(4+t)\mathbf{j}$
d $t = 3$, $49\mathbf{i} - 168\mathbf{j}$

Chapter 14 Differentiation

Discussion point Page 256
The connection is that gravity applies to an apple falling and to the planets in their orbits. Newton realised that the planets were held in their orbits round the Sun by the force of gravity but he needed calculus to work out their equations.

Exercise 14.1 Page 258
1.
   a $4x^3$
   b $6x^2$
   c 0
   d 10
2.
   a $\frac{1}{2}x^{-\frac{1}{2}}$
   b $\frac{5}{2}x^{-\frac{1}{2}}$
   c $\frac{21}{2}t^{\frac{1}{2}}$
   d $\frac{1}{2}x^{\frac{3}{2}}$
3.
   a $10x^4 + 8x$
   b $12x^3 + 8$
   c $3x^2$
   d $1 - 15x^2$
4.
   a $-\frac{2}{x^3}$
   b $-\frac{18}{x^4}$
   c $\frac{2}{\sqrt{x}} + \frac{4}{(\sqrt{x})^3}$
   d $\frac{1}{2}x^{-\frac{1}{2}} + \frac{1}{2}x^{-\frac{3}{2}}$
5.
   a $2x - 1$
   b $4x - 1$
   c $1 - 5x^{-2}$
   d $\frac{3}{2}x^{\frac{1}{2}}$
6. At $(0, 9)$, gradient $= 0$; at $(-3, 0)$, gradient $= -6$; at $(3, 0)$, gradient $= 6$
7. a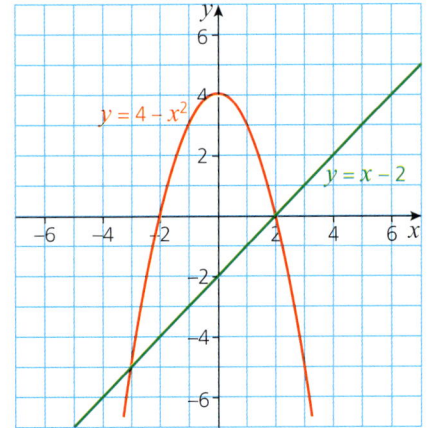
   b $(-3, -5), (2, 0)$
   c $6, -4$

Discussion point Page 261
As the curve does not turn, the gradient is positive both to the left and to the right of D; D is a stationary point of inflexion.

Exercise 14.2 Page 264
1.
   i $\frac{dy}{dx} = 1 - 4x, x = 0.25$
   ii maximum
   iii 1.125
   iv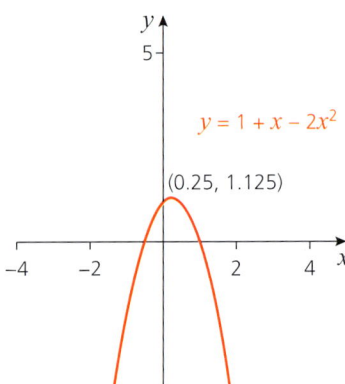

2.
   i $\frac{dy}{dx} = 12 + 6x - 6x^2, x = -1, 2$
   ii minimum at $x = -1$, maximum at $x = 2$
   iii $(-1, -7)$ $(2, 20)$
   iv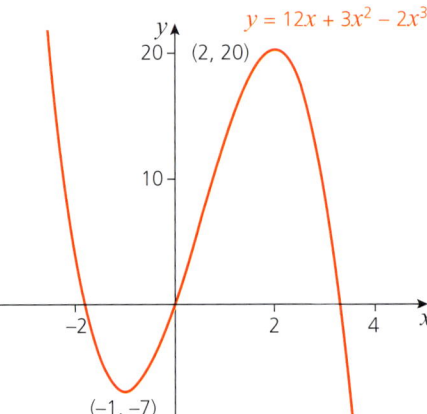

3.
   i $\frac{dy}{dx} = 3x^2 - 8x, x = 0, \frac{8}{3}$
   ii maximum at $x = 0$, minimum at $x = \frac{8}{3}$
   iii $(0, 9), \left(\frac{8}{3}, -\frac{13}{27}\right)$

iv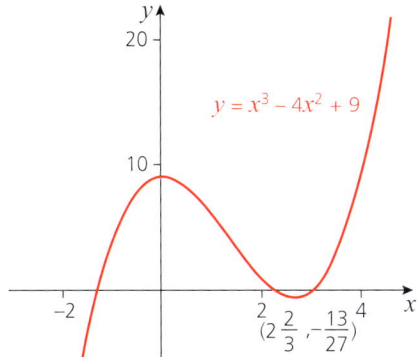

4 i $\frac{dy}{dx} = 4x^3 - 6x^2 + 2x$, $x = 0, 0.5, 1$
   ii minimum at $x = 0$, maximum at $x = 0.5$, minimum at $x = 1$
   iii $(0, 0)$ $(0.5, 0.0625)$, $(1, 0)$
   iv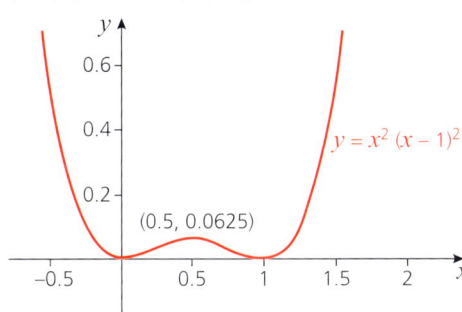

5 i $\frac{dy}{dx} = 4x^3 - 16x$, $x = -2, 0, 2$
   ii minimum at $x = -2$, maximum at $x = 0$, minimum at $x = 2$
   iii $(-2, -12)$, $(0, 4)$, $(2, -12)$
   iv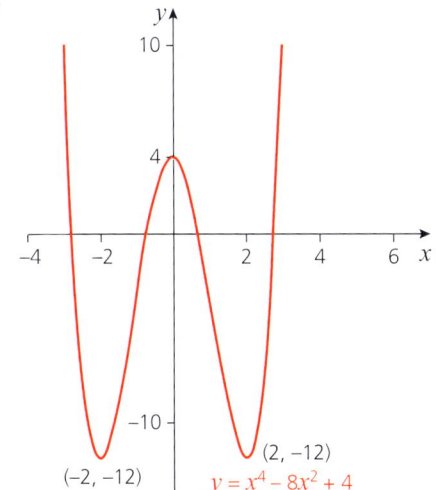

6 i $\frac{dy}{dx} = 3x^2 - 48$, $x = -4, 4$
   ii maximum at $x = -4$, minimum at $x = 4$
   iii $(-4, 128)$, $(4, -128)$

iv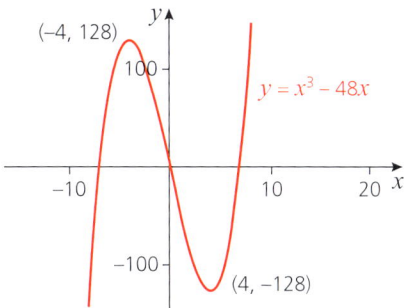

7 i $\frac{dy}{dx} = 3x^2 + 12x - 36$, $x = -6, 2$
   ii maximum at $x = -6$, minimum at $x = 2$
   iii $(-6, 241)$, $(2, -15)$
   iv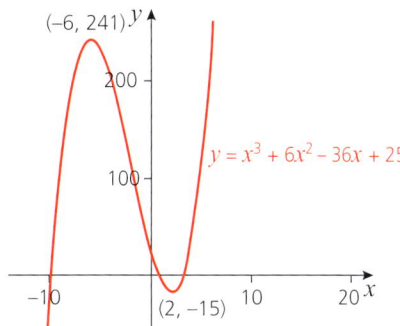

8 i $\frac{dy}{dx} = 6x^2 - 30x + 24$, $x = 1, 4$
   ii maximum at $x = 1$, minimum at $x = 4$
   iii $(1, 19)$, $(4, -8)$
   iv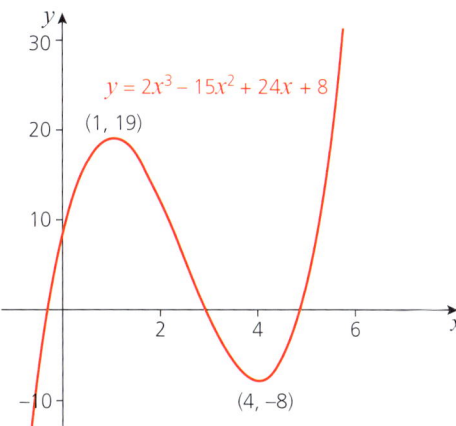

9 a $p = 4$, $q = -3$
   b maximum value $\frac{4}{3}$ when $x = \frac{2}{3}$

10 a minimum at (−0.5, −0.3125)
      maximum at (0, 0), minimum at (1, −2)
   b
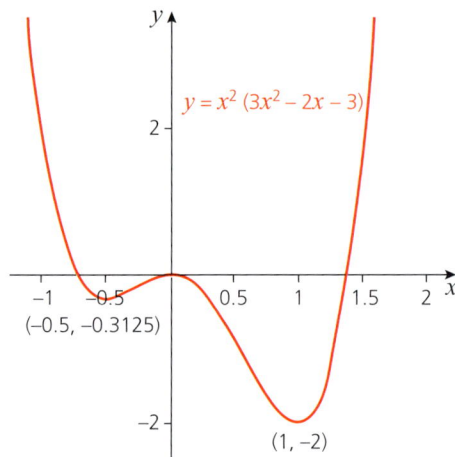

Exercise 14.3 Page 268

1 a $\dfrac{dy}{dx} = 3x^2 - 6x + 2, \dfrac{d^2y}{dx^2} = 6x - 6$

  b $\dfrac{dy}{dx} = 12x^3 - 12x^2, \dfrac{d^2y}{dx^2} = 36x^2 - 24x$

  c $\dfrac{dy}{dx} = 5x^4 - 5, \dfrac{d^2y}{dx^2} = 20x^3$

2 a minimum at (0.75, 2.875)

  b maximum at $(\tfrac{1}{3}, 6.148)$, minimum at (1, 6)

  c minimum at (−0.5, 0.75), maximum at (0, 1), minimum at (0.5, 0.75)

  d maximum at (−1, 4), minimum at (1, −4)

3 a $\dfrac{dy}{dx} = 6x^2 - 6x - 36$, $x = -2, 3$

  b at $x = -2$, $\dfrac{d^2y}{dx^2} = -30$, maximum

     at $x = 3$, $\dfrac{d^2y}{dx^2} = 30$, maximum

  c (−2, 48), (3, −77)

  d
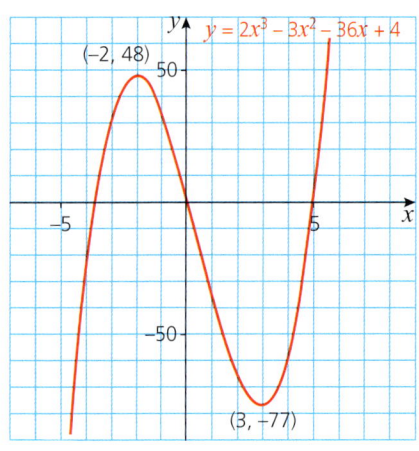

4 The maximum area is equal to 3200 m². It occurs when the enclosure is 40 m wide (perpendicular to the barn wall) and 80 m long (length parallel to the barn wall).

5 a $V = \pi r^2 (3 - r)$

  b $V_{max} = 4\pi$; this occurs when $r = 2$ and $h = 1$

6 a $x = 6$, $y = 6$, $A = 36$ cm² and $\dfrac{d^2A}{dx^2} = -2$ so maximum

  b $x = 6$, $y = 6$, $P = 24$ cm² and $\dfrac{d^2P}{dx^2} = \dfrac{2}{3}$ so minimum

Discussion point Page 270

At the point (1, 2), an increase in x of 0.001 results in an increase in y of 0.001 998 (4 s.f.).

The gradient at $x = 1$ is 2.

$\dfrac{\text{small change in } y}{\text{small change in } x} = \dfrac{0.001\,998}{0.001} = 1.998$

1.998 is approximately equal to the gradient at the point (1, 2).

i At the point (2, 0), an increase in x of 0.001 results in a decrease in y of 0.008 008 (4 s.f.). The gradient at $x = 2$ is −8.

$\dfrac{\text{small change in } y}{\text{small change in } x} = \dfrac{-0.008\,008}{0.001} = -8.008$

−8.008 is approximately equal to the gradient at the point (2, 0).

ii At the point (0, 0), an increase in x of 0.001 results in an increase in y of 0.000 003 998. The gradient at $x = 0$ is 0.

$\dfrac{\text{small change in } y}{\text{small change in } x} = \dfrac{0.000\,003\,998}{0.001} = 0.003\,998$

0.003 998 is approximately equal to the gradient at the point (0, 0).

Exercise 14.4 Page 272

1 a $\dfrac{dy}{dx} = 5 - 2x$

  b −1

  c $y = 9 - x$

  d $y = x + 3$

2 a i $\dfrac{dy}{dx} = 6x - 3x^2$

     ii 0

     iii $y = 4$

     iv $x = 2$

  b i   (3, 0)

     ii −9

     iii $y = 27 - 9x$

  c $y = 0$

3 a (1, 0)
   b $y = 2x - 2$
   c $2y = 1 - x$ or $2y = -x + 1$
   d $Q(0, -2)$, $R(0, 0.5)$
      The area of the triangle is equal to 1.25.
4 a $f'(x) = 3x^2 - 6x + 4$
   b i 5
      ii $y = 4x - 3$
      iii $4y + x = 22$
   c $-1, 3$
5 a $\frac{dy}{dx} = 3x^2 - 18x + 23$
   b -1
   c $y = -x + 5$
   d $Q(4, -3)$
   e $y = -x + 1$
6 a $2p - q = 16$
   b $\frac{dy}{dx} = 3x^2 - p$
   c 12
   d $(-2, 24)$
   e $(0, 8)$
   f $12y = x + 96$
7 a $y = 3x - 5$
   b $\frac{1}{3}$
   c $\left(\frac{1}{3}, -\frac{11}{9}\right)$
8 a $10 - 2x$
   b $y = -2x + 15$
   c $2y = x$
   d $2y = x$ (The equation of the normal.)
9 b $(0, 0)$, tangent $y = 2x$, normal $2y + x = 0$
      $(1, 0)$, tangent $2y + x = 1$, normal $y = 2x - 2$
   c Opposite sides are parallel and adjacent sides are perpendicular so it is a rectangle.
10 a -36
   b -27; y decreases by approximately 0.036
11 4.00125
12 3.0037 (4 d.p.)

Exercise 14.5 Page 285

1 a $3\cos x - 2\sec^2 x$
   b $5\cos\theta$
   c $-2\sin\theta - 2\cos\theta$
   d $\frac{4}{x}$
   e $\frac{1}{x}$
   f $3e^x$
   g $2e^x - \frac{1}{x}$
2 a $x\cos x + \sin x$
   b $-x\sin x + \cos x$
   c $x\sec^2 x + \tan x$
   d $e^x \cos x + e^x \sin x$
   e $-e^x \sin x + e^x \cos x$
   f $e^x \sec^2 x + e^x \tan x$
3 a $\frac{x\cos x - \sin x}{x^2}$
   b $\frac{\sin x - x\cos x}{\sin^2 x}$
   c $-\frac{x\sin x + 2\cos x}{x^3}$
   d $\frac{2x\cos x + x^2 \sin x}{\cos^2 x}$
   e $\frac{\tan x - x\sec^2 x}{\tan^2 x}$
   f $\frac{x\sec^2 x - \tan x}{x^2}$
4 a $4(x + 3)^3$
   b $8(2x + 3)^3$
   c $8x(x^2 + 3)^3$
   d $\frac{1}{2}\sqrt{(x + 3)}$
   e $\frac{1}{\sqrt{(2x + 3)}}$
   f $\frac{x}{\sqrt{(x^2 + 3)}}$
5 a $\frac{1}{(1 + \cos x)}$
   b $\frac{-\cos x - 1}{\sin^2 x}$
   c $\cos 2x + \cos x$
   d $\cos 2x - \sin x$
   e $\cos x + 2\cos^2 x + \cos^3 x - 2\sin^2 x - 2\sin^2 x \cos x$
   f $2\cos^2 x + 2\cos^2 x \sin x - \sin x - 2\sin^2 x - \sin^3 x$
6 a $e^x \ln x + \frac{e^x}{x}$
   b $\frac{e^x \ln x - \frac{e^x}{x}}{(\ln x)^2}$
   c $\frac{1 - x\ln x}{e^x x}$
7 a $e^{-x}(\cos x - \sin x)$
   b $-\frac{e^{-x}\sin x + e^{-x}\cos x}{\sin^2 x}$
   c $\frac{\cos x + \sin x}{e^{-x}}$
8 b At $(0, -1)$ $\frac{dy}{dx} = \cos(0) + \sin(0)$
      The equation of the tangent is $y = x - 1$
      The equation of the normal is $y = -x - 1$
      At $(\pi, 1)$ $\frac{dy}{dx} = \cos\pi + \sin\pi = -1$
      The equation of the tangent is $y = -x + \pi + 1$
      The equation of the normal is $y = x - \pi + 1$

9 b $\dfrac{dy}{dx} = 2\sec^2 x$

At $(0, -1)$

$\dfrac{dy}{dx} = 2\sec^2(0) = 2$

The equation of the tangent is $y = 2x - 1$.

The equation of the normal is $y = -\left(\dfrac{1}{2}\right)x - 1$

At $\left(\dfrac{\pi}{4}, 1\right)$ $\dfrac{dy}{dx} = 2\sec^2\left(\dfrac{\pi}{4}\right) = 2(\sqrt{2})^2 = 4$

The equation of the tangent is $y = 4x - \pi + 1$

The equation of the normal is $y = -\dfrac{1}{4}x + \dfrac{\pi}{16} + 1$

10 b The equation of the tangent is $y = \dfrac{2x}{e} - 1$

The equation of the normal is $y = \dfrac{-ex}{2} + \dfrac{e^2}{2} + 1$

11 c The equation of the tangent is $y = ex - x + 1$

The equation of the normal is

$y = \dfrac{-x}{e-1} + \dfrac{1}{e-1} + e$

12 a $2\pi r$ cm² per second

b Surface area is increasing at a rate of 22 cm² per second (2 s.f.).

Past-paper questions Page 286

1 ii $\dfrac{dA}{dx} = 8x - \dfrac{216}{x^2}$

When $\dfrac{dA}{dx} = 0$, $x = \sqrt[3]{27} = 3$

Dimensions are 3 by 6 by 4

iii Change in $A = -38p$, decrease

2 i $\dfrac{1}{3}\cos 2x \cos\left(\dfrac{x}{3}\right) - 2\sin 2x \sin\left(\dfrac{x}{3}\right)$

ii $\sec^2 x$ and $\dfrac{1}{x}$

$\dfrac{(\sec^2 x)(1 + \ln x) - \dfrac{1}{x}(\tan x)}{(1 + \ln x)^2}$

3 i $\dfrac{dy}{dx} = 12x^2 \ln(2x+1) + \dfrac{8x^3}{2x+1}$

Chapter 15 Integration

Discussion point Page 289

A marathon. She had clearly finished before 3 hours. You can estimate the distance she ran in each half hour, using the average speed. That suggests that by 2.5 hours she had run 41 760 metres and still had 435 metres to go. So an estimate is that she took a bit over 2 hours and 31 minutes.

Exercise 15.1 Page 293

1 a $y = 2x^2 + 2x + c$

b $y = 2x^3 - \dfrac{5}{2}x^2 - x + c$

c $y = 3x - \dfrac{5}{4}x^4 + c$

d $y = x^3 - 2x^2 - 4x + c$

2 a $f(x) = \dfrac{5}{2}x^2 + 3x + c$

b $f(x) = \dfrac{1}{5}x^5 + \dfrac{1}{2}x^4 - \dfrac{1}{2}x^2 + 8x + c$

c $f(x) = \dfrac{1}{4}x^4 - \dfrac{4}{3}x^3 + x^2 - 8x + c$

d $f(x) = \dfrac{1}{3}x^3 - 7x^2 + 49x + c$

3 a $5x + c$

b $\dfrac{5}{4}x^4 + c$

c $x^2 - 3x + c$

d $\dfrac{3}{4}x^4 - 2x^2 + 3x + c$

4 a $9x - 3x^2 + \dfrac{1}{3}x^3 + c$

b $\dfrac{2}{3}x^3 - \dfrac{5}{2}x^2 - 3x + c$

c $\dfrac{1}{3}x^3 + x^2 + x + c$

d $\dfrac{4}{3}x^3 - 2x^2 + x + c$

5 a $y = x^2 - 3x + 6$

b $y = 4x + \dfrac{3}{4}x^4 - 210$

c $y = \dfrac{5}{2}x^2 - 6x - 18$

d $f(x) = \dfrac{1}{3}x^3 + x + 9$

e $f(x) = \dfrac{1}{3}x^3 - \dfrac{1}{2}x^2 - 2x - 44$

f $f(x) = \dfrac{4}{3}x^3 + 2x^2 + x - \dfrac{16}{3}$

6 a $y = \dfrac{4}{3}x^{\frac{3}{2}} - x + \dfrac{2}{3}$

b $y = \dfrac{1}{2}x^2 - \dfrac{2}{3}x^{\frac{3}{2}} - \dfrac{2}{3}$

7 a $x^2 + 3x + c$

b $y = x^2 + 3x + c$

c $y = x^2 + 3x - 11$

8 a $y = x^3 - 2x^2 + x + 9$

9 a $2x^2 - x + c$

b $y = 2x^2 - x + c$

c $y = 2x^2 - x + 1$

d Curve passes above the point.

10 -1

11 a $\dfrac{dy}{dx} = kx(x-2)$ for any $k \neq 0$

b $y = \dfrac{1}{3}x^3 - x^2 + 2$

Exercise 15.2 Page 296
1. 7
2. 255
3. 4
4. 16
5. $\frac{62}{3}$
6. 30
7. 591
8. $\frac{76}{3}$
9. $-\frac{4}{3}$
10. 12
11. 27
12. $\frac{16}{3}$
13. 92
14. $-\frac{4}{3}$
15. $\frac{27}{2}$
16. 0
17. 28
18. $\frac{80}{3}$

Exercise 15.3 Page 297
1. 9
2. 36
3. 2
4. $\frac{20}{3}$
5. 0.25
6. $\frac{3125}{6}$
7. 36
8. 13.5
9. $\frac{128}{15}$
10. $\frac{64}{3}$

Exercise 15.4 Page 303
1. 0.25
2. a area of P = $\frac{5}{12}$, area of Q = $\frac{8}{3}$
   b $3\frac{1}{12}$
3. a $(\sqrt[3]{2}, 0)$
   b 0.95244
4. $\frac{253}{12}$

5. a and b
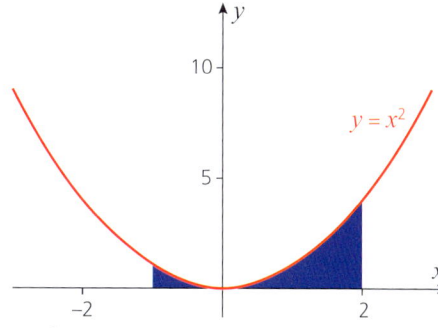
   c 3
6. a
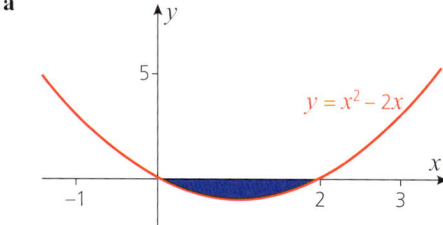
   b $0 < x < 2$
   c $\frac{4}{3}$ below the axis
7. a and b
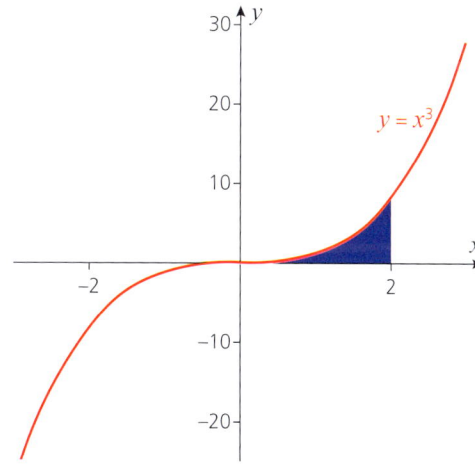
   c 4
   d 0
8. a
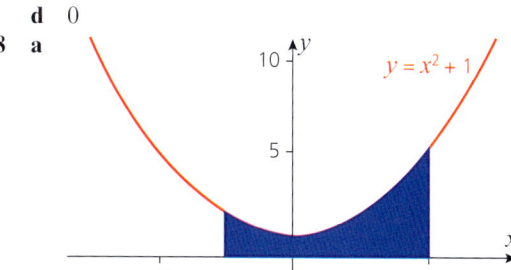
   b 6

9 a 18
b Area between the lines $y = 2x + 1$, $x = 1$, $x = 4$ and the x-axis

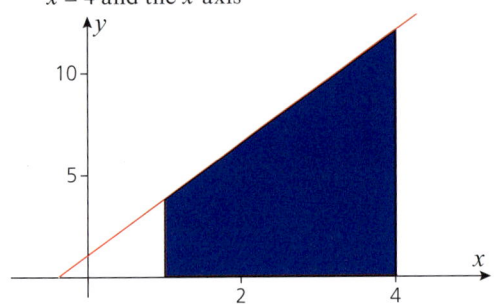

10 a $(-1, 1)$ and $(2, 4)$
b 4.5 units²

11 a

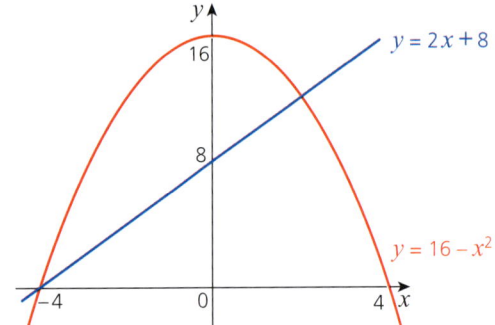

b 36 units²

12 a $(0, 9)$ and $(3, 0)$
b 9 units²

Exercise 15.5 Page 308

1 a $\frac{1}{3}\ln|3x+1| + c$

b $\frac{1}{15}(3x+1)^5 + c$

c $\frac{1}{3}e^{3x+1} + c$

d $(-\cos\frac{3x+1}{3}) + c$

e $(\sin\frac{3x+1}{3}) + c$

f $3\ln|x-3| + c$

g $\frac{1}{8}(2x-1)^4 + c$

h $2e^{2x-3} + c$

i $-\cos(3x) + c$

j $8\sin\left(\frac{x}{2}\right) + c$

k $\frac{2}{5}(x-2)^{\frac{5}{2}} + c$

l $\frac{(2x-1)^{\frac{5}{2}}}{5} + c$

2 a $\frac{1}{3}\ln\left(\frac{13}{7}\right)$

b 23 632.4

c 147 105.59 (2 d.p.)

d $\frac{1}{3}$

e $-\frac{1}{\sqrt{3}}$

f $4\ln 3$

g 5904.8

h $5(1 - e^{-4})$

i $\frac{1}{\sqrt{2}}$

j $\frac{1}{\sqrt{2}}$

k $\frac{\sqrt{3}-1}{3}$

Past-paper questions Page 309

1 a i $2 = a - 3$, $a = 5$
  ii $y = -5e^{1-x} - x^3 + c$
  $c = 10$
  $y = -5e^{1-x} - x^3 + 10$

b i $28(7x+8)^{\frac{4}{3}} + c$

  ii $\left[\frac{3}{28}(7x+8)^{\frac{4}{3}}\right]_0^8$

  $= \frac{180}{7}$ or 25.7

2 ii When $\frac{dy}{dx} = 0$, $(3x-1)(x-3) = 0$

$x = \frac{1}{3}$, $x = 3$

iii Area $= \frac{1}{2}(10+19)3$

$-\left(\int_0^3 x^3 - 5x^2 + 3x + 10 \, dx\right)$

$= \frac{87}{2} - \left[\frac{x^4}{4} - \frac{5x^3}{3} + \frac{3x^2}{2} + 10x\right]_0^3$

$\frac{87}{2} - \left(\frac{81}{4} - 45 + \frac{27}{2} + 30\right)$

$= 24.75$

3 i $x = 2, 4$

ii $\int 3x - 14 + \frac{32}{x^2} dx = 1.5x^2 - 14x - \frac{32}{x}(+c)$

Area $= \left[1.5x^2 - 14x - \frac{32}{x}\right]_2^4$

$= (-)2$

Chapter 16 Kinematics

Discussion point Page 312
No. One component of its motion will be in orbit round the Sun. Another component will be at right angles to it. Given the distance to Jupiter, it is likely

that the spacecraft will complete several orbits of the Sun before it reaches Jupiter.

Discussion point Page 316

The particle is in motion, but has slowed down while still travelling towards *O*.

Exercise 16.1 Page 316

1 a $v = 12t^2 - 12t$
$a = 24t - 12$
b
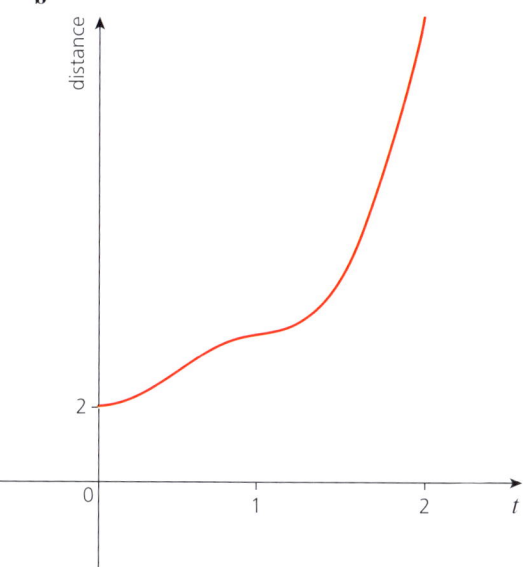
c i
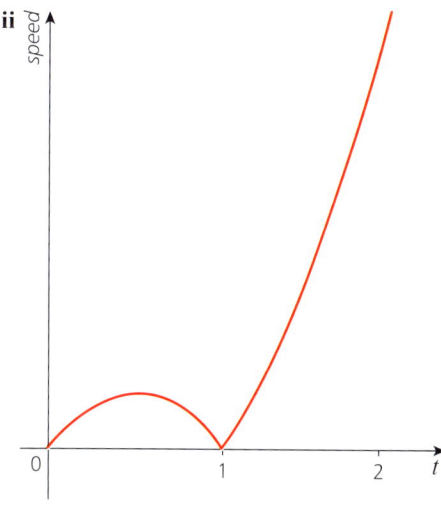
ii
iii
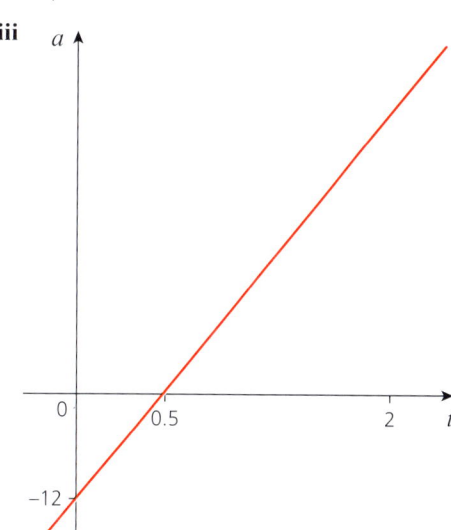

2 a i $v = 10t - 1$, $a = 10$
ii 3, −1, 10
iii when $t = 0.1$, $s = 2.95$
b i $v = 3 - 3t^2$, $a = -6t$
ii 0, 3, 0
iii when $t = 1$, $s = 2$
c i $v = 4t^3 - 4$, $a = 12t^2$
ii −6, −4, 0
iii when $t = 1$, $s = -9$
d i $v = 12t^2 - 3$, $a = 24t$
ii 5, −3, 0
iii when $t = 0.5$, $s = 4$
e i $v = -4t + 1$, $a = -4$
ii 5, 1, −4
iii when $t = 0.25$, $s = 5.125$

3   a $v = 6t - 3t^2$, $a = 6 - 6t$
    b $t = 0, 2$ s
    c 4 m
    d -24 m s^{-1}; travelling in the negative direction
    e 6 m s^{-2}
4   a 1 m
    b $v = 4 - 10t$
    c 0.4 s
    d 1.8 m
    e 1 s
    f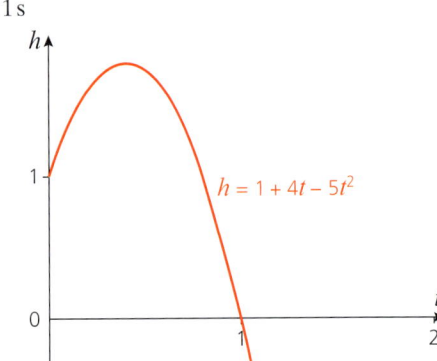
    g -6 m s^{-1}
5   a $v = \frac{2}{3}t^3$, $a = 2t^2$
    b 6 s
    c 216 m, 144 m s^{-1}
6   a 2.5 s, 5 s
    b 5 m s^{-2}, -5 m s^{-2}
    c 3.125 m s^{-1}
    d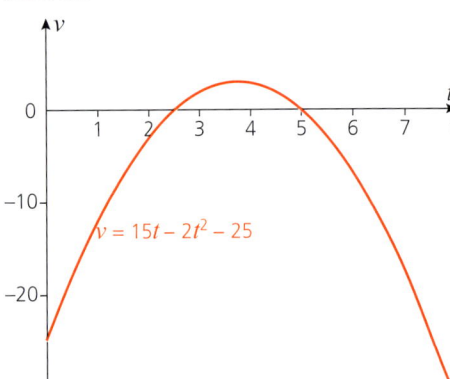

Discussion Point page 320

If the velocity is $v = 3t^2 + 2$, the acceleration is given by $a = \frac{dv}{dt} = 6t$. Since this involves t it is not constant. So you cannot use the constant acceleration formula and must use calculus instead. You can also use calculus if the acceleration is constant.

Exercise 16.2 Page 320

1   a $v = 2t - 3t^2 + 1$, $s = t^2 - t^3 + t$
    b $v = 2t^2 + 4$, $s = \frac{2}{3}t^3 + 4t + 3$
    c $v = 4t^3 - 4t + 2$, $s = t^4 - 2t^2 + 2t + 1$
    d $v = 2t + 2$, $s = t^2 + 2t + 4$
    e $v = 4t + \frac{1}{2}t^2 + 1$, $s = 2t^2 + \frac{1}{6}t^3 + t + 3$
2   a $v = 3t^2 - 12t + 9$, $s = t^3 - 6t^2 + 9t$
    b 3 s
3   a -3 m s^{-1}
    b 2 s, -4 m
4   a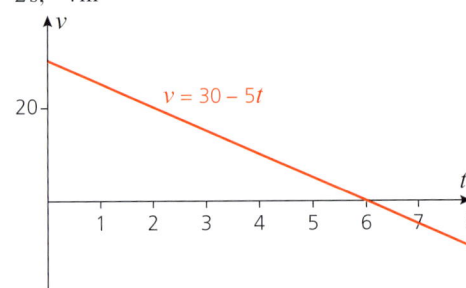
    b 6 s
    c 90 m
5   a $v = 4t + 6t^2$
    b 48 m
6   a $a = 3t^2 - 8t + 4$
    b $\frac{2}{3}$, 2 s; particle is slowing down
    c 8.25 m

Past-paper questions Page 321

1   i $\frac{dy}{dt} = 36 - 6t$
        When $\frac{dy}{dt} = 0$, $t = 6$
    ii When $v = 0$, $t = 12$
    iii $s = 18t^2 - t^3 \, (+c)$
        When $t = 12$, $s = 864$
    iv When $s = 0$, $t = 18$
        $v = -324$
        So speed is 324 m s^{-1}
2   i   Velocity = 9 m s^{-1}
    ii Acceleration = -7.8 m s^{-2}
    iii Distance $OP = 11.1$ m
3   i $t = \sqrt{e^5 - 1}$ or $t^2 + 1 = e^5$
        $t = 12.1$
    ii Distance = $\ln 10 - \ln 5$
        = $\ln 2$ or 0.693
    iii $v = \frac{2t}{t^2 + 1}$, $v = 0.8$

iv $a = \dfrac{(t^2+1)2 - 2t(2t)}{(t^2+1)^2}$

When $t = 2$, $a = -\dfrac{6}{25}$, or -0.24

Review exercise 5

1 a $4\ln(2x+1)\cos 4x + \dfrac{2\sin 4x}{2x+1}$

   b $\dfrac{xe^x - 3e^x}{x^4}$

2 a $\dfrac{3x - 24}{(x-4)^{\frac{3}{2}}}$

   b $5x - 9y + 169 = 0$

3 439 cm² (3 s.f.)

4 a $\dfrac{4}{3}x^3 + 6x^2 + 9x + c$

   b 57 units²

5 $y = -2\sin 2x + \dfrac{x^2}{2} + 2x + 6$

6 i Using $1 + \tan^2 A \equiv \sec^2 A$

   $5 + 4\left(\sec^2\left(\dfrac{x}{3}\right) - 1\right)$

   $= 5 + 4\sec^2\left(\dfrac{x}{3}\right) - 4$

   $= 4\sec^2\left(\dfrac{x}{3}\right) + 1$

   ii $3\tan\left(\dfrac{x}{3}\right) + c$

   iii $8\sqrt{3} + \dfrac{\pi}{2}$ units²

7 a i 0 ms⁻²

      ii 110 m

   b i 2.5 ms⁻¹

      ii $t = \dfrac{\pi}{4}$ s or 0.785s (3 s.f.)

Glossary

absolute value Any real number in its positive size. May also be referred to as the modulus or magnitude.

acceleration The rate of change of velocity with respect to time.

amplitude The amplitude of a curve is the largest displacement from the central position (i.e. the horizontal axis).

arbitrary constant A constant in an expression or an equation that can take any value. Its value does not depend on the variable(s) in the expression or equation.

arithmetic sequence A sequence in which the terms increase or decrease by a fixed amount (the common difference). Also known as an arithmetic progression (A.P.).

arithmetic series When the terms of an arithmetic progression are added together.

ascending When the powers of x are increasing from one term to the next.

asymptotes These may be shown as dotted lines on a graph. The branches of a curve get increasingly close to them but never actually touch them.

base (geometry) The base of a triangle (or any other 2D shape) is one of its sides. Any side can be the base, but the height must be measured perpendicular to the chosen base.

base (logarithms) In $\log_2 8 = 3$, the number 2 is referred to as the base of the logarithm. Read this as 'log to base 2 of 8 equals 3'.

binomial expansion An expression where a binomial expression such as $(x + 1)$ is raised to a power.

chain rule A rule used in differentiation to find the derivative of an expression that is a function of a function.

circumference The distance around a circle.

combination A selection where the order is not important.

common difference When d is a fixed number in an arithmetic sequence, $u_k + 1 = u_k + d$.

common ratio When r is a fixed number in a geometric sequence such that $u_k + 1 = ru_k$.

composition function When two functions are used one after the other.

completing the square Writing a quadratic expression in the form $(x + b)^2 + c$ or $a(x + b)^2 + c$

convergent The terms in a geometric series become progressively smaller in size.

cubic expression An expression of the form $ax^3 + bx^2 + cx + d$ where $a \neq 0$.

definite integral An integral with two limits.

descending When the powers of x are decreasing from one term to the next.

differentiation The process of finding the derivative or gradient function.

differential equation An equation which includes one or more derivatives of a function.

discriminant The discriminant discriminates between quadratic equations with no roots, quadratic equations with one repeated root and quadratic equations with two real roots.

displacement The distance from a starting point.

divergent The terms in a geometric series become progressively larger in size.

domain The domain of a function $f(x)$ is the set of all possible inputs.

exponential decay In an exponential function, when $x < 0$, it is referred to as exponential decay.

exponential function The inverse of the logarithm function, in the form $y = a^x$.

exponential growth In an exponential function, when $x > 0$, the function $y = a^x$ is referred to as exponential growth.

factor theorem This theorem states that if $(x - a)$ is a factor of $f(x)$, then $f(a) = 0$ and $x = a$ is a root of the equation $f(x) = 0$.

flow chart A flow chart can be used to show the individual operations within a function in the order in which they are applied.

function A rule that associates each element of one set (the input) with only one element of a second set (the output).

function notation A way of expressing the relationship between a function and its variable(s). For instance, if f is a function of x, it is common to write $f(x)$.

general solution A solution which is given in terms of a constant. This constant may be arbitrary or it may satisfy one or more conditions.

geometric sequence The terms of this sequence are formed by multiplying the previous term by a fixed number (the common ratio). Also known as geometric progression (G.P.).

geometric series The sum of the terms of a geometric sequence.

gradient function The gradient of the curve at the general point (x, y).

identity Unlike the solution of the equation, an identity is true for all values of the variable.

image set The set of output values for a given function.

indefinite integral An integral which has no limits.

infinite sequence A sequence that has an infinite number of terms.

infinite series The corresponding series of infinite sequences.

integration Using the rate of change of a quantity to find its total value at the end of an interval.

inverse function An inverse function reverses the effect of the function.

limits In integration, the limits are the values between which an integral is evaluated.

linear equation The graph of a linear equation is a straight line. The highest power of the variable is 1.

logarithm Another word for index or power.

magnitude Another word for the modulus (of a quantity).

major sector A sector of a circle where the angle at the centre is between 180° and 360°.

minor sector A sector of a circle where the angle at the centre is less than 180°.

modulus The modulus of a number is its positive value even when the number itself is negative. The modulus is denoted by a vertical line on each side of the number (e.g. $|x|$). It is also referred to as the magnitude of the quantity.

one-one function A function where every object has a unique image and every image comes from only one object.

order The highest power of a polynomial.

parallel Two lines that have the same gradient.

Pascal's triangle The binomial coefficients presented as a triangular array of numbers. It is named after Blaise Pascal.

perfect square An expression in the form $(px + q)^2$.

periodic A function that repeats itself at regular intervals.

permutation An ordered arrangement of people, objects or operations.

perpendicular Two lines that intersect at an angle of 90°.

point of inflexion The point where a curve changes from concave up to concave down, or vice versa.

polynomial An expression in which, with the exception of a constant, the terms are positive integer powers of a variable.

position vectors These start at the origin and are the vector equivalent of coordinates.

product rule The method of differentiating two functions that are being multiplied together.

quadratic expression Any expression of the form $ax^2 + bx + c$, where x is a variable and a, b and c are constants with $a \neq 0$.

quadratic formula The quadratic formula is $x = \dfrac{-b \pm \sqrt{b^2 - 4ac}}{2a}$. It is used to solve a quadratic equation in the form $ax^2 + bx + c = 0$.

quartic A quartic has a term in x^4 as its highest power.

quintic A quintic has a term in x^5 as its highest power.

quotient rule The method of finding the derivative of a function that is the ratio of two differentiable functions.

radian A unit of measure of angles. π radians $= 180°$

radius A line which joins the centre of a circle to a point on the circumference.

range All possible output values of a function, also known as the image set.

remainder theorem This theorem states that, for any polynomial $f(x)$, $f(a)$ is the remainder when $f(x)$ is divided by $(x - a)$.

roots In a quadratic equation, roots are the x-coordinates of the points where the curve either crosses or touches the x-axis.

scalar A quantity with magnitude but no direction.

sector The shape enclosed by an arc of a circle and two radii.

series The sum of the terms of a sequence.

sequence A set of numbers in a given order.

solution The values for which an equation is true.

speed The rate of change of distance with respect to time.

stationary point A point on a curve where the gradient is zero.

summation The process of adding terms in a sequence together.

term When referring to sequences, each of the numbers in a sequence are called terms.

turning point A turning point is a type of stationary point. This is where a curve changes from increasing to decreasing and vice versa.

unit vector Any vector of length 1.

vector A quantity that has both magnitude and direction.

velocity The rate of change of displacement with respect to time.

Index

A

abacuses 41
absolute value *see* modulus
acceleration 313
    variable 315–19
acceleration–time graphs 316, 319
amplitude, of an oscillating graph 177, 183–4
angles
    of any size 175–7
    of a circle 157, 159–63
    positive and negative 171–3
    right-angled triangles 166–71
    see also trigonometry
arc length 158–9, 161–3
area
    between a graph and the x-axis 296–305
    of a circle 157
    of a sector 158–9, 161–3
    under the curve 296–300, 302
arithmetic progressions 224–30
    sum of terms 226–30
arithmetic series 225
arrangements 207–11, 214–15
ascending powers of x 223–4
asymptotes 92, 179, 190, 279

B

base of a logarithm 91, 96–100, 106, 126
bearings 249
binomial coefficients 218–24
    formula for 219–22
    tables of 219
binomial expansion 217–19, 221–4
    of $(1+x)^n$ 222–4
binomial expressions 218, 219–20, 222–4
binomial theorem 217–24

C

calculus 315–19, 322
    see also differentiation; integration
CAST rule 176
chain rule 283–5
chords 257–8
    common 148–9
circles

arc length 158–9, 161–3
    with centre (a,b) 137–41
    coordinate geometry of 136–56, 354–6
    equation of 137–55
    intersections 142–54
    measurement 157–65, 356–7
    sector area 158–9, 161–3
    segments 163
    tangents to 144–8
circular functions 329
circumference 157–8, 160
coefficients 20, 24–7, 32–3, 42, 70, 79–80, 239
combinations 210–14, 363–4
common difference 224–5
common ratio 231–6
completed square form 25–6
completing the square 21, 25–9
composite functions 13–14, 283–5
constants 129, 138
    arbitrary 290
    quadratic functions 24–6
convergent series 235–7
cosecant (cosec) 174–5, 195–6, 199
cosine (cos) 167–70, 198, 308
    angles of any size 175–6
    $\cos x$ 274–7, 306
    differentiation 274–7
    integration 306–9
    positive and negative angles 172
    reciprocal trigonometrical functions 174
    trigonometrical equations and graphs 179–80
    trigonometrical identities 192–6, 199
cosine graphs 177–8, 180, 183, 186–7, 190, 192, 277
cotangent (cot) 174–5, 195–6, 199
cubic equations 43–50
cubic expressions 41, 47
cubic functions, modulus of 70–1
cubic inequalities 70–4

D

definite integrals 294–6, 307–9
derived functions 258
    see also gradient functions

descending powers of x 223–4
differentiation 256–88, 290, 369–73
    composite functions 283–5
    differentiation rule 258–9, 290
    equations of tangents and normals 268–74
    e^x and $\ln x$ 277–80
    gradient function 256–61
    product rule 280–1, 284
    quotient rule 281–2, 284
    second derivatives 265–8
    stationary points 261–5
    sums and differences of functions 259–61
    trigonometrical functions 274–7
discriminants 30–1, 34, 143–4, 151–3
displacement 312–24
displacement–time graphs 313–14, 316–18
distance 315
distance–time graphs 312–13, 316–18
divergent series 235
division, polynomials 42–3
domain of a function 4–5, 11–13
$\frac{dy}{dx}$ *see* differentiation

E

e^x 106–10, 277–80
    graphs of 106–10
    see also exponential functions
elimination method 79, 82
equations 55–76, 339–45
    of a circle 137–55
    cubic 43–50
    differential 290
    and logarithms 99–101
    modulus 57–63
    of a straight line 31, 119–23
    of tangents and normals 268–74
    trigonometrical 179–82, 192–9
    see also quadratic equations; simultaneous equations
exact form 174

exponential functions 105–14, 277, 350, 437–50
   bases 107
   graphs of 105–10, 112–13
exponential growth/decay 110–14
exponential relationships 126, 128–9

F

factor theorem 45–9
factorials 206–9, 221
factorisation 99
   exponential functions 109
   quadratic equations 69–70
   quadratic functions 23–5, 30–1, 36
factors, of polynomials 41–52, 336–8
family trees 89–90
feasible region of an inequality 67
flow charts (number machines) 4
functions 3–19, 330–3
   composite 13–14, 283–5
   derived 258
   domains 4–5, 11–13, 27
   inverse 9–13
   mappings 5–8
   modulus function 15–18, 56–7
   range 4–5, 11–13, 27
   sums of 259–60
   see also exponential functions; gradient functions; logarithms; quadratic functions

G

Gauss, Carl Friedrich 226–7
general solution 290–1
geometric progressions 90, 231–9
   infinite 235–7
   sum of terms 233–4
geometric series 235–7
Global Positioning System (GPS) 136
gradient functions (derivative)
   differentiation 256–61, 269, 272–4
   integration 291, 293–4
gradients 144–9, 312–13, 322
   stationary points 261–5
   of a straight line 119–20, 122–5, 127, 130–1
graphs 55–76, 89, 322, 339–45
   area under the curve 296–300, 302
   displacement and velocity 312
   of exponential functions 105–10, 112–13
   gradient functions 257–8, 260–1, 269, 272–4
   identities and equations 195, 197–9
   of logarithms 92–5, 97, 100, 103–4
   of modulus functions 56–7
   of motion in a straight line 314
   quadratic functions 21–34, 36–8
   solving cubic equations 43–6, 48
   solving cubic inequalities 70–4
   solving modulus equations 57–63
   solving modulus inequalities 65–9
   solving simultaneous equations 78, 81, 83–4, 86
   stationary points 261–6
   of trigonometrical functions 177–92
   of vectors in two dimensions 245–6
   $y = Ab^x$ 126, 128–9
   $y = ax^n$ 126–7
   see also quadratic curves; straight line graphs

I

i (unit vector) 242–52
identities, trigonometrical 192–9
images 4–7
indefinite integrals 294, 306–8
indices (singular: index) 96
   see also logarithms
inequalities 55–76, 339–45
   cubic 70–4
   modulus 64–9
   quadratic 36–8
   using logarithms 101–5
infinite geometric progressions 235–7
infinite sequences/series 217
inflexion, points of 70
integration 289–311, 322, 373–6
   area between a graph and the *x*-axis 296–305
   definite integrals 294–6, 307–9
   indefinite integrals 294, 306–8
interchange 10, 14, 105

J

j (unit vector) 242–52

K

kinematics 312–24, 376–8

L

limits (series) 217
linear simultaneous equations 78–83
lines
   length of a line 121–2, 140, 151
   line segments 36
   motion in a straight line 313–15
   parallel 122–3
   perpendicular 123–5, 144–5, 269
   see also straight line graphs
ln *x* 277–80
logarithms 91–106, 114–15, 250, 437–50
   bases 91, 96–100, 106, 126
   and exponential functions 105–6, 109
   geometric progressions 232
   inverse 106
   laws of 95, 96–7
   solving equations with 99–101
   solving inequalities with 101–5
   and $y = ax^n$ 126–7
long division, algebraic 49–50

M

magnitude *see* modulus
mappings 5–8
measure, circular 157–65, 356–7
method of squaring 61
midpoint of a line 121, 140
modulus 15–18, 56–7
   of cubic functions 70–1
modulus equations 57–63
modulus inequalities 64–9
motion
   in a straight line 313–15
   with variable acceleration 315–19

N

Newton, Isaac 136, 256
normals, equations of 268–74, 277, 279–80
notation 4, 327–9

arithmetic progressions 225–6
circles 157
geometric progressions 231–2
gradient function 258
series 216–17
vectors 241–5, 329

O
objects 5–7
operations, notation 327
outputs 4–7

P
parallel lines 122–3
particles 313
particular solutions 290
Pascal's triangle (Chinese triangle) 218–20
perfect squares 26
perimeters, sector 161–3
period of an oscillation 177, 179
periodic functions 177–9
permutations 209–10, 363–4
perpendicular bisectors 124–5
perpendicular lines 123–5, 144–5, 269
points of inflexion 70
polynomials 21, 218
    division 42–3
    factor theorem 45–9
    factors of 41–52, 336–8
    multiplication 42–3
    order of 21, 41
    remainder theorem 49–50
    solving cubic equations 43–5
    see also quadratic equations; quadratic functions
position vectors 242
powers see indices; logarithms
principal values of trigonometrical functions 180
probability 211
product rule 280–1, 284
Pythagoras' theorem 121–2, 137–8, 169, 193, 242–3

Q
quadrants 176–7
quadratic curves 21, 29–34, 36–8, 269
quadratic equations 21–2, 24–6, 31–5, 41, 142–4, 149–50, 152–3

completing the square 21, 25–6
discriminant 30–1, 34
factorisation 23–5
problem-solving with 35
roots 29–30, 32–4
solution with substitution 69–70
quadratic expressions 21–2, 24–6, 28, 41
quadratic formula 29–30, 144
quadratic functions 20–40, 333–6
    completing the square 21, 25–8, 29
    factorisation 23–5, 36
    intersection of a line and a curve 31–4
    lines of symmetry 21, 22–3, 25, 27
    maximum and minimum values 21–8
    problem-solving 35
quadratic inequalities 36–8
quadrats 55
quartic expressions 41
quintic expressions 41
quotient rule 281–2, 284

R
radians 159–63, 169–71, 275, 307
radius 137–40, 144, 151, 153, 157, 159–61
range of a function 4–5, 11–13
reflections 9–10
    and graphs of exponential functions 107
    of logarithmic graphs 92–4
remainder theorem 49–50
resultant vectors 245–6
roots 143–4, 150–3
    cubic equations 43–6, 49
    principal value 180

S
Same Signs, Subtract (and Opposite Signs, Add) rule 79
scalars 241, 245
scale factors 97, 108, 183–8, 190–1
secant (sec) 174–5, 195–9, 307–8
second derivatives 265–8
sectors of a circle, area 158–9, 161–3
segments of a circle 163

sequences 216–17, 224
    arithmetic 224–30
    geometric 90, 231–9
    infinite 217
series 216–40, 364–5
    arithmetic 224–30
    and binomial theorem 217–24
    convergent, divergent 235–7
    geometric 231–9
    infinite 217
sets 224
simultaneous equations 77–88, 142–3, 345–6
    elimination method 79, 82
    graphical solution 78
    linear, non-linear 78–87
    substitution method 80–6
sine (sin) 167–70, 172, 174–6, 179–80
    angles of any size 175–6
    differentiation 274–7
    integration 307–8
    $\sin x$ 274–7
    trigonometrical identities 192–7, 199
sine curve 177
sine graphs 177–8, 180–92, 275–6
'sohcahtoa' mnemonic 167
speed 241, 313–15, 322–3
speed–time graphs 316–19
spheres, volume 284–5
square numbers
    method of squaring 61
    perfect 26
square roots 30
    and quadratic equations 69–70
stationary points 261–6, 292
    of inflexion 261
straight line graphs 118–35, 352–4
    length of a line 121–2
    midpoint of a line 121
    relationships of the form $y = Ab^x$ 126, 128–9
    relationships of the form $y = ax^n$ 126–7, 129
    $y = mx + c$ 119–26, 129–33
stretches 183–8, 190–1
substitution method 80–6
summation 217
symbols 327
symmetry
    lines of 21–3, 25, 27, 168
    rotational 300

T

tangent (tan) 167–73, 174–7, 179–80
　angles of any size 175–7
　differentiation 274–6
　tan x 274–6
　trigonometrical identities 192–5, 197, 199
tangent graphs 179–81, 183, 190–1
tangents to a curve 32, 257–8, 261, 275, 277
　and circles 142–3, 144–8
　equations of tangents and normals 268–74
terms of a sequence 217–18, 220–35
time 312–23
transformations
　of graphs of exponential functions 107–9
　of logarithmic graphs 92–5
　of trigonometrical graphs 183–92
translations
　and graphs of exponential functions 109
　of logarithmic graphs 92–3, 95
　of trigonometrical graphs 184–5, 187, 189–90
triangles
　adjacent side 166–7
　hypotenuse 166–7
　length of a line 121–2
　opposite side 166–7
　and quadratic equations 25
　types of 166–71
trigonometry 166–203, 357–62
　angles of any size 175–7
　CAST rule 176
　and differentiation 274–80
　equations 179–82
　identities and equations 192–9
　reciprocal functions 174–5, 196
　in right-angled triangles 166–71
trilateration 136
turning points 21, 23–8, 70, 261–6, 276, 278, 292
　maximum and minimum 23–8, 261–6, 276–8

U

undefined 171
unit vectors 242, 247–51

V

variables 77
vectors 241–53, 365–9
　addition and subtraction 245–6
　column 242–4, 247–8, 250
　component form 242, 244
　magnitude/modulus 242–3
　multiplication by a scalar 245
　resultant 245–6
　terminology and notation 241–5
　in two dimensions 241–53
　unit 242, 247–51
　zero 247–51
velocity 241, 312–24
　determination of displacement from 320–3
　determination from acceleration 320–3
　initial 320
velocity–time graphs 313, 316–19
vertical line of symmetry 22–3, 27
Viète, François 20

Y

$y = Ab^x$ 126, 128–9
$y = ax^n$ 126–7, 129
$y = mx + c$ 119–26, 129–33

Z

zero vectors 247–51